THE PHYSICS OF
RAINCLOUDS

THE PHYSICS OF RAINCLOUDS

BY

N. H. FLETCHER

Professor of Physics at the University of New England
Formerly at the Radiophysics Laboratory, C.S.I.R.O.

WITH AN INTRODUCTORY CHAPTER

BY

P. SQUIRES

AND A FOREWORD BY

E. G. BOWEN

CAMBRIDGE
AT THE UNIVERSITY PRESS
1966

PUBLISHED BY
THE SYNDICS OF THE CAMBRIDGE UNIVERSITY PRESS

Bentley House, 200 Euston Road, London, N.W.1
American Branch: 32 East 57th Street, New York ! N.Y. 10022.
West African Office: P.M.B. 5181, Ibadan, Nigeria

©

CAMBRIDGE UNIVERSITY PRESS
1962

First Published 1962
Reprinted 1966

First printed in Great Britain at the University Press, Cambridge
Reprinted by photo-lithography by Lowe & Brydone (Printers) Ltd., London

FOREWORD

WATER is one of the basic commodities on which our civilization depends and it is no accident that the highly developed regions of the world are those which are endowed with good supplies of water. Practically the whole of our usable water comes in the form of precipitation from the atmosphere, and few studies could be more important than those which lead to a complete understanding of how it is stored in the atmosphere and how it precipitates out.

The study of cloud physics has been a comparatively neglected branch of meteorology, and an understanding of the relatively simple physics of the precipitation process has come very late in that particular science. The main reason for this seems to have been the difficulty of measuring, in cloud, many of the quantities involved. However, in the past ten or fifteen years, some of these difficulties have been overcome and a vast extension has taken place in our knowledge of the subject.

Many factors have combined in bringing about this renaissance. In the first place, aircraft and radar have proved to be invaluable tools for investigating what goes on in cloud. The former provide a platform from which measurements may be made, and the latter gives a direct insight into the processes at work in even the most complex cloud systems. A second reason has been the possibility of modifying clouds to produce additional rainfall. This gave a tremendous stimulus to research into the mechanisms at work in clouds and incidentally revealed serious gaps in our knowledge of the precipitation process.

Lastly, the new knowledge would not have been obtained without the efforts of a few dedicated groups who have pursued the subject in different parts of the world. One of these exists at the Radiophysics Division of C.S.I.R.O. in Australia and for many years Dr N. H. Fletcher was a distinguished member of this group. The present book gives a comprehensive treatment of the recent developments in cloud physics and stands as a monument to the work which has been carried out to a common end in many countries of the world.

E. G. B.

DIVISION OF RADIOPHYSICS, C.S.I.R.O.
January 1961

CONTENTS

Preface		*page* ix
1	The Dynamics of Clouds	1
2	The Microphysics of Clouds	31
3	The Condensation of Water-vapour	37
4	Condensation Nuclei in the Atmosphere	64
5	The Microstructure of Non-freezing Clouds	103
6	Theory of the Development of Non-freezing Clouds	122
7	Rain from Non-freezing Clouds	160
8	Nucleation of the Ice-phase	197
9	Ice-forming Nuclei in the Atmosphere	229
10	The Growth of Ice-crystals	259
11	Rain from Sub-freezing Clouds	283
12	Artificial Modification of Clouds	293
13	Production and Properties of Silver Iodide Nuclei	309
14	Large-scale Rain-making Experiments	327
Appendix—Recent Advances		348
References and Author Index		351
Supplementary References		379
Subject Index		381

PREFACE

THE physics of rainclouds is a subject of which our knowledge and understanding is expanding at an impressive rate, largely because of the possibility that by suitable intervention on our part a measure of control may be exercised over the weather. If this intervention is to be effective then it must be based upon a sound understanding of the physical processes involved in cloud evolution and in the development of precipitation. It is in an attempt to summarize and place in perspective the developments which have taken place in this field in the last few years that this book has been written.

The book has been written with two main classes of readers in mind. In the first place it is designed to serve as an up-to-date introduction to cloud physics for physicists and meteorologists new to the subject, and for this reason the basic physical principles underlying each process have been dealt with in considerable detail. Secondly, by including copious references to recent original papers it is hoped that it may provide a useful reference volume for workers in the various branches of cloud physics.

The emphasis throughout the book is on microphysical processes in cloud development which lead ultimately to the production of precipitation. This treatment leads to only one major omission—the physics of cloud electrification and its associated phenomena. We have been content with brief mention of these topics for two reasons: first, the role of electrification in precipitation is not yet clear and, secondly, a discussion of the rival theories of electrification and of the phenomena associated with thunderstorms could well occupy a whole book in itself.

This volume was originally planned to be written in collaboration with Dr E. G. Bowen, Chief of the Division of Radiophysics of the C.S.I.R.O., but unfortunately the pressure of other phases of the Division's research effort made this impossible. The general plan of treatment, however, owes much to his advice, and without his help and encouragement the book would never have been written. In its final form it also owes much to the results of numerous discussions with members of the Cloud Physics group at the Radiophysics Laboratory, and to them it is a pleasure to record my gratitude. In particular I should like to thank Pat Squires for writing the introductory chapter on the 'Dynamics of Clouds', Sean Twomey who made many helpful suggestions about the earlier chapters, Miss Robin Rawson and her assistants in the Radiophysics publications office for typing the manuscript, and Miss V. W. Rowntree for drawing the diagrams.

UNIVERSITY OF NEW ENGLAND
ARMIDALE, N.S.W.
July 1960

N.H.F.

Note for reprint of 1966

THE reprinting of this book has given the opportunity to correct an error in equation (3.34) and in the discussion immediately following equation (8.4). An appendix has also been prepared giving a summary of advances in the field over the past four years and a selection of important references.

N.H.F.

December 1965

CHAPTER I

THE DYNAMICS OF CLOUDS

(*by* P. SQUIRES)

WATER IN THE ATMOSPHERE

WATER-VAPOUR is an ever-present, though minor constituent of the atmosphere. Unlike the major components oxygen and nitrogen, its concentration varies greatly in space and time; it is added to the lowest layers by evaporation from the oceans and land surfaces, and removed from the mid-troposphere by the formation of rain and snow which falls to the surface. On the average, therefore, water-vapour must diffuse upwards through the lower troposphere; since this transport is effected by turbulent rather than molecular diffusion, this evidently implies that on the average the concentration of water-vapour relative to the other constituents must decrease upwards through this layer.

The flux of water-vapour from the earth's surface—evaporation—depends on the nature and temperature of the surface, the wind strength and the humidity of the air. The flux of water back from the atmosphere to the earth is mostly due to precipitation, direct condensation—dew and rime—being relatively negligible. Precipitation is zero most of the time, but rises to very high values for short periods during heavy rain. Obviously these two fluxes must balance over the earth's surface over a sufficiently long time but, because of their great variations, a complex transport of water-vapour occurs on all scales of motion up to the general circulation of the atmosphere. Thus, tropical regions are on the whole a source-region for water-vapour, and polar regions a sink; the oceans as a whole are a source-region, and obviously the continents, or at least those parts of them which drain to the oceans, must be sinks.

Because of its high latent heat of condensation, water-vapour plays an important role in the transport of energy. The upward diffusion of water-vapour which has evaporated at the surface accounts for a large part of the total flux of energy from the surface to the atmosphere. On a global scale, the poleward flow of water-vapour transports a large fraction of the net radiative energy income of tropical regions to higher latitudes, which suffer a net radiative loss. Finally, the radiative properties of the air itself depend strongly on the concentration of water-vapour, and clouds radiate much more efficiently than clear air, so that the distribution of water-vapour and clouds determines the location of the most effective energy sinks in the troposphere, and hence affects the general circulation.

For these reasons the water-cycle of the atmosphere is an important topic in meteorology, and indeed extends outside this subject since it involves not only atmospheric events but also the movement of water on and below the earth's surface. The purpose of this book is to consider only one link in this complex chain—the processes by which clouds give rise to precipitation. In this chapter, the formation and general properties of the clouds themselves will be briefly discussed by way of introduction to the main subject-matter.

THE FORMATION OF CLOUDS

Clouds form in the free atmosphere almost entirely as a result of the expansion and consequent cooling of ascending air. The process may be studied in its simplest form by considering the history of a parcel of air which is lifted adiabatically, although, as will appear later, matters are often complicated by the fall-out of precipitation, and by mixing between air-parcels of different properties. As is shown in general textbooks of meteorology, it follows from the conservation of energy that in such an adiabatic parcel of cloud-free air, the temperature decreases with height at a constant rate of $0.98°$ C per 100 m (the dry adiabatic lapse-rate). At the same time, the partial pressure of water-vapour in the air is reduced, in proportion to the total pressure, so that the dew-point of the air also decreases —although much more slowly than the temperature (about $0.15-0.2°$ C per 100 m depending on the temperature and pressure). If the lifting continues far enough the temperature will eventually overtake the dew-point, and condensation will begin. From this point on, the situation is more complex, since condensation releases a large amount of latent heat. In tropical air, which is rich in condensable water-vapour, the subsequent rate of cooling (the saturated or wet adiabatic lapse-rate) is only about one half of the dry adiabatic lapse-rate. In polar regions, or in the upper parts of the troposphere, there is little difference between the two lapse-rates.

Evidently, the stability of the atmosphere depends on its vertical gradient of density or temperature. Thus, if the temperature of a cloud-free layer decreases less than $0\cdot98°$ C per 100 m (the dry adiabatic lapse-rate), a parcel displaced upwards adiabatically will be colder and denser than its new surroundings, and will tend to fall back to its level of origin. Such an ambient lapse-rate is therefore stable; one exceeding the dry adiabatic lapse-rate is unstable, and one equal to it, neutrally stable. Similarly, a cloudy air-parcel is stable (unstable) if the local vertical gradient of temperature is less (greater) than the wet adiabatic lapse-rate appropriate to the temperature and pressure of the parcel.

These events are most conveniently followed by the use of one of the many aerological diagrams described in several textbooks and in full detail in the publication of the International Meteorological Organization *Über Aerologische Diagrammpapiere*. On these diagrams are plotted dry and wet adiabatics, and constant saturation mixing-ratio lines, so that the variation of temperature and mixing-ratio in the air can be found as a function of height. By comparing the water-vapour saturation mixing-ratio of the air at the condensation level with that at a higher level, it is a simple matter to find the concentration of liquid water which must appear in a parcel of air which is lifted adiabatically. The rate of release of liquid water in such a parcel varies from around 2 g per kg of air per kilometre of rise in tropical air to less than 1 g kg^{-1} km^{-1} in cold air as shown in Fig. 1.1.

On reaching the so-called freezing-level (0° C), the cloud droplets usually remain supercooled until the temperature falls considerably lower. If the cloud is eventually transformed to ice, an additional release of latent heat occurs, amounting to about 13 % of that resulting from condensation.

CLOUD FORMS

The International Classification of clouds used in synoptic meteorology is based on that proposed by Howard early in the nineteenth century. A full description of the many genera, species and varieties is given in the International Cloud Atlas (1956) of the World Meteorological Organization. Excellent photographs of the various types can be found in this Atlas, or in the collection published by Ludlam and Scorer (1957).

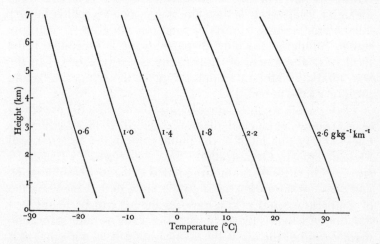

Fig. 1.1. Isopleths of the rate of release of liquid water (g/kg of air/km of lifting) in an adiabatically ascending air-parcel, as a function of height and temperature.

The International Classification evolved from the experience of ground observers over many years, and depends primarily on the appearance of clouds as seen from the ground. From a physical point of view, the important distinctions to be made are those arising from the character of the vertical motions which give birth to the clouds, and their microphysical properties—the presence or absence of ice, and the size and concentration of droplets or ice-crystals. Since the microphysical properties of clouds will form a large part of the subject-matter of this book, they will not be discussed in this chapter.

Most types of clouds are formed as a result of vertical motions which arise in one of the following ways:

Widespread, gradual lifting

Upglide motion of warm air at a frontal surface occurs in the cyclones of temperate latitudes, and gives rise to extensive sheets of deep and, commonly, layered alto-stratus and nimbo-stratus. In tropical and subtropical regions this type of cloud is far less common, arising from convergent flow over wide areas associated with airstream boundaries and with tropical storms and cyclones. The vertical component of air-velocity, which is responsible for the formation of the cloud, may be estimated from the convergence of the low-level flow, and has been found to be of the order of a few to a few tens of centimetres per second. Alto-stratus is one of the pre-eminent rain- or snow-forming clouds, and gives rise to long-continued, steady precipitation.

The formation of widespread layer-clouds of this kind is an aspect of the dynamics of synoptic scale weather disturbances, far too extensive and long-lasting to be adequately observed with the means commonly available. There is, moreover, no known way of directly measuring these slow, long-continued upward movements. For these reasons, experimental cloud physics has concentrated very largely on other types of clouds (in particular, on convective cloud), despite the importance of layer-clouds, especially in temperate regions, in the formation of precipitation.

Convection currents

Convection currents give rise to heaped-up or 'cumuliform' clouds, ranging from small fair-weather cumuli a few hundred metres deep to huge cumulo-nimbi and thunderstorms sometimes reaching to the tropopause. The speeds of the vertical motions in these clouds can be estimated by observing the rate of climb of the cloud-top, and have been measured during aircraft penetrations. They are typically a metre a second in cumuli and as much as 20 m a second in cumulo-nimbi. Convective clouds arise from the diurnal heating of the land, or when a cold air-mass moves over a relatively warm surface. A less obvious cause is the release of 'convective instability' in the free air. This can occur when the concentration of water-vapour in the air falls off rapidly with increasing height. Then, if the air as a whole is lifted by any agency, it may happen that condensation first commences in the lower layers of the lifted column. With continued lifting these cloud-filled layers cool at the wet adiabatic lapse-rate, while the clear air above cools more rapidly at the dry adiabatic lapse-rate. If this steepening of the lapse-rate in the column continues long enough, it can cause strong and widespread convection. The agency which gives rise to the lifting of the air-mass may be general convergence, orographic lifting, or undercutting of warm air by colder air.

Of limited horizontal extent and lasting only some tens of minutes in the case of cumuli and some hours in the case of cumulo-nimbi, convective clouds yield showers rather than continuous rain. However, because of their considerable depth and the strength of the upcurrents which form them, these showers can at times be quite intense. Most of the rain of tropical regions falls from these clouds, and even in temperate latitudes they are important precipitation sources. Together with alto-stratus and nimbo-stratus, therefore, they play a significant role in the water-cycle of the atmosphere, and because of the importance of latent heat in the energy balance, in its thermal cycle also.

Orographic lifting

When an air-mass moves against a mountain barrier, some of the air is forced to rise and this can result in an extensive sheet of deep stratiform cloud. If the mountain range is high enough, the lifting of the airstream may release convective instability and so give rise also to cumuliform clouds. Even mountain ranges of modest height can notably influence the formation of cloud and rain by accentuating the lifting associated with moving weather disturbances.

Lenticular wave-clouds form at times high above mountains, and in their lee, the air reaching saturation for a short time as it passes through the crest of a wave. The clouds may remain stationary although the air and cloud droplets move through them at the speed of the wind, which may be very strong.

Turbulence

The random stirring by ground friction of a layer of air trapped below an inversion of temperature often results in the formation of widespread sheets of strato-cumulus. Continued stirring results in a dry adiabatic lapse-rate and uniform moisture-content below the cloud, condensation occurring in randomly upward moving air-parcels which pass beyond the condensation level. Within the cloud the lapse-rate is wet adiabatic. This type of cloud is typically less than a kilometre in depth, although it may have very great lateral extent and may last for days at a time. The upcurrents in strato-cumulus are weak, usually less than a metre a second, and this cloud is of virtually no importance in the production of rain, rarely yielding more than a very light drizzle.

THE LIQUID WATER CONTENT OF CLOUDS

Comparison of various cloud types

It is evident that one of the most important properties of a cloud is the concentration of condensed material in it; this must directly affect its capacity to release precipitation. The first measurements of the concentration of liquid water in clouds were made by A. and H. Schlaginweit in 1851 in cumuli over the Alps (Hann, 1889). Diem (1942) and Mazur (1943) made the first measurements in the free air from aircraft by capturing the cloud droplets in a layer of oil. Soon after, a lengthy series of measurements was undertaken in the United States in connexion with studies of the icing of aircraft. Icing is caused by the impingement of supercooled droplets which freeze on impact and not, as a rule, by ice or snow, and consequently these measurements aimed at determining only the liquid water content of supercooled clouds. Rotating cylinders were exposed to the airstream and the amount of ice formed on them after a given time was measured. Since 1950, several series of measurements have been undertaken in the study of the physical processes occurring in clouds, but up to the present no method of measuring ice-concentrations in cloud has been devised. In view of the great importance of the ice-phase in the release of precipitation, especially in temperate latitudes, this is a serious gap in our knowledge of the basic properties of clouds; however, the instrumental problem appears difficult.

In measurements concerned with aircraft icing the aim was to sample all kinds of clouds, in order to provide adequate data for aircraft-design purposes. Later work directed towards the physical study of clouds has been more eclectic and attention has largely been concentrated on convective cloud; these results will be considered separately. Thus, in making comparisons between different cloud types, we must lean heavily on the icing investigations. The data from these, and such other series of measurements as were not exclusively concerned with convective cloud, are summarized in table 1.1. Liquid water content (w) is expressed in grams per cubic metre. All these measurements were made either by droplet impaction or by means of rotating cylinders, with the exception of those of Neiburger, who used a method due to Vonnegut (1949a) in which the droplets are impacted on to a porous plug which is connected to a glass capillary tube, the whole system being filled with water. The impacted water droplets are drawn into the porous plug by capillary forces and the movement of the water-column along the tube gives a direct measure of volume of water absorbed by the porous plug.

The values of liquid water content given in table 1.1 may of course lie significantly below the total concentration of condensed material: Lewis (1951a), states that, in the icing investigations undertaken in the United States, alto-stratus was normally found to consist almost entirely of ice-crystals. Again, Ludlam (1951a) has pointed out a source of error in methods which depend on the deposition of ice: with high liquid water content, and at temperatures not far below freezing, the air-current may not remove heat from the cylinders fast enough to carry away the latent heat of freezing of all the impacted water. If this happens the surface temperature of the ice-coating will rise to $0°$ C and some of the water will remain unfrozen and run off. With a given airspeed, cylinder diameter and temperature there is, therefore, a maximum possible rate of accretion of ice, and this sets an upper limit (the 'Ludlam limit') to the measured value of the liquid water content. In the rotating cylinder measurements of table 1.1, the measured values of w

Table 1.1. *Average and maximum liquid water contents observed in various types of clouds in the free air*

Author	Method	Stratus and strato-cumulus			Alto-stratus and alto-cumulus			Cumulus			Cumulo-nimbus		
		Average	Max.	No. of observations	Average	Max.	No. of observations	Average	Max.	No. of observations	Average	Max.	No. of observations
Diem, 1942	Droplet sampling	0·17	0·75	34	0·11	0·24	17	0·19	0·21	4	—	—	—
Mazur, 1943	Droplet sampling	0·19	0·38	—	0·13	—	—	0·22	0·41	—	—	—	—
Lewis, 1947	Ice accumulation on rotating cylinders	0·26	0·70	7	0·16	0·70	5	0·38	0·90	3	0·40	1·10	2
Lewis, Kline and Steinmetz, 1947	Ice accumulation on rotating cylinders	0·22	0·57	99	0·16	0·34	34	0·32	0·94	19	—	—	—
Lewis and Hoecker, 1949	Ice accumulation on rotating cylinders	0·23	0·36	21	0·15	0·33	98	0·64	1·71	81	0·51	1·72	76
Kline, 1949	Ice accumulation on rotating cylinders	0·23	0·57	41	—	—	—	—	—	—	—	—	—
Neiburger, 1949	Capillary collector	0·23	0·67	61	—	—	—	—	—	—	—	—	—
Kline and Walker, 1951	Ice accumulation on rotating cylinders	0·31	0·88	119	0·15	0·24	5	0·52	1·30	12	—	—	—
Frith, 1951	Droplet sampling	0·38	0·88	36	—	—	—	—	—	—	—	—	—

Table 1.2. *The liquid water content of convective clouds in the free air*

Author	Method	Cloud type	Liquid water content (w) g/m³ Average	Liquid water content (w) g/m³ Max.	Ratio w/w_a Average	Ratio w/w_a Max.	No. of observations	Averaging distance	Locality
Diem, 1942	Droplet sampling	Cu	0·19	0·21	—	—	4	1 m	Germany
Mazur, 1943	Droplet sampling	Cu	0·22	0·41	—	—	—	1 m	England
Lewis, 1947	Rotating cylinders	Cu	0·38	0·90	—	—	3	Some km	U.S.A.
		Cb	0·40	1·10	—	—	2	Some km	U.S.A.
Lewis et al. 1947	Rotating cylinders	Cu	0·32	0·94	—	—	19	Some km	U.S.A.
Lewis and Hoecker, 1949	Rotating cylinders	Cu	0·64	1·71	—	—	81	Some km	U.S.A.
		Cb	0·51	1·72	—	—	76	Some km	U.S.A.
Zaitsev, 1950	Filter-paper	Cu and Cu cong.	0·48	4·10	—	—	600	1 km	Russia
Kline and Walker, 1951	Rotating cylinder	Cu	0·52	1·30	—	—	12	Some km	U.S.A.
Weickmann and aufm Kampe, 1953	Visibility	Cu	1·0	3·00	0·65	1·00	9	—	New Jersey, U.S.A.
		Cu cong.	3·9	6·50	0·96	1·60	9	—	New Jersey, U.S.A.
		Cb	2·5	10·00	0·89	3·50	16	—	New Jersey, U.S.A.
Day and Murgatroyd, 1953	Icing disk and heated cylinder	Cu cong. and Cb	0·33	2·10	0·14	0·52	Continuous record	Probably some hundreds of metres	England
Warner, 1955	Paper tape	Cu and Cu cong.	—	—	0·26	—	Continuous record	—	Eastern Australia

Table 1.2 (cont.)

Author	Method	Cloud type	Liquid water content (w) g/m³		Ratio w/w_a		No. of observations	Averaging distance	Locality
			Average	Max.	Average	Max.			
Day, 1955	Refrigerated icing disk	Cb	—	4·20	—	—	Continuous records	Probably some hundreds of metres	England
Day, 1956	Icing disk and heated cylinder	—	0·53	2·00	0·15	> 1·0	Continuous records	Probably some hundreds of metres	England
Battan and Reitan, 1957	Droplet sampling	Cu	0·18	0·28	0·18	0·34	5	1 m	Central U.S. and Caribbean Sea
	Paper tape	Cu	0·33	0·37	0·31	1·30	—	10 m	Caribbean Sea
Squires, 1958c	Droplet sampling	Cu	—	—	0·20	1·0	321	1 m	Eastern Australia and Hawaii
Warner and Squires, 1958	Droplet sampling	Cu	0·27	0·80	—	—	155	1 m	Eastern Australia and Hawaii
	Paper tape	Cu	0·30	0·95	—	—	—	30 m	Central U.S.
Draginis, 1958	Paper tape	Cu and Cu cong.	0·60	3·60	0·23	2·60	Continuous records	Some metres	Caribbean Sea
	Droplet sampling	Cu		4·00	0·24	2·40			
Durbin, 1959	Paper tape	Cu	0·43	'Almost 5'	< 1	> 1	150	1 m	England
Ackerman, 1959	Heated wire	Cu	0·54	3·36	0·35	About 2·00	Continuous records	Tens of metres	Caribbean Sea

were mostly less than half the Ludlam limit. However, since rotating cylinders must be exposed for some tens of seconds to collect a measurable amount of ice, the resulting values of w are averaged over some kilometres. Several series of observations have shown that w often undergoes wide fluctuations over tens or hundreds of metres. For this reason it seems likely that some truncation of peaks has occurred; both the maximum and mean values of w found using rotating cylinders are probably somewhat too low, even when the measured value is less than the Ludlam limit.

Measurements made from aircraft in the orographic cloud which forms on the windward slope of the island of Hawaii gave values similar to those in table 1.1 (Squires and Warner, 1957). The mean value found by droplet sampling was 0·35 g/m³. A continuous record obtained by measuring the electric resistance of a paper tape on which the droplets impinged (Warner and Newnham, 1952) gave a mean value of 0·28 g/m³, and showed that w was less than 1 g/m³ nearly all the time, rising briefly as high as 4 g/m³ in heavy rain in the lower reaches of the cloud sheet, which, unlike the upper portion, was turbulent and showed some convective structure.

Observations on mountain-tops have also yielded results which are consistent with the free air data of table 1.1. Thus Learnard (1953) found a mean value of 0·46 g/m³ from some 3500 rotating cylinder measurements on Mount Washington (New Hampshire, 6300' MSL); Bricard (1953) observed various types of clouds on the Pic du Midi and found mean values ranging from 0·41 to 0·64 g/m³; Levin and Starostina (1953) in measurements made on Mount Elbruz in several cloud types, found a maximum value of 1·5 g/m³; Rittenberger (1959), at the Feldberg Observatory in the Black Forest (4900' MSL), found a mean value of 0·22 g/m³ and a maximum of 1·54 g/m³ in 370 observations. The data of table 1.1 are consistent also with the results of a very extensive icing study carried out over the United States, and over the North Atlantic, Pacific and Arctic Oceans, described by Perkins (1959). Over three thousand icing-cloud encounters were recorded, and liquid water contents averaged over a period of one minute were derived. These measurements, like those made with rotating cylinders, depend on the accumulation of ice and are

Liquid Water Content of Clouds

equally subject to the error pointed out by Ludlam. The mean value found was 0·25 g/m^3; only 10% of the measurements yielded values above 0·6 g/m^3. The maximum value found was 1·34 g/m^3, but this value may be too small, since the Ludlam limit at the time of the measurement appears to have been rather less than the measured value.

The data of table 1.1 makes it possible to compare the values found in different cloud types using the same equipment. The lengths of the paths over which the liquid water concentration was averaged varied from about ·1 m in the case of droplet sampling methods to several kilometres in the rotating cylinder method. Considering that these results were obtained by means of a variety of techniques, their consistency is surprising: it would seem that all these cloud types have comparable liquid water contents, the lower stratiform clouds being slightly wetter than those at middle levels, and convective clouds being wetter still. Except in convective clouds, no observation appears to have yielded a value exceeding 1 g/m^3.

Convective cloud

As mentioned earlier, recent cloud-physics studies have tended to concentrate on convective cloud, to the relative exclusion of other important types such as alto-stratus. This is of course largely a matter of scale. In dealing with a convective cloud a few kilometres across and lasting perhaps an hour or even less, it is reasonable to take the synoptic situation as constant and given. It is possible also, at least in principle, to measure the more important factors governing the formation, growth and dissipation of the cloud and to formulate a theory to explain its observed properties. On the other hand, it would be impossible to discuss the formation and evolution of the middle-level cloud system of an extra-tropical cyclone without reference to the dynamics of the atmosphere on a synoptic scale and without detailed observations of the cloud sheet extending several hundreds of kilometres and lasting for several days. This is clearly a task of a higher order of difficulty.

The mean and maximum values of liquid water content found in convective cloud are shown in table 1.2. For completeness, the data on these clouds from table 1.1 is repeated here.

The table indicates, where possible, the mean and maximum values of the ratio of the observed liquid water content (w g/m^3) to that which would appear in a parcel of air saturated at cloud base and lifted adiabatically to the observation level (w_a). The techniques used in obtaining the data of table 1.2 were similar to those already described except for three investigations: Zaitsev (1950) impacted the droplets on a filter-paper impregnated with a dye and measured the area of the resulting stain; by means of a calibration, this yielded the volume of water received by the filter-paper. Weickman and aufm Kampe (1953) recorded the attenuation of light caused by the cloud and used the relation given by Trabert (1901) between visibility and w. Ackerman (1959), by means of a calibration, deduced w from the cooling of a heated wire exposed to the cloud droplets.

Considering the variety of cloud types and sizes, and the wide range of geographical location, the values of mean liquid water content are surprisingly consistent. The values for the maximum liquid water are, however, widely disparate. Similarly, there is reasonable agreement that, in the mean, w is about a third or a quarter of w_a but the maximum values found for the ratio w/w_a are not in agreement. Some of the discrepancies in maximum values may be due to differences in the averaging distance of various methods; fast-responding instruments will naturally yield higher maximum values. However, this does not explain all the discrepancies. The values of w/w_a exceeding unity found by Weickmann and aufm Kampe, Battan and Reitan, Draginis and Ackerman can be explained only by supposing that sedimentation of droplets has enriched some parts of the cloud at the expense of others. Warner and Squires (1958), summarizing the observations on convective cloud up to 1956, concluded that 'the full adiabatic value is realized, if at all, only in regions which are of negligible size in relation to the cloud as a whole. In most cases the liquid water in the main body of the cloud is less than a quarter of the adiabatic value, and often considerably less.' This view is consistent with the later results of Draginis and Ackerman, who both emphasize that in their observations the regions where w exceeded w_a were very restricted.

There has been an understandable reluctance to accept the

implications of these results. Since the falling velocities of nearly all cloud droplets are a few centimetres per second or less, it is clear that sedimentation cannot remove liquid water from most of the volume of a cloud extending some kilometres in height and breadth, and lasting probably less than an hour. It is generally agreed that the measured values of liquid water can be explained only if it is supposed that dry air from outside mixes into the cloud. It is of course possible that the results so far obtained are not representative of all convective clouds. Ludlam and Scorer (1953) have shown some evidence that in the particular case of polar outbreaks over the North Atlantic, the largest clouds, provided they exceed 7000 ft in depth, are effectively undiluted. The data of table 1.2 includes many observations from clouds of this, and greater, depth from other places, and there seems little doubt that in most parts of the world, and in particular over the oceans in the trade-wind zone, clouds of this size are usually influenced strongly by mixing. There is indeed a dearth of observations in cumulo-nimbi, and although such observations as have been made do not show that the largest convective clouds are on the average any wetter than moderate-sized cumuli, it seems intuitively likely that, if large volumes of adiabatically ascending cloud air are to be found anywhere, it will be in the violent upcurrents which form these clouds.

CONVECTION AND CONVECTIVE CLOUD

The reasons mentioned earlier for a certain concentration of observational effort on convective cloud to the relative exclusion of other important cloud types have naturally affected theoretical discussion as well. A considerable literature exists on atmospheric convection and, although much of it is not immediately relevant to the physics of clouds, it appears logical to give here a brief outline of some observations and theories of convection in the clear air layer below cloud base before discussing cloudy convection itself.

The roots of cumuli

(a) *Observational data*. In the air below cloud base, such properties as temperature and water-vapour content are much easier to measure than in cloud, but against this, the clear air gives us no such indication of its state of motion as may be gleaned by the careful observation of clouds. Most of our knowledge of the pattern of convection below cloud is derived from observations of soaring birds and the experience of glider pilots. Woodcock (1940) was the first to observe birds carefully for this purpose. In an elegant paper he has described how the convection pattern is revealed by the soaring routines of herring-gulls off the east coast of the United States in autumn and winter when the air is colder than the sea surface. At low wind speeds the birds soar in circles in a manner which clearly shows that rising air is concentrated in continuous updraughts which drift with the wind. At higher wind speeds their routines indicate that it is concentrated in vertical sheets aligned along the wind. When the wind is stronger than 13 m/sec, or the air is warmer than the water, the birds do not soar.

Woodcock and Wyman (1947) have used smoke to study convection over the ocean in a trade-wind zone. They found indications of the presence of Bénard type polygonal convection cells and longitudinal roll vortices in the air near the sea surface. Bunker, Haurwitz, Malkus and Stommel (1949) have described in detail the temperature and humidity structure of the clear air below cloud in the same region. Up to within 100–300 m below cloud base the lapse-rate is dry adiabatic and the air well-mixed. A shallow layer of rather more stable air in which the humidity mixing-ratio decreases with height separates the well-mixed layer from the cloud bases. The trade-wind cumuli apparently form when an unusually buoyant parcel of air in the well-mixed layer breaks through the shallow stable layer. On these grounds, and especially on the basis of observations of turbulence in and below the clouds, it was concluded that oceanic trade-wind cumuli have no 'roots', in contrast to diurnal landward cumuli where clouds form as a result of convective motions which are driven by the heating of the surface. James (1954) has also observed a shallow stable layer immediately below cloud-base level over land.

(b) *Plume theory.* Steady state, buoyant plumes rising from heat sources on the ground have been discussed by Priestley and Ball (1955), Morton, Taylor and Turner (1956) and Morton (1957). The Reynolds numbers being very high, the motion is

Convection and Convective Cloud

fully turbulent, and independent of molecular viscosity. The surrounding fluid is steadily entrained into the plume, which it dilutes, as a result of turbulent mixing. The theory involves only one empirical constant; in the formulation of Priestley and Ball, this appears as a coefficient in an assumed quadratic relationship between shearing stress and relative velocity, which the theory shows may be directly obtained experimentally from the rate of lateral spread of the plume upwards; in the formulation of Morton, Taylor and Turner, a different constant appears as the ratio between the horizontal velocity of the environmental fluid entering the plume at its periphery and the vertical velocity of the plume. A comparison of the two approaches has been given by Morton et al. (1956). Laboratory studies of turbulent plumes in air and water have yielded a value close to one-tenth for this constant (in both formulations). When this value is applied to atmospheric plumes, the agreement with observation is satisfactory. The absence of disposable constants is the great strength of plume theory, and sharply distinguishes it from the theory of lateral entrainment into convective clouds to be discussed later where no corresponding position has been reached and in consequence theoretical treatments have at times tended to become rather formalistic.

Although plume theory originated in connexion with studies of atmospheric pollution rather than convection, it seems likely that there would be situations in nature where, in a calm or light wind, variations of ground slope, vegetation, albedo and thermal conductivity of the soil would give rise to local hotspots and upward-moving plumes. Woodcock's observations (*loc. cit*) indicate that plume-like convection occurs sometimes over the sea. The theory deals in the first place with pointsources of buoyant fluid, but is extended to consider finite sources by the postulate that the flow some distance above such a source is the same as would arise from a virtual point-source some distance below the surface. This yields satisfactory agreement with observation. However, in practice a major difficulty in applying plume theory to natural convection problems is that usually no well-defined source-region can be identified.

(*c*) *The bubble theory of convection.* Scorer and Ludlam (1953), drawing on the experience of glider pilots, formulated a theory which aimed at presenting a unified picture of convection from the ground to cloud-top level. The basic tenet is that atmospheric convection is carried on primarily by the agency of buoyant bubbles of air. In the original qualitative formulation of the theory, these bubbles were supposed to be steadily eroded as they rose, the eroded material mixing with environmental air and lagging behind the undiluted bubble to form a trailing wake; in this form, the theory was developed quantitatively by Malkus and Scorer (1955), using observations of cumulus tops. However, Scorer and Ronne (1956) reported that experiments with slightly buoyant parcels or 'bubbles' of water in water-tanks showed that erosion occurred only when the surroundings were stably stratified; when stability was neutral, corresponding to a dry parcel of air ascending through a dry adiabatic layer, or a cloudy parcel moving through a wet adiabatic layer, erosion did not occur; in fact the bubbles grew by entraining environmental fluid. Scorer (1957) has developed a dimensional theory for entraining bubbles in a neutrally stable environment resulting in equations of the form

$$u = C(gBr)^{\frac{1}{2}}, \qquad (1.1)$$

$$h = nr, \qquad (1.2)$$

where u is the upward velocity, r the radius and h the height above the virtual source of the bubble, and $B = (\rho_0 - \rho)/\rho_0$, where ρ_0 is the density of the environment and ρ the mean density of the bubble. The dimensionless constants C and n were found from water-tank experiments to be approximately 1·2 and 4 respectively. These values were confirmed by Woodward (1959). The validity of the quantitative theory of bubbles in a neutrally stable environment hinges on the correctness of the postulate, supported by observation of buoyant bubbles in water, that a bubble or blob of buoyant fluid tends after a time to take on a typical shape and circulation pattern irrespective of its mode of release; once this has happened, all bubbles obey the same laws. By observing the motion of suspended particles in water-tank experiments, Woodward (1959) was able to deduce the velocities in the various parts of the bubble, and found that while the core is rising faster than the bubble-cap, the edges are actually descending; thus the bubble resembles a spherical vortex. Mixing with the environment occurs primarily over the cap, but also at the rear of the bubble.

Since atmospheric convection usually occurs in neutrally stable or unstable conditions, it seems likely that the entraining bubble model discussed by Scorer and Ronne (1956), Scorer (1957) and Woodward (1959) may apply in general rather than the eroding bubble model first suggested. This would be so both in the layer below cloud base when the lapse-rate is dry adiabatic, or inside convective cloud, where it is usually steeper than wet adiabatic. At the tops of these clouds, on the other hand, rising towers often penetrate into a stable environment, and here the eroding bubble model may be nearer the truth. In addition, Scorer and Ronne (1956) and Ludlam and Saunders (1956) have suggested that evaporative cooling of the exposed outer surfaces of such turrets may tend to accentuate erosion. As a result of gliding experience Woodward (1959) considers that the compact bubbles do not form at the surface of the earth, but first appear at a height of some hundreds of metres. The suggested mechanism is that the heated air rises from the surface at first as a plume: the head of the plume being strongly diluted by mixing, it is slowed down and overtaken by the less dilute lower portions so that a compact buoyant bubble finally results.

Cloudy convection

(a) *The structure of convective cloud.* During the Wyman–Woodcock expedition to the Caribbean Sea (Bunker *et al.* 1949) many measurements were made of the vertical temperature distribution in and around trade-wind cumuli. Although the interpretation of cloud-temperature measurements made from aircraft is rather doubtful (Byers and Braham, 1948) these results left no doubt that the lapse-rate in these cumuli was notably steeper than the wet adiabatic, and gave the first clear observational evidence that the adiabatic model of cumulus development is inadequate.

By far the most extensive investigation of convective cloud structure was that undertaken in the Thunderstorm Project (Byers and Braham, 1948, 1949). Vertical motions and air temperature were recorded by five aircraft simultaneously.

Some ten ground radars were used to follow balloons in order to determine the horizontal wind field around the storms. Byers and Braham found that thunderstorms usually consist of a number of convection units or cells, each of which passes through a characteristic life-cycle; at any moment, a storm may contain several cells in different stages of development. Three stages of cell life were distinguished: the cumulus stage, characterized by updraughts throughout the cell; the mature stage in which rain reaches the surface and downdraughts begin to appear at least in the lower half of the cell; and the dissipating stage, where weak downdraughts prevail throughout the cell. The whole process occupies a period of the order of an hour. Unicellular storms seldom grow as tall, or contain such vigorous updraughts and precipitation as cells which grow adjacent to others; they last on the average only 20 min. The well-known cold squally wind associated with thunderstorms is the result of cell downdraughts reaching the ground and spreading out; the formation of the downdraught was explained by Byers and Braham in terms of the theory of lateral entrainment, and will be discussed later. Despite the difficulties of interpreting cloud temperature measurements these authors concluded that lapse-rates in both the updraught and downdraught regions of thunderstorms are steeper than the wet adiabatic and indeed approximate that of the surroundings.

Zaitsez (1950) made many measurements of liquid water content (w g/m^3) and droplet concentration (n cm^{-3}) in cumuli. He found a maximum of w a little above the centre of the cloud, with a steady decrease upwards, downwards and outwards. The droplet concentration, after rising to a maximum in the first hundred metres or so decreased steadily upwards. In the peripheral region, 50–100 m wide, the mean drop size and w were both smaller than in the rest of the cloud. The tendency for droplet concentration to decrease upwards was found also by Weickmann and aufm Kampe (1953) and Squires (1958c).

Warner (1955) has made extensive measurements of liquid water content using a continuously recording device in cumuli up to about 2 km deep. These records showed the distribution of water to be far more complex than indicated by Zaitsev. During a horizontal traverse, w varied widely and often rapidly. In general, there seemed to be no steady rise to a central maximum, as found by Zaitsev; w often increased rapidly to its peak value quite close to the cloud edge. The peak value of w found in a traverse increased from the base upwards and usually reached a maximum some hundreds of metres below cloud-top after which it quickly fell away to zero. The ratio of the observed liquid water to the adiabatic value, w/w_a, decreased steadily from about 0·5 a few hundred metres above cloud base to about 0·2 at 2 km above it. Except in very small clouds of diameters less than 500 m, cloud width seemed to be unrelated to the value of w/w_a. From droplet samples taken every few seconds during each cloud traverse, Squires (1958c) found a variation of w/w_a with height similar to that reported by Warner. He also emphasized the sharpness of many cloud edges, and by combining many cloud traverses, without regard to direction, to give a mean variation of n and w on entering a cloud, concluded that beyond the first sample taken about 100 m inside the cloud there was on the average no further increase in either n or w. However, Ackerman (1959) has combined forty-seven liquid water records taken at one level on one day in a similar way, but taking account of the direction of the traverse; analysed in this way, that is in two dimensions, the data revealed that in the 'average cloud' there was a region of maximum w somewhat to the west of the cloud centre, with a steady gradation out to the cloud edge. As the upper winds reports on this day indicated a westerly shear, this asymmetrical distribution of water is consistent with the ideas of Malkus (1949) to be discussed later. Ackerman also found that w increased with cloud depth up to $2\frac{1}{2}$ km and also with cloud width up to 3 km. The latter result is contrary to Warner's finding; however, the data confirmed the conclusion of Warner that w/w_a decreases steadily with increasing height. Draginis (1958) found evidence of a similar trend. Like Ackerman, Draginis found regions in many clouds where w exceeded w_a; these regions were in fact characteristic of clouds which produced a radar echo at some stage of their development.

Many of the authors quoted here emphasized the rapid fluctuations in the value of w during an aircraft traverse. Weickmann and aufm Kampe (1953), using an optical transmissometer, found large variations over distances of a few metres.

The measurements of microwave refractive index, temperature and liquid water content made by Cunningham, Plank and Campen (1956) also show that convective clouds often possess a very detailed fine structure. Frith (1951) observed droplet spectra in strato-cumulus and found erratic variations from one sample to the next; however n and w tended to rise and fall together. Squires (1958c) obtained a similar result in cumuli, and attributed the fluctuations of n and w to the mixing into the cloud of varying amounts of dry air from the surroundings.

(b) *Lateral entrainment.* Almost simultaneously Schmidt (1947) in Holland and Stommel (1947) in the United States proposed that the adiabatic model of a cumulus should be modified to take account of interaction between the cloud and its environment. Schmidt attempted to evaluate the form drag on a developing cumulus on the one hand, and the friction drag on a cumulus regarded as a steady-state jet on the other. He made use of the solution obtained by Tollmien (see Goldstein, 1938) for a steady-state jet entraining fluid from its surroundings by turbulent mixing. Using Tollmien's profile for the distribution of temperature and moisture, he deduced, in a hypothetical situation, that the visible cloud would be roughly cylindrical in shape, occupying only a fraction of the cross-section of the conical jet.

Stommel set out to explain the measurements of temperature obtained in trade-wind cumuli during the Wyman–Woodcock expedition to the Caribbean Sea (Bunker *et al.* 1949) and, using the model of an entraining jet as studied by Tollmien and as used by Rossby (1936) in the theory of ocean currents, postulated lateral entrainment of outside air into the cloud. The treatment was purely empirical, and involved no theories concerning the dynamics of entrainment. Using observed vertical distributions of temperature and specific humidity both inside and outside the cloud, Stommel deduced from the conservation of heat and moisture how much environmental air would have to be entrained over each small height interval in order to result in a cloud with the observed temperature. Unfortunately, the liquid water content of the clouds had not been measured, but Stommel went on to deduce it from the computed entrainment, and found values much less than the adiabatic and indeed quite comparable with those which have been observed since in trade-wind cumuli.

This model clearly implies an upward mass flux steadily increasing with height. In most of the cases treated, the entrainment required was such that the flux roughly trebled over the cloud's depth, which ranged from 500 to 1400 m. This would imply on a steady-state model that either the updraught speed or the cloud's horizontal cross-section increased markedly with height, and is probably not realistic. However, since trade-wind cumuli are often rather transient, this does not amount to a criticism of Stommel's basic suggestion that the observed properties of clouds might result from the lateral entrainment of dry environmental air. In addition, this difficulty may be obviated by postulating a suitable rate of 'detrainment' from the cloud to the environment, as first suggested by Malkus (1949). Since the 'detrained' cloud air has the same temperature and humidity as the cloud at that level, its loss in no way affects the specific properties of the cloud.

Stommel's idea provided the framework for all discussion of the properties of convective cloud until the appearance of Scorer and Ludlam's (1953) bubble theory of convection, and because of its greater theoretical simplicity, for much since then. Theoretical discussions and development of the notion of lateral entrainment have been given by several authors. Like Stommel's original paper, they treat a steady-state cloud, homogeneous in any horizontal plane, and do not inquire in detail into the dynamics of the entrainment process. Austin (1948) discussed graphical methods for evaluating the effects of mixing and showed from synoptic data that the growth of cumuli depended on the relative humidity of the air into which the clouds grew as well as on its lapse-rate. This would be expected if entrainment occurred, since the drier the entrained air, the greater is its efficacy in cooling and stabilizing the ascending cloud stream. Austin and Fleischer (1948) gave a more detailed thermodynamic analysis, and concluded that provided the environmental lapse-rate is less than the dry adiabatic, as is practically always the case, the upward growth of a cloud terminates because of loss of buoyancy rather than by drying out. Stommel (1951) extended his original empirical treatment of trade-wind cumuli observed over the Caribbean Sea to take account of the conservation of vertical momentum as well as of heat and moisture, and in this way deduced vertical velocities for the updraughts. Houghton and Cramer (1951) examined theoretically the consequences of the conservation of vertical momentum. Postulating a cylindrical shape for the upward-moving stream they found it necessary to further postulate lateral entrainment in order to satisfy the condition of continuity. This they called 'dynamic entrainment'. Malkus (1952b) has identified 'dynamic' with 'net' entrainment, the difference between 'gross' entrainment as computed by Stommel from the temperature and humidity distributions and a postulated 'detrainment'. Haltiner (1959) has carried out elaborate calculations on a model similar to that of Houghton and Cramer, but allowing, in addition to the entrainment needed for continuity, a mass exchange as a result of general turbulence. This is equivalent to a gross entrainment greater than the dynamic or net entrainment, balanced by a detrainment equal to their difference.

Although the mechanism considered in theories of lateral entrainment into clouds is similar to that treated in plume theory, the approach to the problem is quite different. Morton (1957) has made the only attempt so far to apply the physical principles of plume theory to clouds. He considered a plume rising from a fire and computed the vertical variation of humidity mixing-ratio in it. From this, it was possible to compute the height at which condensation would begin. He further extended the theory to take account of the release of latent heat above the condensation level, assuming that entrainment occurs here in the same manner as in the clear air below. However, recognizing the possibility that different phenomena may intervene as a result of cloud formation, he limited his treatment to small clouds. The theory shows that, if the condensation level is close to the level where the plume would have ceased to rise in the absence of water-vapour, the only effect is to form a small cloud the top of which lies slightly above the level of the undisturbed plume top. If, however, condensation begins below a certain critical level, the behaviour of the upper part of the plume is changed completely: because of the release of latent heat, the plume buoyancy begins to increase and the upward growth of the cloud can be limited only by factors not considered in the theory, such as a strong temperature inversion at a higher level.

The entrainment model has been used in the interpretation of several studies of convective clouds. During the Thunderstorm Project, Byers and Hull (1949) measured the proportional rate of change of the area of triangles formed by balloons around thunderstorms. From this they were able to compute the horizontal divergence of the air-flow at various levels and deduced a typical entrainment rate into the cloud of 100 % in 4 km of height. This is indeed the only direct measurement which has been made of entrainment. Other studies mentioned below have not produced independent evidence of entrainment, although they leave no doubt that the cloud is far from adiabatic, and that in some way dry air is entering and mixing with the rising cloudy air. Byers and Braham (1948) explain the formation of the downdraughts which they observed in thunderstorms, mentioned in section (a), by the effects of entrainment. During the cumulus and mature stages, dry air is entrained by the updraught, so that the lapse-rate in the cloud becomes steeper than moist adiabatic. If then a parcel of saturated air inside the cloud is for any reason displaced downwards far enough it will become unstable, since it warms at the wet adiabatic lapse-rate, and so must eventually become colder than the environment. If dry air is entrained by the downward moving parcel, it will become still colder and denser, relative to its surroundings. The weight of precipitation in the air may provide the initial downward impulse.

Malkus (1949) first pointed out the importance of wind shear in the growth of convective clouds. Earlier discussions had assumed an environment at rest, despite the well-known fact that wind shear is rarely absent and produces marked and easily observed slopes in cumuli. Assuming that the transfer of horizontal momentum to the cloud resulted solely from the entrainment of environmental air with a horizontal velocity different from that of the cloud, Malkus deduced the cloud slopes which would result with a linear variation of wind speed with height, and went on to point out that one result of vertical wind shear would be that clouds would be asymmetrical, new vigorous towers formed by strong updraughts building on the up-shear side of the cloud, while spent cloud material which had lost its buoyancy and upward speed would be blown towards the down-shear side. Measurements of turbulence in and around trade-wind cumuli during the Wyman–Woodcock expedition had shown that vertical gust speeds in clouds were on the average nearly an order of magnitude larger than those in the clear air at the same level. However, cloud-scale turbulence could often be found in the clear air just outside clouds, and the area of such turbulence was asymmetrically distributed with respect to the cloud. In the examples given, this area extended further up-shear than down-shear, and Malkus explained this as being associated with the greater turbulence of the vigorous updraughts on the building, up-shear side. However, Ackerman (1958), from a study of similar trade-wind cumuli, found an even more asymmetrical distribution of clear air turbulence, but extending much further down-shear than up-shear. Since turbulence was still found in areas where clouds had evaporated some two to seven minutes earlier, Ackerman suggested that the turbulence characteristic of cloud outlived the visible cloud, and that this explained the down-shear extension of the turbulent region. Ackerman considered that the discrepancy between these results and those of Malkus might perhaps be explained by a difference in the definition of turbulence: Malkus was concerned with 'effective gust velocities' exceeding a certain value, Ackerman with all turbulence large enough to cause a measurable aircraft response. As mentioned in section (a) above, Ackerman (1959) has also found evidence of a concentration of liquid water on the up-shear side of the centre of cumuli, that is, in the region where new active turrets would be expected to build.

Byers and Braham (1949) described the inhibitory effects of vertical wind shear on the growth of thunderstorms and developed a theory similar to that of Malkus (1949) using the conservation of horizontal momentum in an entraining updraught to compute the change in horizontal velocity with height. The shearing rates in cloud computed in this way were smaller than those found from measurements of the displacement of radar echoes in thunderstorms, and this was taken as evidence that entrainment alone cannot account for the observed transfer of horizontal momentum to the cloud, but that aerodynamic forces due to the flow of the environmental air around the cloud must also be important. Malkus (1952a) solved the equation of horizontal momentum assuming a linear law for the horizontal form drag on the updraught due to the environment; the drag on a solid cylinder was taken as giving an upper bound for the form drag on the updraught. This approach gave an indication that the momentum transfer due to form drag on clouds a kilometre in diameter was of the same order as that due to entrainment.

Malkus (1954) used measurements of cloud slopes and of temperature, humidity and updraught speed in and around trade-wind cumuli to test whether the observations could be consistently interpreted in terms of the entrainment model. This is perhaps the most elaborately instrumented and ambitious attempt of this kind which has been made. Even so, it seems doubtful whether sufficient data were available to unravel the complicated phenomena which occur even in these relatively small and simple clouds; for example, the net or dynamic entrainment, that is the difference between gross entrainment and detrainment, was estimated from the speed and extent of updraughts found from one aircraft traverse at each level. This is evidently a very approximate estimate only, since the traverse may or may not have passed through the centre of the updraught. In the experimental investigation of convective clouds by means of a single aircraft, this typifies a recurring problem: if observations are sought at a number of levels in order to study the cloud as a whole, there is rarely time for more than one pass at each height if the observations are to be regarded as even quasi-simultaneous on the scale of the lifetime of such clouds.

Stommel's original use of the idea of lateral entrainment, suggested by examples from hydrodynamics and oceanography, was purely empirical and implied no specific mechanism. Within its assigned limits, this approach was completely successful and in the hands of Stommel and later in the discussions of the effects of wind shear on convective clouds by Malkus (1949) and Byers and Braham (1949) gave for the first time some quantitative ideas of the degree of interaction between cumuli and their environment. Later and more ambitious formulations which have aimed at presenting a full dynamical description of the steady-state, jet-like convective cloud have not been quite so successful. On the one hand it has proved impossible to rid the theory of arbitrary assumptions, and on the other it may be questioned whether the model used displays enough of the essential physical features of convective clouds to justify the erection of an extensive theoretical superstructure. It is of interest to compare the rates of entrainment measured by Byers and Hull, and those deduced from the measurement of cloud properties, with those implied by plume theory. According to the latter, the inflow velocity around the periphery of the cloud would be approximately one tenth of the updraught speed. If then α is the mass flux upward through a horizontal cross-section of a circular cloud of radius r, the entrainment rate, as normally defined, is given by:

$$\frac{1}{\alpha}\frac{d\alpha}{dh} \approx \frac{0 \cdot 2}{r} \qquad (1.3)$$

at all levels. The large clouds investigated by Byers and Hull had radii of the order of 5 km, giving a computed entrainment rate of 4×10^{-7} cm^{-1}; the measured value was 2×10^{-6} cm^{-1} approximately. The cumuli investigated by Stommel (1947) and by Malkus (1949, 1952a, 1954) had radii of the order of 0·5 km, giving a computed entrainment rate of 4×10^{-6} cm^{-1}; the values deduced by both authors from cloud properties were a few times 10^{-5} cm^{-1}. From liquid water measurements in similar clouds, Ackerman (1959) deduced an entrainment rate of about 10^{-5} cm^{-1}. These comparisons suggest that the mixing-process is rather more vigorous than would be expected from simple turbulent entrainment. Two possible explanations suggest themselves: (i) the steady-state jet model underestimates mixing, in particular because it ignores mixing across the top of the cloud mass; and (ii) the presence of the cloud itself may affect the dynamics of the mixing-process. These possibilities have both received consideration, as will be seen in the following sections.

(c) *Bubble theory as applied to cloudy convection.* Stommel (1951) and Malkus (1952b) both recognized certain deficiencies in the lateral entrainment theory: it neither took account of the life-cycle of cumuli, nor provided any adequate dynamical description of the mixing-process. A step towards meeting both these needs was taken by Scorer and Ludlam (1953) with the presentation of their bubble theory of convection described above. The early studies of Malkus and Scorer (1955) and of Malkus and Ronne (1954) made use of time-lapse photographs of cumulus tops and were based on the eroding bubble model. However, according to the indications of later experimental work, this model probably does not apply to other aspects of atmospheric convection such as that below cloud base or within cloud.

Ludlam (1958) has recently given a very clear description of cumulus clouds in terms of the bubble theory. A cumulus consists largely of the remnants of spent bubbles, containing little liquid water or buoyancy; nevertheless, being saturated, this mass serves to shelter new active bubbles or thermals which enter its base from evaporative cooling and consequent erosion and loss of buoyancy. In this way it enables them to rise to the level of its top. Arriving here, where they become exposed to the dry outside air, they quickly lose their buoyancy, and usually rise at most a distance equal to their diameter on emergence from the cloud mass. The pyramidal shape of cumuli is explained by the better protection of thermals rising in its centre rather than near its edges. Applying Scorer's theory of bubbles or thermals to this model of a convective cloud, Ludlam found that, on reasonable assumptions, the mean temperature excess and liquid water content of active thermals should both be about one third of the adiabatic value. This conclusion seems consistent with observation, although it has not as yet been directly checked.

Levine (1959) has approximated the buoyant cloudy bubble by a spherical vortex. The solution of the equations involves two parameters which need to be determined by observation: one relating to the drag on the rising bubble, the other to the rate at which mass is exchanged with the surroundings as a result of turbulent mixing. It is shown that, with a suitable choice of these parameters, the model can be made to fit measurements of cloud properties in particular situations.

(*d*) *Penetrative downdraughts.* Like the lateral entrainment theory, the bubble theory was suggested by phenomena occurring in liquids or in dry convection. Phase changes do not occur in such cases; as a result, they are not considered in either theory as affecting the dynamics of the mixing between the cloud and its clear air environment. However, a phase change is the distinguishing characteristic of cumulus convection, and Squires (1958c, d) has suggested that the presence of liquid water in cloud may provide a local source of energy available to drive small-scale motions which would promote mixing and that this factor should be taken into account in addition to those considered in the bubble and lateral entrainment theories. The mechanism envisaged is termed 'penetrative downdraughts'; it is supposed that parcels of unsaturated environmental air become engulfed or immersed in the cloud, are cooled by the evaporation of liquid water which diffuses into them, and consequently subside.

Some illustrative calculations designed to show the potency of this mechanism were carried out. These indicated that a dry air-parcel of a suitable size, estimated to be some tens to some hundreds of metres in radius, could subside several kilometres through a cloud against a steady updraught of 1 m sec^{-1} before becoming saturated. The speed and depth of penetration both increased with increasing cloud liquid water content, as would be expected, since it is the liquid water alone which provides the energy to drive the motion. This mechanism therefore introduces a negative feed-back loop into the cloud-environment system which will tend to limit the liquid water content in a manner which seems consistent with the observations summarized in table 1.2. Obviously, also, any significant temperature excess of the cloud-top over the environment would promote the initiation of the kind of mixing considered here, so that this mechanism would also help to keep the cloud lapse-rate close to that of the environment, as is observed.

It is evident that this theory leads to a picture of the final cloud structure little different from that resulting from the bubble theory; in both cases, the cloud consists of an intimate mixture of moist air from below condensation level with dry air which originally occupied approximately the space now filled with cloud. The significant difference is that the downdraught mechanism assigns an active, penetrative role to the dry air, evaporation being called upon to play a part in the dynamics of mixing just as condensation plays a dynamical part in cloud growth, with the consequence that, except in a saturated environment, such as occurs in tropical cyclones, the liquid water content of convective clouds tends to be self-limiting.

CHAPTER 2

THE MICROPHYSICS OF CLOUDS

INTRODUCTION

In chapter 1 we discussed some of those features of clouds which are immediately apparent to an observer, features ranging in scale from a few metres to a few kilometres and encompassing many of the important macroscopic attributes of clouds.

From this starting-point it is possible to proceed in two directions; towards a study of cloud systems and of the general circulation patterns which give rise to them on the one hand, or towards an investigation of the individual droplets, crystals and dust-particles which make up an individual cloud on the other. The first field belongs essentially to the province of meteorology, while the second is more akin to the fields of physics and chemistry.

It is this second, or microphysical approach which we shall adopt in this book, the overall programme being to explain the behaviour of a cloud as a macroscopic entity in terms of interactions between the microscopic entities which compose it. It must be admitted at the outset that this programme is still far from complete, but as we shall see it is sufficiently advanced in most aspects to give at any rate a semi-quantitative description of the processes involved.

In a detailed study of interactions between microscopic entities in the atmosphere it is all too easy to lose sight of the general pattern generated by these interactions through concentration on the details of a single process. In order to avoid this as far as possible, the present chapter is devoted to a general survey of the microphysical processes involved in cloud formation and evolution. With this general pattern in mind the detailed studies of the various interactions which follow can then be placed in proper perspective and related to the development of the cloud as a whole.

CONDENSATION

The existence of clouds in the atmosphere is due, as we have seen, to the fact that when moist air is expanded adiabatically its relative humidity is increased. Such approximate adiabatic expansions are common in the atmosphere when air moves from one altitude to another, and in an uplift of sufficient magnitude the relative humidity may be expected to exceed saturation. This does not, however, occur, and once saturation is exceeded by an almost negligible amount condensation begins.

Condensation is a process which does not occur easily in a pure environment. For pure water-vapour at room temperature the vapour-pressure must be about four times its saturation value before appreciable condensation occurs. However, the atmosphere is not a pure environment but contains numerous small dust-particles which may be neutral or electrically charged, droplets of various solutions and soluble crystals. Upon this suspended material condensation occurs readily at super-saturations ranging from a few hundredths of one per cent up to some tens of per cent.

After the basic physics of this condensation process has been discussed the study must extend into the atmosphere itself to determine the origins and characteristics of these 'condensation nuclei' suspended in it. Various instruments have been designed to measure the total concentration of condensation nuclei, and more recently it has been found possible to determine an activity spectrum giving the number of nuclei active below a certain supersaturation level. On the basis of such measurements air-masses can be classified into various types and it is found that the nucleus content is closely related to the recent history of the air-mass considered.

When an air-mass is lifted above the condensation level it becomes supersaturated and the most active condensation nuclei begin to grow into droplets. The growth of these droplets tends to oppose further increase in the supersaturation, which passes through a maximum at a value usually less than one per cent and then decreases. The number of droplets in the cloud thus formed depends upon the number of nuclei which have been activated and hence on the maximum value of the supersaturation. This in turn depends upon the nucleus spectrum, the temperature and the uplift speed, and variations in these parameters are the cause of most of the variations in the droplet populations of clouds.

A study of this condensation process in the atmosphere is clearly basic to our understanding of cloud microphysics, and for this reason considerable space has been devoted to its development in the chapters which follow. Unsolved problems still remain, but the basis of our understanding seems firm and many observed cloud structures are fairly well described.

SUBLIMATION AND FREEZING

As a cloud rises through the atmosphere it is cooled by its nearly adiabatic expansion and its summit temperature may often fall below the freezing-point. We are led then to consider the ice-phase as another important feature of the development of clouds.

Just as a pure vapour when made supersaturated does not condense spontaneously until a very large supersaturation is reached, so a pure liquid will not freeze spontaneously until supercooled well below its equilibrium freezing-temperature. In the case of water the supercooling required for such homogeneous freezing is in the neighbourhood of 40° C. Again, as in the case of condensation, suspended particles may act as nuclei for the freezing-process and if such nuclei are present freezing may occur with only a few degrees of supercooling.

There is another way in which ice-crystals can form in the air, that is by direct sublimation from the vapour in the same way as droplets are formed by condensation. This can only occur spontaneously under very special conditions, but when suitable nuclei are present it is probably an important process. Since sublimation is so closely related to condensation they are often competing processes and close consideration is necessary to determine the conditions under which one or the other may occur.

Ice-forming nuclei (leaving aside the question of whether they act by sublimation or freezing) occur naturally in the atmosphere, but their origin and composition are not yet clearly established. In comparison with condensation nuclei which exist in hundreds per cubic centimetre they are rare, and nuclei producing ice-crystals at temperatures warmer than $-20°$ C typically occur only in concentrations of the order of one per litre. For this reason ice-crystals do not usually occur in appreciable numbers in cloud-tops until their temperature has fallen to about $-20°$ C.

STABILITY OF CLOUDS

A cloud, as formed, is an assembly of tiny droplets numbering perhaps one hundred or so per cubic centimetre and having radii of about 10 μ. This structure is remarkably stable as a rule, the droplets showing little tendency to come together or to change their sizes except by a general growth of the whole population.

For a long time this stability was ignored, and it was assumed in general terms that continued condensation would lead to the development of precipitation by some mechanism which was not discussed. It is now recognized that the stability of clouds is a question of fundamental importance to cloud physics. A study of the means by which instability and ultimately precipitation develops has uncovered many important features of cloud structures, as well as paving the way for man's first scientific attempts to influence the weather.

There are two different mechanisms by which the microstructure of a cloud may become unstable. The first mechanism involves the direct collision and coalescence of water droplets and may be important in any cloud. The second involves the interaction between water droplets and ice-crystals and is confined to those clouds whose tops penetrate above the freezing-level.

Consideration of the relative motion of two small droplets involves complex aerodynamic analysis and it is only recently that accurate solutions have been obtained. Very small droplets are unable to collide with each other, no matter what their original trajectories. However, when the radius of one droplet exceeds about 18 μ, collisions with a limited range of smaller droplets become possible; for larger drops the collision efficiency increases sharply.

It is thus to be expected that clouds containing negligible numbers of droplets larger than 18 μ will prove stable as far as coalescence is concerned, whilst clouds containing appreciable numbers of large droplets may develop precipitation. This critical radius lies within the range of normal large cloud drop sizes and clouds belonging to both categories exist. It is found that their behaviour is in fact as predicted, and it is possible to trace theoretically the expected development of precipitation. Whilst many features of this discussion remain to be put on a more firm quantitative basis, the general behaviour predicted is in good agreement with experiment.

When an ice-crystal exists in the presence of a large number of supercooled water droplets the situation is immediately unstable. The vapour-pressure over ice is less than that over water (by about one per cent for every degree below freezing) and as a consequence the water droplets tend to evaporate whilst the ice-crystal grows. This direct vapour transfer is most efficient at temperatures near $-15°$ C, where the absolute vapour-pressure difference is greatest. Growth is most rapid when the ice-crystal is small because the diffusion gradient is then very sharp; as growth continues the growth-rate decreases.

Once the ice-crystal has grown appreciably larger than the water droplets, however, it begins to fall relative to them and collisions become possible. If these collisions are primarily with other ice-crystals then snowflakes form, whilst if water droplets are collected graupel (sleet) or hail may result. Once the ice-structure falls below the freezing-level melting may occur, and upon emerging from the cloud base the raindrop may be indistinguishable from one formed by coalescence. In cold weather on the other hand, or when large hailstones are involved, the precipitation may reach the ground unmelted. Again this process is fairly well understood in essence, but many important aspects remain to be examined in detail, and many quantitative details are still undetermined.

Because these two mechanisms of precipitation development differ greatly in their early stages they have been treated separately in the discussion to follow. It should be remembered, however, that they are merely two different forms of instability and, though it is probably not very usual, both mechanisms may be operative to comparable extents in the one cloud at the same time.

INDUCED INSTABILITY

Once man has attained some detailed knowledge of some phenomenon of practical importance, the obvious step is to try to exert some measure of control over it. The physics of rainclouds is a field where this has happened. In the paragraphs above we have outlined those topics in cloud microphysics which will concern us in this book, and shown that we have at least a semi-quantitative understanding of most of the processes involved in the growth of clouds and in the development of precipitation in them.

Perhaps the most important area of study concerns the stability of clouds and the processes which tend to destroy this stability in nature. Once this is understood the next step is to anticipate natural effects and to endeavour to provoke instability in apparently stable clouds. Armed with the knowledge gained from close study of cloud microstructures this has proved possible, though the overall effects of such intervention are still cloaked to some extent in statistical uncertainty. A review of the methods used and progress achieved in this field will bring our discussion to a close.

CHAPTER 3

THE CONDENSATION OF WATER-VAPOUR

PHASE CHANGES

In cloud physics we are concerned with three principal changes of phase—the condensation of water-vapour to form droplets, the freezing of water droplets to form ice-crystals and the direct formation of ice-crystals from the vapour by sublimation. These three phase-changes have an important property which is characteristic of phase-changes in any material whatever, namely that they do not begin in a continuous manner, but require nucleation.

Nucleation processes can be divided into two main classes. Suppose we have a quantity of pure water-vapour and cause it to become supersaturated either by cooling it, or by compressing it isothermally. To attain the state of lowest energy the vapour should now condense, but this condensation process must commence with the formation of small droplets, and these are unstable under the buffetings of thermal agitation. The vapour thus remains in its metastable gaseous state and no condensation takes place. As the supersaturation is increased small droplets become more stable, and the likelihood of their survival increases, until at a certain critical supersaturation the survival-rate becomes appreciable and condensation to a fog takes place. Such a nucleation process occurring within the bulk of a pure substance is referred to as homogeneous nucleation, and is characterized by the rather high degree of supersaturation required for it to occur.

On the other hand, if the supersaturated vapour is confined in some sort of container, then the walls of this container may provide sites where condensation occurs more easily than in the bulk of the vapour. Similarly, if the vapour is contaminated with small particles of airborne dust these may serve as nuclei for the condensation of water droplets. In either case, where condensation occurs on foreign substances, the nucleation is called heterogeneous. Because heterogeneous nucleation generally requires much less supersaturation than does homogeneous nucleation, and because in most situations with which we deal many foreign particles are present, we shall mostly concern ourselves with nucleation by these particles.

In the present chapter we shall discuss the formation of water droplets from a supersaturated vapour, and later we shall deal with the formation of ice-crystals. A brief discussion of homogeneous nucleation will be given first because it makes more explicit some of the ideas involved, and then we shall turn our attention to foreign particles.

Before becoming involved in the details of nucleation, it may be as well to say a few words about the thermodynamics of phase-changes in general. We shall not go into details since excellent treatments may be found in the many standard texts on chemical thermodynamics (for example, Glasstone, 1947, ch. 10).

Phase-equilibria and changes of phase may be conveniently discussed with the help of two thermodynamic free-energy functions. The first of these, which is called variously the Helmholtz free energy, the work function, or simply the free energy, is defined by the equation

$$F = U - TS, \qquad (3.1)$$

where U is the internal energy, T the absolute temperature and S the entropy of the system under consideration. Similarly, the function G called the Gibbs free energy, the Gibbs function or again simply the free energy† is defined by

$$G = U - TS + pV, \qquad (3.2)$$

where p is the pressure and V the volume of the system. From the first law of thermodynamics we can derive differential forms of (3.1) and (3.2) for a system in equilibrium, and these are

$$dF = -SdT - pdV; \qquad (3.3)$$

$$dG = -SdT + Vdp. \qquad (3.4)$$

From these equations it follows that if a system is in equilibrium at constant temperature and volume, then F is stationary, and

† The symbols used for these quantities, as well as their names, are unfortunately rather confused. Instead of F for the Helmholtz free energy we sometimes find A or ψ, and instead of G for the Gibbs free energy American authors often use F.

Phase Changes

can in fact be shown to be a minimum. Similarly, for a system in equilibrium at constant temperature and pressure, G is a minimum. In any spontaneous process therefore, since the final state must be nearer to equilibrium than is the initial state, F or G, whichever is appropriate, must decrease.

In the systems we shall be considering the pressure is usually held constant, so that we shall only be concerned with the Gibbs free energy G, which we shall usually simply call the free energy.

One final thermodynamic concept is required for our later use. Suppose that a system contains n_1 molecules in phase 1, n_2 molecules in phase 2, etc., and that the total free energy is G. Then the partial molar quantities

$$\mu_i = \left(\frac{\partial G}{\partial n_i}\right)_{T,\,p,\,n_j}, \qquad (3.5)$$

which are just the free energies of single molecules in the various phases, are called the chemical potentials of those phases. If two phases i and j are in equilibrium, then from the fact that $dG = 0$ it follows that

$$\mu_i = \mu_j. \qquad (3.6)$$

Thus, for example, the chemical potential of a molecule in the saturated vapour over a plane liquid surface is the same as that of a molecule in the liquid.

We shall use these thermodynamic ideas constantly in the discussion which follows.

HOMOGENEOUS NUCLEATION OF WATER-VAPOUR CONDENSATION

In any phase near the transition point there are fluctuations in density and structure which Frenkel (1939, 1946) has discussed under the descriptions 'homophase' and 'heterophase' fluctuations. A homophase fluctuation is a microscopic variation in parameters like density, temperature and pressure, and concerns us only indirectly. A heterophase fluctuation, on the other hand, is a microscopic variation in structure corresponding to the transient appearance of an embryo of the neighbouring phase. It is from a study of these fluctuations that an understanding of the nucleation process can be gained.

In this section we shall discuss the nucleation process in the absence of any foreign particles. This homogeneous nucleation process is not, of itself, of any interest from the point of view of cloud physics, but we shall discuss it in detail for two reasons. The first of these is that the basic processes are the same, whether or not foreign particles are present, and the basic physical mechanisms are more easily understood for the simpler homogeneous case. The second reason is that all the concepts and explicit results developed for homogeneous nucleation carry over to the heterogeneous case with only minor modification, so that our detailed examination will be worth while.

Returning now explicitly to consideration of the condensation of water-vapour, let μ_V denote the free energy of a water molecule in the vapour. This free energy will depend upon the pressure and temperature in a way which we shall investigate presently. Similarly, let μ_L denote the free energy of a water molecule in the liquid state. Suppose now that within the vapour-phase molecules come together to form a liquid droplet of volume V and surface area A. Then the free energy of the whole system has been increased by an amount

$$\Delta G = n_L(\mu_L - \mu_V)V + \sigma_{LV}A, \qquad (3.7)$$

where n_L is the number of molecules per unit volume of liquid and σ_{LV} is the interfacial free energy per unit area between liquid and vapour. This quantity σ_{LV} is, in a liquid, equivalent to the surface-tension, and we shall for the moment neglect any possible variation with surface-curvature.

Now consider the form of μ_L and μ_V. If two phases are in equilibrium, then the chemical potential is the same for molecules in each phase. In the present case a liquid will be in equilibrium with a saturated vapour with pressure p_∞ over a plane liquid surface. The vapour we are considering is at an arbitrary pressure p, so that

$$\mu_L - \mu_V = \mu_V(p_\infty) - \mu_V(p). \qquad (3.8)$$

From (3.4) applied to a single molecule, it follows that for an isothermal process

$$d\mu_V = V dp, \qquad (3.9)$$

where V is the volume occupied by one vapour molecule at pressure p and temperature T. Assuming that the vapour

Homogeneous Nucleation of Water-vapour Condensation 41

behaves as a perfect gas, (3.9) can be immediately integrated to give

$$\mu_L - \mu_V = -kT\ln(p/p_\infty), \qquad (3.10)$$

where k is Boltzmann's constant, and we shall call p/p_∞ the saturation ratio. It can immediately be seen that the free-energy change $\mu_L - \mu_V$ is positive when the vapour is unsaturated and negative for a supersaturated vapour.

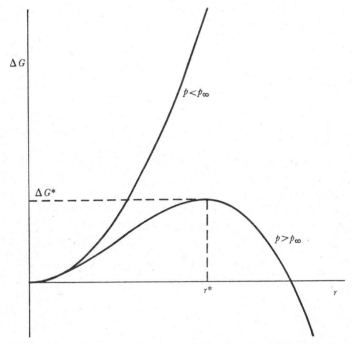

Fig. 3.1. The free energy ΔG required to form a liquid droplet of radius r in a vapour with partial pressure p.

If we assume that the embryo droplet is spherical, then V and A in (3.7) can be written in terms of its radius r, and using (3.10) we get

$$\Delta G = -\tfrac{4}{3}\pi r^3 n_L kT \ln(p/p_\infty) + 4\pi r^2 \sigma, \qquad (3.11)$$

where for simplicity we have dropped the subscripts on σ.

The variation of ΔG with r is shown in fig. 3.1 for the cases of both an unsaturated and a supersaturated vapour. Looking first at the unsaturated case $p < p_\infty$, we see that the free energy required to form droplets increases sharply with their size. Because of thermal fluctuations we expect the number of droplets of a given size to be given by a Boltzmann distribution. If we assume that the total number of embryos is negligibly small in comparison with the number of unassociated molecules, then this distribution is given by

$$n(r) \approx n(1)\exp[-\Delta G(r)/kT], \qquad (3.12)$$

where $n(r)$ is the number of embryos of radius r, and $n(1)$ is the number of unassociated molecules per unit volume. This distribution is shown in fig. 3.2. It should be noted that even though the vapour is unsaturated there are appreciable numbers of small water droplet embryos present. This is the significance of the heterophase fluctuation phenomenon discussed by Frenkel. The distribution is in dynamic equilibrium, and there is no tendency towards condensation, because large droplets are unstable.

The case of greatest practical interest is, of course, that of supersaturated vapour. From fig. 3.1 we see that for $p > p_\infty$ the free energy involved in the creation of an embryo has a maximum value ΔG^* at a radius r^*. Embryos with radii $r < r^*$ are unstable and tend to disappear under thermal agitation. On the other hand, embryos which have exceeded the critical radius r^* tend to grow without limit, and become macroscopic droplets. Droplet embryos having the critical radius r^* are in unstable equilibrium with the supersaturated vapour.

To find the value of r^* we set $\partial \Delta G/\partial r = 0$ in (3.11), and this immediately gives

$$r^* = \frac{2\sigma}{n_L kT \ln(p/p_\infty)}. \qquad (3.13)$$

Since droplets of radius r^* are in equilibrium with vapour of pressure p, (3.13) can be rewritten to yield the vapour-pressure over a curved surface in the form

$$\ln(p/p_\infty) = \frac{2\sigma}{n_L kTr^*}, \qquad (3.14)$$

which is Kelvin's classical formula† (Kelvin, 1870; Gibbs, 1928).

† Also sometimes called the Thomson or Gibbs–Thomson formula.

Homogeneous Nucleation of Water-vapour Condensation

Substituting (3.13) back into (3.11), the critical free energy can be written

$$\Delta G^* = \tfrac{4}{3}\pi r^{*2}\sigma \qquad (3.15)$$

or

$$\Delta G^* = \frac{16\pi\sigma^3}{3[n_{\rm L}kT\ln(p/p_\infty)]^2}. \qquad (3.16)$$

It is now, of course, our main interest to evaluate the rate at which embryos of this critical size are generated within the vapour. This calculation was first made by Volmer and Weber (1926), and their approach was subsequently improved by Becker and Döring (1935) and by more recent workers. We shall not go into the argument in detail, but sketch its outline.

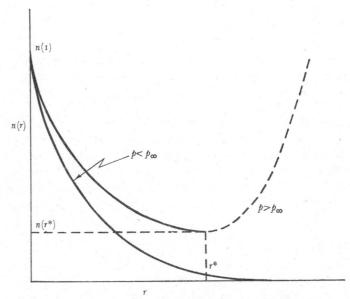

Fig. 3.2. The concentration $n(r)$ of embryo droplets of radius r in a vapour with partial pressure p.

If we try to set up an equilibrium distribution of embryos as we did for the unsaturated case, then we have the situation shown by the curve for $p > p_\infty$ in fig. 3.2. For $r > r^*$ the population of embryos increases without limit, corresponding to growth of the condensed phase, and clearly violates the initial assumptions of the model. However, since we are interested in the onset of nucleation as supersaturation is increased, a reasonable model may be constructed by breaking off the distribution at $r = r^*$ and supposing all embryos which grow past this point to be broken up and returned to the system as individual molecules. In this way the system is maintained in a steady state, with a steady flux of embryos growing through all values of r less than r^*.

Substitution of this steady-state model for a true equilibrium system has some effect on the distribution of embryos $n(r)$ for $r \leqslant r^*$, but to a first approximation we may take the value of $n(r^*)$ to remain unaltered at

$$n(r^*) \approx n(1)\exp(-\Delta G^*/kT), \qquad (3.17)$$

and the flux of embryos through the value r^* is then

$$J \approx Bn(1)\exp(-\Delta G^*/kT), \qquad (3.18)$$

where B is the net rate at which an embryo of critical size gains one molecule through collisions with surrounding vapour molecules.

Volmer and Weber (1926) obtained an approximate value for B by neglecting the evaporation process, and computing the rate at which molecules of the vapour collide with a critical embryo on the basis of kinetic theory. The collision rate, per unit area, can be shown to be $p/(2\pi mkT)^{\frac{1}{2}}$, where p is the pressure of the vapour and m the mass of a vapour molecule. B is therefore $4\pi r^{*2}$ times this quantity, so that (3.18) becomes

$$J \approx [p/(2\pi mkT)^{\frac{1}{2}}]4\pi r^{*2}n(1)\exp(-\Delta G^*/kT). \qquad (3.19)$$

Becker and Döring (1935) included the evaporation process in their calculation and obtained an equation differing from (3.19) by a factor

$$\left[\frac{\ln(p/p_\infty)}{8r^{*3}n_\text{L}}\right]^{\frac{1}{2}}, \qquad (3.20)$$

which is of order 10^{-1}. Including this correction the coefficient $Bn(1)$ of (3.18) has the approximate value 10^{25} cm^{-3} sec^{-1} for water-vapour under the sort of conditions we are considering. A more accurate value is not required since, as we shall see later, the nucleation process is not at all sensitive to the exact value of B.

Before looking at some of the practical implications of this treatment we should note some of the weaknesses of the theory.

Homogeneous Nucleation of Water-vapour Condensation

These are caused almost entirely by the fact that the critical embryo, which is the basis of the theory, typically consists of less than 100 molecules. A cluster of this size is too small to be approximated well by a spherical droplet, and yet too large for its molecular configurations to be considered in detail. The division of free energy into bulk and surface terms is thus somewhat ambiguous, and the validity of using macroscopic values of surface free energy may be questioned. These points have been discussed by Buff and Kirkwood (1950) and by Benson and Shuttleworth (1951) who conclude that the procedure is probably not greatly in error even for droplets containing only a few tens of molecules. More recently, Reiss (1952) has attempted to formulate a new approach to the whole calculation on the basis of statistical mechanics, but this is not yet complete.

PRACTICAL IMPLICATIONS—COMPARISON WITH EXPERIMENT

To calculate the nucleation behaviour of supersaturated water-vapour it is simply necessary to substitute appropriate numerical values into the general formulae (3.16) and (3.18). At 0° C, $\sigma = 75 \cdot 6 \text{ erg cm}^{-2}$ and $n_\text{L} = 3 \cdot 3 \times 10^{22} \text{ cm}^{-3}$. The kinetic coefficient $Bn(1)$ has the approximate value $10^{25} \text{ cm}^{-3} \text{ sec}^{-1}$ and we may take an appreciable nucleation rate to be say $J \approx 1$ droplet per cm³ per sec. Equations (3.16) and (3.18) then show that the saturation ratio required for this nucleation rate is 4·4. Fig. 3.3 shows the variation of nucleation rate with saturation ratio in the vicinity of this critical value. Clearly, had we chosen an 'appreciable' nucleation rate to be as high as 10^3 or as low as 10^{-3} droplets cm⁻³ sec⁻¹, we should still have arrived at a critical saturation ratio between 4·0 and 4·8. A corresponding uncertainty of even a few orders of magnitude in the value of the coefficient B in (3.18) similarly has very little effect upon the final result.

Translated into physical terms, fig. 3.3 implies that as the supersaturation is increased the nucleation rate remains essentially zero until the critical saturation ratio is approached, at which point it becomes very large over an extremely short supersaturation range, and a fog forms in an almost discontinuous manner.

The critical saturation ratio has a considerable temperature variation, both because of the temperature variation of the surface free energy and from the T^3 term in the exponent. Fig. 3.4 shows the calculated variation with temperature over most of the range of interest.

In principle an experimental verification of this behaviour is quite straightforward. Air saturated with water-vapour is made

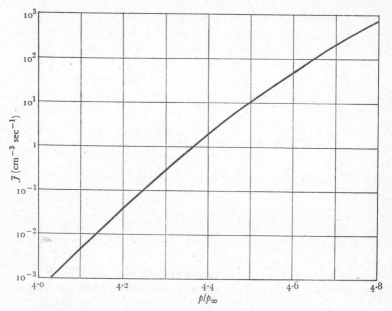

Fig. 3.3. The nucleation rate J as a function of saturation ratio near the critical value.

completely free of foreign particles and ions. It is then expanded adiabatically to give a supersaturated state, and the lowest expansion ratio for which a fog is produced defines the critical saturation ratio.

In practice particles can be removed by a combination of careful filtering and successive expansions, and ions can be removed by a strong electric field. Exact comparison with theory requires, however, that the course of the expansion near its end point be known with considerable accuracy, deviations from true adiabatic behaviour must be allowed for, and the rate

Practical Implications—Comparison with Experiment 47

of droplet production must be estimated to within one or two orders of magnitude. Additional problems arise because the supersaturation due to the expansion is only transient, and because the first droplets to form upset the supersaturation by removal of water-vapour and by release of latent heat. The extent of the agreement between theory and experiment is shown in fig. 3.4. It is evident that whilst the general agreement is quite good, particularly at fairly warm temperatures,

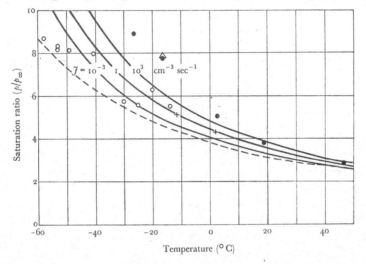

Fig. 3.4. Various experimental determinations of the critical saturation ratio for homogeneous nucleation of water-vapour, compared with theoretical curves for different nucleation rates J. △, Wilson (1897). ●, Powell (1928). +, Volmer and Flood (1934). – – –, Sander and Damköhler (1943). ○, Pound, Madonna and Sciulli (1951).

there is a considerable scatter of results below about $-20°$ C. This scatter reflects the difficulty experienced in deciding exactly what supersaturation has been attained in a given expansion. Pound, Madonna and Sciulli (1951) interpret their results and those of Sander and Damköhler (1943) as implying that the surface-tension of small droplets decreases with increasing surface-curvature, as suggested by the theories of Tolman (1949), of Kirkwood and Buff (1949), and of Benson and Shuttleworth (1951). Powell's results (Powell, 1928) however, suggest the opposite tendency, and the point has yet to be finally settled.

In addition to their work on water, Volmer and Flood (1934) measured homogeneous nucleation thresholds for a variety of other vapours, and found good agreement with theory for all except methyl alcohol. We shall not discuss these experiments in detail since homogeneous nucleation is not of direct importance in cloud physics. The general good agreement of these experiments with the predictions of theory is, however, of great importance. It shows that, despite some fundamental uncertainties about the application of macroscopic concepts to small clusters of molecules, the theory can be relied upon to produce numerical results of sufficient accuracy for most practical purposes. With this assurance we can go on to apply the theory to more complex nucleation problems with reasonable confidence that the results will be not only qualitatively, but even quantitatively correct.

CONDENSATION ON IONS

As we remarked in the introduction to this chapter, condensation may be aided by the presence of ions and foreign particles in the condensable vapour. As the first, but by no means the simplest example of this heterogeneous nucleation we shall discuss condensation upon ions. Wilson (1897, 1899) was among the first to examine this phenomenon in detail. In his classic series of experiments, to which we have already referred, he found that a rain of droplets formed at a saturation ratio of about 4·2, in contrast with the dense fog formed for saturation ratios above 7·9. He attributed the formation of these droplets to the presence of naturally produced ions in the vapour, and subsequently verified this by producing ionization with an X-ray beam. In later experiments he separated the positive and negative ions by means of an electric field, and established that negative ions are active nuclei with $(p/p_\infty)^* \approx 4$ at about $-6°$ C, while positive ions are active with $(p/p_\infty)^* \approx 6$ at a slightly lower temperature.

These experiments have since been repeated by many other workers with substantially the same results. Minor disagreements can be ascribed to differences in the final temperature and in the fog density detectable under the conditions of the experiment. By analogy with the case of homogeneous nucleation, as shown in fig. 3.4, we can expect these variations to have quite an appreciable effect on the observed critical saturation ratios.

No really satisfactory theory has yet been proposed to explain these observations, but that of Tohmfor and Volmer (1938) supplies a basis for further work. We shall discuss this briefly and point out the remaining gaps in our understanding.

Just as the Kelvin equation in the form (3.7) gives a relation between the radius of a droplet and the saturation ratio of a vapour with which it is in unstable equilibrium, so we can find a corresponding equation for the case of a droplet bearing a charge Q. Consider a vapour with partial pressure p, and suppose that it contains an ion of charge Q and radius r_0. Then, if a droplet of radius r and dielectric constant ϵ condenses on the ion, the change in the free energy of the system is, analogously with (3.11),

$$\Delta G = -\tfrac{4}{3}\pi r^3 n_L kT \ln(p/p_\infty) + 4\pi r^2 \sigma + \frac{Q^2}{2}\left(1 - \frac{1}{\epsilon}\right)\left(\frac{1}{r} - \frac{1}{r_0}\right). \tag{3.21}$$

The last term represents the change in the electrostatic energy of the ion, and for simplicity the dielectric constant of the surrounding vapour has been taken as unity.

To find the radius of a droplet in equilibrium with a given supersaturation we require $\partial \Delta G/\partial r = 0$ as before, and obtain the result

$$\ln(p/p_\infty) = \frac{2\sigma}{n_L kTr} - \frac{Q^2}{8\pi n_L kTr^4}\left(1 - \frac{1}{\epsilon}\right), \tag{3.22}$$

which was first given (for a conducting sphere with $\epsilon = \infty$) by J. J. Thomson (1888). Comparing this with the Kelvin equation (3.14), it can be seen that the saturation ratio is smaller in the case of a charged droplet. The dependence of vapour-pressure upon radius is, however, quite different from the case of an uncharged droplet, as is shown in fig. 3.5. Some writers have taken the peak of this curve as defining the critical saturation ratio at which nucleation occurs, but this procedure is no more valid than using the Kelvin equation to predict the onset of homogeneous nucleation (a method which predicts that an infinite saturation ratio is required). To determine the true nucleation threshold, which is considerably lower in both cases, the theory of fluctuations must be used.

Consider a saturation ratio corresponding to the horizontal line which we have drawn on fig. 3.5. The point A is a condition of metastable equilibrium, and all ions in the vapour will take up water molecules and form droplets of this radius. The nucleation process now consists of some of the droplets making a transition, under the influence of statistical fluctuations, from

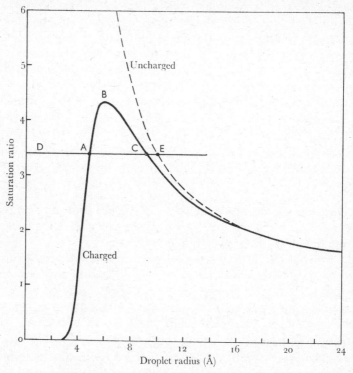

Fig. 3.5. Saturation ratio as a function of radius for charged and uncharged water droplets.

this point A to the state of unstable equilibrium represented by the point C. This process will clearly take place when the line AC lies at some distance below the peak, B, of the curve.

Comparing this process with that for homogeneous nucleation at the same saturation ratio, which is represented by transitions from the point D, giving the radius of a single molecule, to the point E, the critical radius on the Kelvin curve for uncharged droplets, it is obvious that ions assist the nucleation process.

Returning to a calculation of the nucleation rate for ions, the theory involves the critical free energy ΔG^*, which is simply given by (3.21) with r_A corresponding to point A substituted for r_0, and the critical radius r_C corresponding to the point C substituted for r. The nucleation rate is then given as before by an expression like (3.19) with the appropriate ΔG^*, and with N, the number of ions per cm³, substituted for $n(1)$. Tohmfor and Volmer derive a more accurate expression by the method of Becker and Doring and, substituting numerical values, find

$$J \approx 10^7 N \exp(-\Delta G^*/kT) \qquad (3.23)$$

at 265° K, assuming the ions to have unit electronic charge. A calculation along the lines indicated above then gives $r_A = 4.8 \times 10^{-8}$ cm, $r_C = 10.2 \times 10^{-8}$ cm and $(p/p_\infty)^* = 3.2$.

Whilst of the right order of magnitude, this critical saturation ratio is significantly lower than the observed value of about 4·1 at this temperature. Tohmfor and Volmer proposed that this disagreement was due to using a value of 80 for the dielectric constant of water, whereas in the strong field near the ion its value is likely to be much less than this. They suggested a value $\epsilon = 1.85$ as giving agreement with experiment. This dielectric saturation effect is supported in a general way by the work of Debye (1945), but it seems unlikely that ϵ could decrease below the optical value of about 3, and this value is not low enough to give good agreement with experiment.

At this stage, however, the major shortcoming of the theory becomes evident, namely that it makes no distinction between positive and negative ions. This is not meant to imply that the theory is not correct in its general approach, but merely that it is insufficiently detailed. The additional detail required is clearly of a molecular nature. Water is a highly polar liquid, and there is strong evidence that molecules in the surface-layers are preferentially oriented (Henniker, 1949; Good, 1957). In the case of water the preferred dipole direction is with the negative (oxygen) end outwards, and this orientation is preferred in the first two or three layers from the surface. When condensation is on a negative ion, the free energy of these surface dipoles is lowered and condensation is made more easy, whilst the reverse is true of positive ions. There may also be an orientation effect in vapour molecules approaching the surface, caused by the strong electric field (Loeb, Kip and Einarsson, 1938), and this may help or hinder the attachment of these molecules to the embryo droplet. No detailed theories along these lines have yet been proposed, but it appears that at least a partially molecular approach of this nature will be required if the results of experiment are to be understood.

NUCLEATION BY INSOLUBLE SURFACES

In natural condensation processes in the free atmosphere the air is never sufficiently clean for condensation to occur homogeneously, or upon ions. It is, therefore, very important that

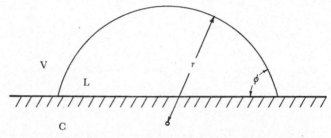

Fig. 3.6. A spherical-cap embryo of liquid (L) in contact with its vapour (V) and a nucleating surface (C).

we know something about the behaviour of small dust-particles, hygroscopic crystals and other naturally occurring contaminants as condensation nuclei. In this section we shall discuss the basic theory of nucleation by foreign surfaces, which leads naturally to a treatment of small insoluble particles. We shall then examine the properties of small soluble particles.

Volmer (1939, p. 100) developed a modification of the original Volmer–Weber nucleation theory to include the case of condensation on a plane surface, and this approach, like the original theory, has been the basis of all the work which has followed. Because this is the simplest case, we shall first discuss nucleation by plane surfaces, and then go on to consider the effect of particle-size and surface-structure.

The geometry of a small droplet of liquid in contact with a solid surface can be characterized by two parameters. The droplet takes up the form of a spherical cap as shown in fig. 3.6 and the two parameters are r, the radius of curvature of

Nucleation by Insoluble Surfaces

the spherical surface, and ϕ, the angle of contact of the liquid on the substrate. For convenience we denote the liquid droplet by subscript L, the catalysing substrate by C, and the vapour by V. Then from the mechanical equilibrium of the line common to these three phases we have the following relation

$$\mathfrak{m} \equiv \cos\phi = (\sigma_{CV} - \sigma_{CL})/\sigma_{LV}. \quad (3.24)$$

Suppose that an embryo of this shape forms from a vapour, then, as before, the change in free energy of the system is

$$\Delta G = n_L(\mu_L - \mu_V)V_L + \sigma_{LV}A_{LV} + (\sigma_{CL} - \sigma_{CV})A_{CL}, \quad (3.25)$$

which is just a generalization of (3.7). If the radius of curvature of the liquid surface is r, then we can express the volumes and areas in terms of r and $\mathfrak{m}$ as

$$\left.\begin{aligned} A_{LV} &= 2\pi r^2(1-\mathfrak{m}), \\ A_{CL} &= \pi r^2(1-\mathfrak{m}^2), \\ V_L &= \frac{\pi r^3}{3}(2+\mathfrak{m})(1-\mathfrak{m})^2. \end{aligned}\right\} \quad (3.26)$$

The size of the critical embryo can now be determined as before by setting $\partial \Delta G/\partial r = 0$, which leads to the result

$$r^* = \frac{2\sigma_{LV}}{n_L kT \ln(p/p_\infty)}. \quad (3.27)$$

This is exactly the same result as for homogeneous nucleation (3.13), which is to be expected, since the curved surface of the critical embryo must be in equilibrium with the vapour. Substituting (3.26) and (3.27) in (3.25), the critical free energy is found to be

$$\Delta G^* = \frac{16\pi \sigma_{LV}^3}{3[n_L kT \ln(p/p_\infty)]^2} f(\mathfrak{m}), \quad (3.28)$$

where $\quad f(\mathfrak{m}) = (2+\mathfrak{m})(1-\mathfrak{m})^2/4. \quad (3.29)$

The result (3.28) differs from that for the homogeneous case (3.16), only by the factor $f(\mathfrak{m})$. Since $-1 \leqslant \mathfrak{m} \leqslant 1$, (3.29) shows that $0 \leqslant f(\mathfrak{m}) \leqslant 1$ so that foreign surfaces have the general property of reducing the free energy necessary to form a critical embryo.

Calculation of the nucleation rate follows the same lines as before. If $n'(1)$ is the number of single molecules adsorbed per unit of surface-area, then the number of critical embryos is

$$n(r^*) = n'(1)\exp(-\Delta G^*/kT). \qquad (3.30)$$

To a reasonable approximation the surface-area of a critical embryo on a plane substrate is πr^{*2}, and the rate of impact, per unit area, of molecules from the vapour is $p/(2\pi mkT)^{\frac{1}{2}}$. The nucleation rate per unit area of catalyst surface is thus

$$J \approx [p/(2\pi mkT)^{\frac{1}{2}}]\pi r^{*2} n'(1)\exp(-\Delta G^*/kT), \qquad (3.31)$$

with ΔG^* given by (3.28) and (3.29).

The kinetics of this nucleation process have recently been considered by Pound, Simnad and Yang (1954). They develop a detailed expression for $n'(1)$, the density of adsorbed molecules on the nucleating surface, and consider the addition of molecules to growing embryos to take place primarily by diffusion of adsorbed molecules from the surrounding surface, rather than directly from the vapour. The additional factors introduced by this mechanism are estimated to increase the rate given by (3.31) by a factor between 10^2 and 10^3 in typical cases.

The complete kinetic coefficient of (3.31) clearly depends on the vapour-pressure p and upon the nature of the adsorption on the nucleating surface. If we assume that the adsorption is an appreciable fraction of a monolayer at the supersaturations involved, then near 0° C the kinetic coefficient is of order 10^{24} to 10^{27} cm^{-2} sec^{-1} when an additional correction similar to that of Becker and Döring has been made.

Inserting this and other appropriate numerical values into (3.31), the saturation ratio for a given nucleation rate may be determined as a function of contact angle. It should be noted that the contact angle may depend to some extent upon the saturation ratio, and this must be taken into account in any comparison with experiment. The saturation ratio for a nucleation rate of 1 droplet per cm² per sec at 20° C as a function of contact angle is shown in fig. 3.7. Confirmation of this curve has recently been found by Twomey (1959d), using glass surfaces covered with films of more or less hydrophobic substances in a continuous cloud chamber which we shall describe presently. It should be pointed out that for $\phi = 0$ the surface is wet by water, and behaves as a plane water surface, so far as

Nucleation by Insoluble Surfaces

condensation is concerned, whilst for $\phi = 180°$ there is no condensation on the surface, and nucleation is effectively homogeneous.

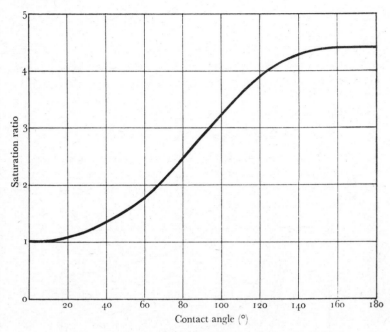

Fig. 3.7. Critical saturation ratio for $J = 1$ cm^{-2} sec^{-1} upon a plane substrate of given contact angle, at 0° C.

NUCLEATION BY INSOLUBLE PARTICLES

A further extension of the Volmer–Weber theory has been made recently by Fletcher (1958a) who has included the effects of particle-size. If the nucleating particle is considered to be a sphere of radius r then the critical free energy can be evaluated just as before, except that the geometry is rather more complicated. The final result is

$$\Delta G^* = \frac{16\pi\sigma_{LV}^3}{3[n_L kT \ln(p/p_\infty)]^2} f(m, x), \qquad (3.32)$$

where

$$x = \frac{r}{r^*} = \frac{rn_L kT \ln(p/p_\infty)}{2\sigma_{LV}} \qquad (3.33)$$

and
$$f(m,x) = \tfrac{1}{2}\left\{ 1 + \left[\frac{1-mx}{g}\right]^3 + x^3\left[2 - 3\left(\frac{x-m}{g}\right) + \left(\frac{x-m}{g}\right)^3\right] \right.$$
$$\left. + 3mx^2\left[\frac{x-m}{g} - 1\right] \right\} \quad (3.34)$$
with
$$g = (1 + x^2 - 2mx)^{\tfrac{1}{2}}. \quad (3.35)$$

If one is interested in the nucleation rate per particle, then the kinetic coefficient is simply proportional to particle area to a first approximation. The nucleation rate per particle is then, analogously to (3.31),
$$J \approx [p/(2\pi mkT)^{\tfrac{1}{2}}] 4\pi^2 r^2 r^{*2} n'(1) \exp(-\Delta G^*/kT). \quad (3.36)$$
The pre-exponential factor in this case has the approximate value $10^{25}\, 4\pi r^2$. This approximation breaks down if $r < r^*$, in which case it is better to substitute r^* for r, but this is of very minor practical importance.

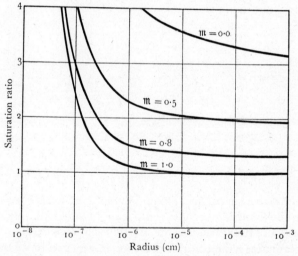

Fig. 3.8. Critical saturation ratio for nucleation of a water droplet in one second upon a particle of given radius and with surface properties defined by $m \equiv \cos\phi$ (Fletcher, 1958a).

These expressions have been used by Fletcher to derive curves showing the nucleation behaviour of insoluble particles as a function of radius and contact angle. These curves are reproduced in fig. 3.8, from which it can be seen that in order to be an efficient condensation nucleus a particle must not only have a reasonably large diameter, but must also have a small contact angle for water. This fact has been overlooked by many writers who simply use the Kelvin equation (3.14) to determine the nucleation efficiency of small particles. This corresponds approximately to considering just the curve for $m = 1$, which is clearly not applicable to particles which exhibit a finite contact angle.

No detailed experimental confirmation of these curves has yet been produced. Experimental confirmation of Kelvin's equation for water droplets (La Mer and Gruen, 1952) is, however, indirect confirmation of the curve for $m = 1$, whilst Twomey's experiments (Twomey, 1959d) on essentially flat surfaces verify the dependence on m for large values of r. An experimental determination of the efficiency of silver iodide smoke as a condensation nucleus is also in agreement with the predictions of the theory based on measurements of contact angle on large particles (Fletcher, 1959a).

THE EFFECTS OF SURFACE STRUCTURE

We have so far dealt only with ideal plane or spherical surfaces, and it is clear that real surfaces will possess imperfections of various kinds. If we exclude chemical impurities, the most important imperfections from the point of view of condensation are pits and steps in the surface. Turnbull (1950) has examined the retention of embryos in conical cavities, and found that once a condensed phase has been formed in such a cavity it may be retained even under unsaturated conditions. The nucleation threshold depends in detail upon the geometry of the cavity, but is always lower than the supersaturation required for nucleation upon a plane or convex surface.

Similar observations apply to steps on otherwise perfect surfaces. Nucleation will occur at a lower supersaturation than required for a plane surface, and the embryo may grow to some extent as a fillet along the step. If the contact angle is small enough and the angle of the step sufficiently acute that the embryo has a concave surface, then the embryo may be retained under unsaturated conditions as for a conical cavity.

These remarks apply to imperfections which are large in

comparison with a critical embryo. Very small-scale features such as the growth steps produced by the emergence of screw dislocations on crystal faces have a much less important effect. The steps in this case may provide favoured nucleation sites, but will not greatly alter the critical saturation ratio. We shall discuss some of these topics in more detail in connexion with the nucleation and growth of ice-crystals, where their importance is much greater.

NUCLEATION BY SOLUBLE PARTICLES

The nucleation of water droplets by small soluble particles is in many ways similar to nucleation by ions, but there is an important difference in that the droplets involved are so large that it is a valid approximation to neglect the effects of statistical fluctuations and treat the droplets from a macroscopic equilibrium point of view. The exact behaviour of an individual particle then becomes completely predictable, and we should perhaps use the phrase 'droplet growth' instead of nucleation.

A discussion of this growth process was given as long ago as 1921 by Köhler (1921a, b; 1925a; 1926) and later developed by Wright (1936). The basis of the growth process is the fact that the water vapour-pressure is less over an aqueous solution than it is over pure water. Water therefore tends to condense towards the solution, and this effect is able to balance to some extent the increase in vapour-pressure caused by the surface curvature of small droplets.

Raoult's law describes the behaviour of the vapour-pressure of solutions through the equation

$$p'/p = \mathbf{m}, \qquad (3.37)$$

where p' is the vapour-pressure over the plane surface of a solution containing a mole fraction $\mathbf{m}$ of water, and p is the vapour-pressure over pure water. In calculating the mole fraction care must be taken to include the van't Hoff factor i, which takes account of the dissociation of inorganic salts. This factor is not a constant, but varies to some extent with concentration (McDonald, 1953b).

Consider a mass m of a substance of gram-molecular weight M dissolved in water of gram-molecular weight M_0, to form a

Nucleation by Soluble Particles

droplet of solution of radius r and density ρ'. Allowing for the van't Hoff factor i, this droplet contains effectively im/M moles of solute and $(\frac{4}{3}\pi r^3 \rho' - m)/M_0$ moles of water. The mole fraction of water is thus

$$\mathbf{m} = \left[1 + \frac{imM_0}{M(\frac{4}{3}\pi r^3 \rho' - m)}\right]^{-1}. \tag{3.38}$$

By (3.37) the vapour-pressure of the droplet is thus decreased from the pure water value by the factor $\mathbf{m}$ given by (3.38).

Now, from the Kelvin equation (3.14), if the vapour-pressure over a flat surface is p_∞, then that over a droplet of radius r is given by

$$p_r/p_\infty = \exp(2\sigma/n_\mathrm{L} kTr), \tag{3.39}$$

where σ is the surface-tension or surface free energy. Combining (3.37), (3.38) and (3.39) we have for the water vapour-pressure over a solution droplet

$$p'_r/p_\infty = \left[\exp\frac{2\sigma'}{n'_\mathrm{L} kTr}\right]\left[1 + \frac{imM_0}{M(\frac{4}{3}\pi r^3 \rho' - m)}\right]^{-1}, \tag{3.40}$$

where the primed quantities all refer to the solution at the concentration involved. This expression differs only in minor details from those given by Köhler and Wright. A more detailed discussion has been given by Dufour and Defay (1953).

A simplified approximate form of (3.40) can be written down for the case when the droplet is a solution sufficiently dilute that m is very much smaller than the mass of water in the droplet. Making this approximation and examining the magnitudes of the terms in an expansion in powers of $1/r$, the dominant terms are found to yield

$$p'_r/p_\infty \approx 1 + \left(\frac{2\sigma'}{n'_\mathrm{L} kT}\right)\frac{1}{r} - \left(\frac{imM_0}{\frac{4}{3}\pi\rho' M}\right)\frac{1}{r^3}. \tag{3.41}$$

Substituting numerical values this can be written

$$p'_r/p_\infty \approx 1 + \frac{a}{r} - \frac{b}{r^3}, \tag{3.42}$$

where $a \approx 3.3 \times 10^{-5}/T$ and $b \approx 4.3 \, im/M$.

If the saturation ratio p'_r/p_∞ for a solution droplet is plotted as a function of radius, a curve is obtained which we shall refer to as a Köhler curve. A set of these curves is shown in fig. 3.9

for various masses of solute. From our point of view it is much more useful to regard these curves as giving the radius of a solution droplet as a function of the saturation ratio. Considered

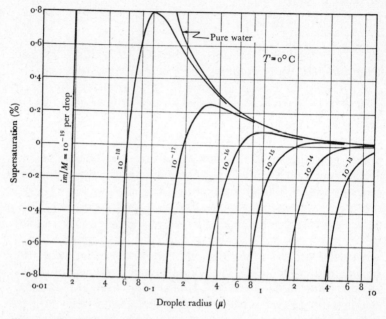

Fig. 3.9. Equilibrium supersaturation as a function of droplet radius for water droplets, each containing mass m of solute of molecular weight M and van't Hoff factor i (after Howell, 1949).

in this way we see that as the saturation ratio is raised, the solution droplet has a definite radius, until a critical value is reached above which the radius increases without limit.

Critical data for a given nucleus are easily calculated from the approximate equation (3.42) which gives

$$(r^*)^2 = 3b/a; \qquad (3.43)$$

$$\left(\frac{p}{p_\infty}\right)^* = 1 + \left(\frac{4a^3}{27b}\right)^{\frac{1}{2}}. \qquad (3.44)$$

The Köhler curves are very similar in shape to the curve shown in fig. 3.5 for a charged water droplet, but it is not necessary to consider the fluctuation mechanism in the present case because of the relatively immense droplet sizes. The peak in the curve for a singly charged droplet occurs for a cluster containing only about 100 molecules, whereas even for a solute particle as small as 10^{-16} g the critical droplet contains about 10^9 molecules. Growth of these solution droplets can therefore be treated as an equilibrium process.

One important phenomenon has been left out of account in deriving these curves. In practice it is not possible to increase the concentration of a solution to an arbitrarily large value. When a critical concentration is reached the crystalline salt begins to separate out, and if at this stage the droplet is in equilibrium with aqueous vapour complete desiccation occurs almost immediately. There may be some hysteresis in this phase change since a nucleation process is involved in the production of the crystalline phase, and a certain amount of supersaturation of the solution is ordinarily required.

Experimental confirmation of this theory for rather large particles has been obtained by several workers using a fine spider web as a particle support. This device was introduced by Dessens (1949) who was able to obtain fibres as thin as 0·01 μ. In these experiments Dessens verified the form of the curves at less than saturation for NaCl and $ZnCl_2$, and found that, instead of crystallizing, these solution droplets could remain liquid to relative humidities as low as 40 %, whereas in the case of NaCl the phase-change is expected at 78 % r.h. A similar effect was found by Junge (1952).

Twomey (1953 b, 1954) has made a detailed study of this phase-transition. He found that whilst considerable supersaturation of the salt could be produced in solution droplets, the phase-change from solid to liquid always took place at a sharply defined and reproducible relative humidity which, with a few exceptions, was in good agreement with the expected value. He therefore proposed this phase-change as a convenient method of identification for large hygroscopic atmospheric particles.

Fig. 3.10 shows typical behaviour for a particle of NaCl, as found by Junge (1952). Note the hysteresis loop due to supersaturation of the solution, the rather large range over which crystallization takes place, and the relatively sharp onset of solution. More recently Orr, Hurd, Hendrix and Junge (1958) have found similar curves for particles in the size range 10^{-6}

to 10^{-5} cm. They observed in addition that phase-transitions occurred at slightly lower relative humidities for very small particles.

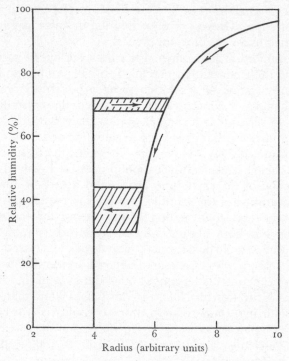

Fig. 3.10. Phase-transition curve for a droplet of NaCl solution (after Junge, 1952).

MIXED NUCLEI

In any real atmospheric situation it is unlikely that the natural nuclei of condensation will be described completely by one of the types we have considered above, though this may well be the case for the purer artificially produced nuclei. The two mixed types of nuclei which we shall consider here are small electrically charged particles (or 'large ions'), and larger particles consisting of a mixture of soluble and insoluble components.

The nucleation efficiency of small charged particles can be calculated by a combination of the methods we have used

Mixed Nuclei

above, but it is scarcely worth while to carry out this rather complicated calculation. From fig. 3.5 it can be seen that if the particle has a radius less than that corresponding to the point A (about 5 Å) it will have no effect on the nucleation efficiency of the ion, which will acquire water molecules until its total radius reaches this value. On the other hand, (3.22) shows that the effect of the electric charge falls off very rapidly, as r^{-4}, with droplet radius. This is borne out by the close approximation between the curves for charged and uncharged droplets in fig. 3.5 for radii greater than 20 Å or so. Thus, if a particle is smaller than about 10 Å in diameter, electrical effects completely determine its nucleation behaviour, whilst if it is greater than a few tens of Ångstroms in diameter, electrical effects can be neglected. Whilst the transition region is of interest in other fields, from the point of view of atmospheric condensation we can dismiss it after this brief discussion.

Of more practical importance are nuclei consisting of a mixture of particles, and we shall discuss in a later chapter the way in which such composite particles may arise by coagulation. Even when all the components of a coagulated nucleus are similar particles, it may possess properties differing from those of a simple particle of the same size. This is due to the fact that its surface will contain cracks and cavities between the particles, and these, as we mentioned before, are favoured condensation sites. Condensation in these crevices effectively reduces the contact angle for water on the particle, and in the limit it may behave as completely wettable, even though its component particles have a finite contact angle for water.

If the nucleus consists of a mixture of soluble and insoluble particles, then nucleation behaviour is dominated by the soluble particles. Above the phase-transition for these components the particle is enveloped in a film of solution, and behaves as a solution droplet in the way we have already considered. The presence of the insoluble components enhances the effectiveness of the soluble salts since less water is required to make a droplet of given size, and the resulting solution is more concentrated. The behaviour of such mixed nuclei has been examined in some detail by Junge (1952). The overall variation of droplet radius with saturation ratio is qualitatively similar to that of pure solution droplets, but differs slightly in magnitude.

CHAPTER 4

CONDENSATION NUCLEI IN THE ATMOSPHERE

INTRODUCTION

In chapter 3 we discussed the condensation of water-vapour from the point of view of the basic physical processes involved. We differentiated between homogeneous nucleation of the vapour within itself and heterogeneous nucleation upon foreign particles, and found that in general the latter process requires the smaller supersaturation.

There are normally large numbers of foreign particles in suspension in the atmosphere, so that we should expect these to be the centres upon which natural condensation occurs, and hence of prime importance in the formation of clouds. This chapter will deal therefore with the material suspended in the atmosphere upon which condensation can take place. We shall first examine the main sources of suspended material in some detail and see in what way they make their contribution to the aerosol. Then after a brief description of some of the experimental methods which have been employed, we shall discuss the results of the many measurements which have been made on the properties of natural aerosols.

SOURCES OF ATMOSPHERIC AEROSOLS

Two main types of process produce particles which may become suspended in the air. The first of these is generally termed a condensation process, and includes, as well as true condensation as we have discussed it, formation of solid particles by sublimation and of all types of products by gas phase-reactions. The second type of process is termed dispersion, and involves the break-up of large particles or drops to form small particles which can remain in suspension in air. We shall discuss in turn particularly important examples of each type of process.

Gas-phase reactions

The origin and concentration of the minor gaseous constituents of the atmosphere are still matters for considerable speculation, and, since the concentrations involved are extremely small, reliable experiments are difficult to perform. It is possible, however, to examine some of the processes which may produce active gases and then to see in what way they may react.

The upper parts of the atmosphere are continually subject to a high intensity of ultraviolet radiation from the sun. This radiation causes ionization of some of the major constituents of the atmosphere and leads to appreciable quantities of ozone together with small amounts of the oxides of nitrogen, NO and NO_2. These reactions are limited to the upper atmosphere because the ultraviolet radiation is rapidly absorbed, and the components which reach the earth's surface are of too long a wavelength to cause ionization.

Ozone itself does not form condensation nuclei but, because of its chemical activity, may react with other substances. An example of this type is a further oxidation of NO or NO_2 in the presence of water-vapour to form nitric acid. It must be remembered, however, that before any such reactions can occur the components must be present in sufficient amounts that the combination reaction proceeds at a greater rate than does subsequent dissociation.

Lightning discharges reproduce these same reactions in the lower atmosphere, with the added activity produced by high temperatures. Since the flash itself is very intense, large initial concentrations of new components may be produced, which facilitates further reactions.

The surface of the earth may yield large quantities of many gases. The decay of animal or vegetable matter releases large amounts of ammonia, while animals in life release carbon dioxide. Plants, particularly at warm temperatures, may release sap components such as terpenes (Went, 1956), whilst hydrocarbons of many kinds are liberated in swamps and marshes.

Photochemical reactions are probably of little importance near the earth's surface, though there is some evidence (Cauer,

1951) that chlorine may be liberated from sea-water droplets in sunlight, and some of Aitken's experiments (Aitken, 1911–12) suggest that sunlight may produce nuclei in air containing gas traces.

Combustion, of course, contributes many gases as well as particles to the atmosphere. We shall treat these in more detail in the next section, but CO_2 and SO_2 can be recognized as appreciable products. Special fuels release unusual combustion products, as for example the iodine liberated during the heating of sea-weed (Cauer, 1951) but we shall not consider these in detail.

Reactions between these components depend, as we stated before, upon there being an adequate concentration of the two reacting components and upon the stability of the reaction product. Possible products of significance from the point of view of condensation nuclei are HNO_2, HNO_3, H_2SO_3 and H_2SO_4 together with their salts with NH_3 and Na. Sodium chloride and other chloride salts are of course important, but they originate primarily from dispersion of sea spray rather than by gas phase-reactions.

Table 4.1. *Surface concentrations of nucleus-forming substances in Europe (Cauer, 1951)*

Substance	Region	Mean ($\mu g/m^3$)	Range ($\mu g/m^3$)
Mg	Mainland and coast	3·1	0·0–65·2
Cl	Mainland and coast	32·2	0·0–964·8
SO_4	Mainland and coast	2·6	0·0–732·0
NH_3	Mainland	7·9	2·5–54·4
NH_3	Island of Norderney	5·5	0·0–25·0
NO_2	Mainland	1·0	0·0–21·6
NO_2	Island of Norderney	0·14	0·0–1·2
H_2O_2	—	0·0	0·0–0·0
HCHO	Mainland	0·5	0·0–16·0

Table 4.1 shows a summary given by Cauer (1951) for various components of the atmosphere in Europe. Unfortunately, most measurements tend to be made near centres of population and industry so that values found may be very far from representative of the atmosphere as a whole.

Combustion

Two processes are important in combustion. The first is the heating and vaporization of the volatile components of the fuel, and the second is the mechanical disruption of the less volatile parts. Either or both these processes may be followed by partial or complete burning of the material involved.

The volatile components, either burnt or unburnt, are released into the atmosphere with a very high vapour-pressure because of the high temperature of the flame. This vapour is chilled very rapidly by mixture with surrounding cold air and a very high supersaturation of the volatile component results. This may typically amount to many hundreds or even thousands of percent, and is sufficient to cause homogeneous nucleation to occur. Because the supersaturation is so high the droplets formed are exceedingly small, and because their concentration is so high the flame is a prolific source of condensation nuclei. A typical size-range for such particles would lie below $0 \cdot 1\ \mu$ dia.

The mechanically disrupted particles, either unburnt or partially burnt, are of much larger size, ranging indeed up to thin sheets some millimetres in diameter, though predominant sizes are probably in the micron range. These particles are mainly carbon for most fuels, and are very much less numerous than the condensation products.

In addition many gaseous components are liberated, their nature depending upon the fuel concerned; CO_2 is always given off by organic fuels, and large amounts of SO_2 are given off when most varieties of coal are burnt.

Coste and Wright (1935) made an important observation in the course of their study of condensation nuclei produced by combustion. They found that even very pure fuels like absolute alcohol gave large numbers of nuclei when burnt, and subsequently also found many nuclei to be given off by carefully cleaned nichrome or platinum wires when heated. After careful investigation they attributed this effect to the formation of oxides of nitrogen by direct reaction of oxygen and nitrogen at high temperatures. This finding has been borne out by later work, and we must therefore recognize nitrous acid formed in this way as a very important and universal condensation nucleus produced by combustion.

Dispersion of solids

Since we have found that almost any solid particle can act as a condensation nucleus, it is evident that the mechanical disruption of solid particles on the earth's surface to produce particles small enough to remain suspended in the air must contribute to the sum of atmospheric condensation nuclei.

The initial stages of the dispersion process are usually carried out by a combination of chemical attack and water erosion. These processes have over long periods of time covered most of the earth's surface with relatively finely divided materials in the form of sands and soils. Wind action now takes over to sift out the finer components of the soil and to further break up the particles by mechanical action. There are, of course, other sources of dusts. We have already mentioned combustion, and many mechanical processes associated with industry produce dust, but all these sources are usually small compared with wind erosion.

The particle-size distribution which can be produced by dispersion processes has been the subject of considerable study (Herdan, 1953). In general it can be seen that it is easy to break up large particles and progressively harder to break up small ones, so that some sort of peaked distribution about a mean particle-size is to be expected. Experiments show that many dispersion processes produce in fact a normal probability distribution when particle-number is plotted against the logarithm of particle-diameter. This implies that ratios of diameters are important rather than absolute diameter differences, and particles of say twice the mean diameter should occur with the same frequency as particles of half the mean diameter. This is physically reasonable.

When such a log-normal distribution is plotted on a linear radius scale an asymmetric curve is obtained as shown in fig. 4.1. This has the appearance of giving a long tail on the large particle end of the distribution and a rather sharp cut-off towards small particle diameters. Since the peak of this distribution usually occurs at a diameter of a few tenths of a micron for the finest dispersed products that can be made, particles smaller than $0 \cdot 1\ \mu$ dia. are very rare indeed.

Dispersion aerosols of solid particles therefore contribute to the atmospheric suspension only in the diameter range greater than about 0·1 μ. In the last section we saw that condensation products usually have diameters less than 0·1 μ, so that the two classes do not overlap greatly.

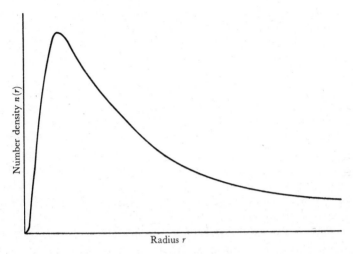

Fig. 4.1. The log-normal size-distribution converted to a linear radius scale.

Dispersion of solutions

Just as solid material can be broken up to give an aerosol, so can a solution be dispersed to give many fine droplets which may either remain liquid or evaporate to give solid particles. The dispersion process for a liquid is, however, quite different from that for a solid. A solid is broken up essentially by repeated impacts which continually reduce the particle-size, whilst a liquid must be dispersed by a single event.

The only important naturally occurring body of solution is, of course, the sea, so that all our remarks will be directed explicitly to processes involving salt droplets produced over relatively large expanses of ocean.

Once again the wind is the main agent responsible for disruption of the liquid. Strong winds raise waves and tear spray plumes from their crests and this obviously supplies some droplets (Köhler, 1941). Droplets produced in this way, however, are quite large as the wind shear is insufficient to cause very fine subdivision.

The crests of large waves tend to break over and trap pockets of air which then rise to the surface as bubbles, and it was first suggested by Aliverti and Lovera (1939) that in this way very small droplets might be produced. Recently Woodcock, Kientzler, Arons and Blanchard (1953) and Knelman, Dombrowski and Newitt (1954) have examined by high-speed photography the bursting of such small bubbles, and found that on collapse of the bubble a column and several droplets of liquid are shot into the air. Bubbles thus produce droplets as expected, and an investigation of aerosol samples near bursting bubbles by Moore and Mason (1954) shows that though many of the droplets produced are so large as to fall back into the liquid almost immediately, numbers of droplets as small as 0·03 μ dia. are produced. Facy (1951) has suggested that the smallest droplets may originate in the bursting of the bubble-film or more particularly in the films in bubble clusters, and this seems very probable.

Whether the solution droplet remains liquid or crystallizes depends upon the humidity of the surrounding air, and in trajectories over land it is likely that solution droplets may pass through this phase-transition several times. Dessens (1949) has observed that salt particles a few microns in diameter sometimes shatter into several smaller fragments if they are caused to dissolve in a droplet which is then evaporated. Dessens's remarks and further experiments by Lodge and Baer (1954), however, make it plain that this effect is confined to very large particles and even then is rare. It does not, therefore, seem to be an appreciable mechanism for further subdivision of hygroscopic nuclei.

Twomey and McMaster (1955) have found a rather different effect. They observed that when a large solution droplet ($\sim 10^{-10}$ g) crystallized, very large numbers of very small nuclei were produced as well as the single large salt particle. They estimated the mass of these small particles to be between 10^{-14} and 10^{-18} g, so that very large numbers could be produced without any appreciable change in the original particle. If established, this mechanism could provide a convenient multiplication process for nuclei over land, both by evolution of very small salt particles from sea-salt suspended in the atmosphere, or by repeated solution and desiccation of hygroscopic salts in the soil. Aitken's observation that the nucleus count on a sea beach increased sharply when the sun came out (Aitken, 1910–11) may be related to this effect.

THE COAGULATION OF AEROSOLS

We have discussed the probable origins of most of the components of the atmospheric aerosol and said something about the approximate size-distributions of particles produced by these various processes. Since suspended matter is constantly being added to the atmosphere there must be mechanisms for its removal, since the total amount in suspension is fairly constant, except perhaps in increasingly industrial areas.

Very large particles raised by strong winds settle out rather quickly under gravity, and most other large particles are removed by precipitation. We shall discuss this in more detail in the next chapter, but we may note here that the largest particles, being generally the most efficient condensation nuclei, are involved preferentially in cloud formation and are hence largely removed when rain falls. Smaller particles may be swept out by falling rain droplets, or even collected by the cloud droplets during their growth. Facy (1957) has discussed this effect, and shown that in a vapour-pressure gradient a small particle experiences a net force which is in a direction to drive the particle towards a condensing droplet. Such a process could cause clouds to remove from suspension a very large number of small particles which are not directly involved in their formation.

In the absence of cloud formation the main agent modifying the distribution of aerosol particles is their coagulation one with another, and it is this process which we shall discuss now. The basic theory was developed by Smoluchowski for liquid suspensions and was extended to aerosols by Whytlaw-Gray and Patterson (1932) in their work on smokes. We shall outline their discussion and conclusions below.

At the outset we make the assumption that when two particles collide they stick to one another. This appears to be fairly valid in practice for uncharged particles. Consider now an isolated sphere of radius r_s in a suspension of particles with initial concentration n_0. When steady-state conditions are attained there will be a distribution $n(r)$ of particles set up such that

$$n(r_s) = 0, \quad n(\infty) = n_0 \qquad (4.1)$$

and from the requirement that the flow of particles be continuous (considering for the moment only collisions with the sphere)

$$\nabla^2 n(r) = 0. \qquad (4.2)$$

The solution to these equations is simply

$$n(r) = r_s n_0 \left(\frac{1}{r_s} - \frac{1}{r} \right), \qquad (4.3)$$

and the total flux of particles by diffusion towards the collecting sphere is

$$J = 4\pi r^2 D \frac{\partial n(r)}{\partial r}, \qquad (4.4)$$

where D is the diffusion coefficient of the suspended particles. Since J is just the rate of removal of particles from suspension, from (4.3) and (4.4), substituting values at $r = r_s$, we have

$$J = 4\pi r_s D n_0. \qquad (4.5)$$

We are interested, of course, in the case where each particle of the suspension acts as a collecting sphere, and the obvious generalization of (4.5) for this case is

$$-\frac{dn}{dt} = 2\pi r_s D n^2, \qquad (4.6)$$

where D is an effective diffusion coefficient given by

$$D = D_1 + D_2, \qquad (4.7)$$

D_1 and D_2 being diffusion coefficients for two individual particles, and

$$r_s = r_{s1} + r_{s2}. \qquad (4.8)$$

n now represents the average particle concentration. The radii r_{s1} can be treated in a general way, so that, if particles stick on touching, the r_{s1} are equal to the geometrical radii r_1, whilst if there are attractive or repulsive forces the r_{s1} will be respectively greater or less than the r_1.

The Coagulation of Aerosols

Diffusion coefficients can be calculated by using Einstein's equation which relates diffusion coefficient D to mobility B by

$$D = kTB, \qquad (4.9)$$

where k is Boltzmann's constant and T the absolute temperature. For particles of moderate size B is simply given by Stokes' law. For particles comparable in diameter with the mean free path l of air molecules a correction term is required, and the complete expression is

$$B = \left(1 + a\frac{l}{r}\right) \Big/ 6\pi\eta r, \qquad (4.10)$$

where η is the viscosity of air and a is a constant approximately equal to unity.

For relatively coarse smokes of mean radius r, in which $l/r < 1$, (4.6) to (4.10) can be combined to give the approximate relation

$$-\frac{dn}{dt} = \frac{2kTs}{3\eta} \frac{(r_1+r_2)^2}{4r_1 r_2} \left(1 + a\frac{l}{r}\right) n^2, \qquad (4.11)$$

where we have written sr_1 for r_{s1}. The term $(r_1+r_2)^2/4r_1 r_2$ is approximately unity for r_1 and r_2 comparable in size, but increases greatly when the size disparity exceeds a factor of 5 or so. This implies, as is indeed found experimentally, that heterogeneous smokes coagulate more rapidly than do uniform smokes. Because from (4.10) the mobility of a small particle is greater than that of a large particle, the coagulation process principally involves collection of the small particles by the large particles, which collide with each other only rarely.

In the air under ordinary conditions the mean free path l is about 10^{-5} cm or $0.1\,\mu$. For particles below this size the Cunningham correction $[1 + a(l/r)]$ in (4.10) becomes large and loss by coagulation is accelerated. As we have pointed out, many of the products of combustion have diameters near 10^{-6} cm, and in addition are produced in very high concentrations. Coagulation will therefore be of considerable importance in modifying the size-distribution of these particles.

To gain an idea of numerical magnitudes we substitute explicit values in (4.11) and find for a reasonably homogeneous aerosol

$$-\frac{1}{n}\frac{dn}{dt} \approx 10^{-6} n \left(1 + \frac{10^{-5}}{r}\right), \qquad (4.12)$$

where r is in cm, n in particles per cm³ and t in hours. The value of this expression gives a measure of the time-scale of the process, and is actually the reciprocal of the time required for the particle concentration to drop to one-half its initial value. This time is of order 500 hr for $0\cdot2$ μ dia. particles of initial concentration 1000 cm⁻³, whilst for typical city air containing perhaps 10^5 particles per cm³ of average diameter $0\cdot01$ μ, the time is only 30 min.

Equation (4.6) can be written in the integrated form

$$\frac{1}{n} - \frac{1}{n_0} = (2\pi r_s D)t \qquad (4.13)$$

if it is desired to follow the course of the coagulation in detail. The decay is not exponential but follows this reciprocal law which has been established experimentally.

Junge (1955, 1957a) has applied this theory to a consideration of the modifications in the size-distribution of an aerosol which would be produced by coagulation alone. His model size-distribution shown in fig. 4.2 is an idealized form of the general distribution found experimentally in contaminated continental air. Shown also in this figure are the calculated size-distributions after various coagulation times. The modification is almost completely confined to particles less than $0\cdot1$ μ in radius, larger particles being almost unaffected except for a slight size-increase from collection of a coating of fine particles. After some time this leads to a distribution rather sharply peaked at a radius of $0\cdot1$ μ. In practice of course additional fine material is continually added, but it is unlikely that appreciable concentrations of particles smaller than about 10^{-7} cm radius can be maintained.

In addition to their importance in the atmosphere, coagulation effects must also be considered whenever aerosol samples are collected for subsequent examination, to be sure that the sample as examined does in fact give a reasonable picture of the size-distribution in the original aerosol.

EXPERIMENTAL METHODS

Experimental investigations of condensation nuclei can be divided into two broad classes—those which deal with individual particles, and those which treat the aerosol as a whole. Indi-

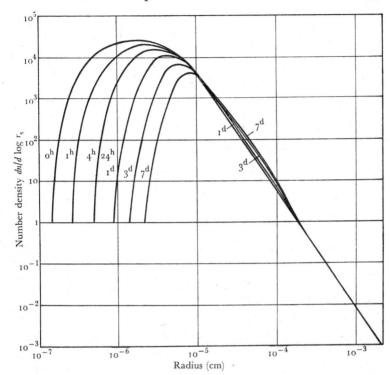

Fig. 4.2. Change in model size-distribution ($t = 0$) caused by coagulation after various periods of time in hours (h) or days (d) (Junge, 1957a).

vidual particle studies are chiefly concerned with size and number counts, and with chemical identification of individual particles so that information can be obtained about their origin. All the more obvious techniques of microchemical analysis, electron microscopy and electron or X-ray diffraction have been employed in this study. Since these methods are not peculiar to cloud physics we shall only mention them in passing and concentrate instead upon the problem of obtaining a representative aerosol sample in a form suitable for examination.

Turning to methods of examining the aerosol as a whole we find that many special techniques have been developed which are for the most part used only in cloud physics, though some are closely related to the cloud-chamber methods of nuclear physics. The quantity of prime importance is the number of condensation nuclei active at a given supersaturation, and most of the apparatus we shall describe is designed to determine this quantity. We shall discuss these bulk methods first, and then return to the sampling of aerosol particles.

The Aitken counter

During his classic researches on atmospheric condensation nuclei, Aitken (1887–8, 1888–9, 1890–1, 1923) developed several counters which have since been very widely used in routine measurements. The principle of operation of all is similar, and they differ mostly in mechanical arrangements.

The air sample to be investigated is drawn into a small chamber and saturated by contact with a piece of wet blotting-paper. When equilibrium has been reached, the sample is suddenly expanded in a more or less adiabatic manner to produce a considerable supersaturation. This causes droplets to grow on all the active nuclei, and these droplets are allowed to fall on to a graticule where they are counted with the aid of a magnifying glass.

Fig. 4.3 shows Aitken's 'portable counter' which has had very general use. The air sample is drawn into the chamber by means of the graduated pump, which also serves later to produce a standard expansion by drawing it quickly down to its fullest extent. An arrangement of stopcocks and a cotton-wool filter are provided so that in the case of very concentrated aerosols small samples diluted with filtered air may be drawn into the chamber. The mirror provides dark-field illumination for the graticule and droplets to make counting more easy.

Since counting is done by eye, only a small number of droplets can be counted at one time, and for an accurate determination of the number of nuclei present several counts must be made. This is rather time-consuming, and of course precludes study of very short-term concentration changes. Pollak (1957), in a review of methods of measuring condensation nuclei, quotes the range of an Aitken counter as 100–150,000 nuclei per cm^3, with an accuracy of 10–20 %.

The saturation ratio which one would expect to achieve under adiabatic conditions for a given expansion ratio may be calculated by application of the perfect gas laws, assuming that

Experimental Methods

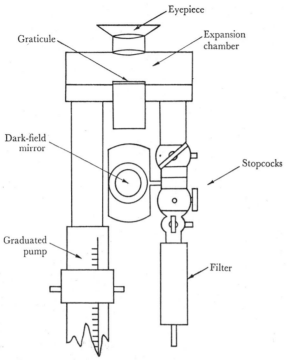

Fig. 4.3. The Aitken Portable Nucleus Counter (after Landsberg, 1938).

negligible condensation takes place during the expansion itself. The result of this calculation is

$$\frac{p}{p_\infty} = \frac{p_1}{p_2}\left(\frac{V_1}{V_2}\right)^{c_p/c_v}, \qquad (4.14)$$

where V_1 is the volume before and V_2 that after expansion, p_1 is the saturation vapour-pressure of water at the temperature obtaining before expansion, and p_2 that at the temperature after expansion. c_p and c_v are respectively the specific heats of air at constant pressure and at constant volume. The theoretical saturation ratio attained by expansion from an initial temperature of 15° C is shown in fig. 4.4.

Standard counts with present-day instruments are ordinarily made with a fixed expansion ratio, so that a single count is obtained, rather than a series of counts as a function or saturation ratio. It is difficult to estimate the maximum saturation ratio attained because of deviations from ideal adiabatic behaviour, but estimates range from values of 2 to 4 (that is, 100% to 300% supersaturation). Such a supersaturation is sufficient to activate almost all the naturally occurring condensation nuclei, with the exception of small ions and particles with very large contact angles. There is some possibility that the

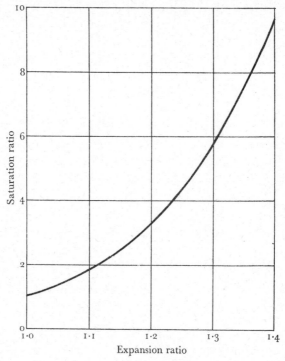

Fig. 4.4. Saturation ratio produced by adiabatic expansion of saturated air from an initial temperature of 15° C.

largest nuclei may be lost by sedimentation before the count is made, but their relative concentration is usually so small that this has negligible effect on the final count. The Aitken count may thus be taken as a measure of the total number of condensation nuclei present in the atmosphere.

Various modifications and improvements have been made to Aitken's original counter from time to time, notable among them being those of Scholz (1931, 1932), who introduced two widely

Experimental Methods

used counters which bear his name. These are described in detail in a review article by Landsberg (1938), to which the reader is referred for further information. More recently several improvements have been made towards reducing the time required for a measurement. Pollak (1952) introduced a photographic counter of the Aitken type which allowed one expansion to yield an adequate drop-count for reliability, and Wieland (1955), working with a Scholz counter, introduced a technique of taking replicas of the droplets falling on the base plate in a film of varnish, and later scanning and counting these impressions with a television-like photo-electric system. Verzar (1953) in turn has constructed a completely automatic counter which can take continuous records of nucleus concentration. We shall not consider these modifications in detail, since they leave the basic operation of the counter unchanged.

The photo-electric counter

A somewhat different type of counter introduced by Pollak and Morgan (Nolan and Pollak, 1946) has proved very useful in recent years. Again supersaturation is produced by adiabatic expansion of moist air as in the Aitken counter, but instead of counting individual droplets, the density of fog produced is evaluated by measuring the extinction of a beam of light passing through the expansion chamber.

Referring to fig. 4.5, which shows a typical counter in a somewhat schematic manner, the expansion chamber is a brass tube about 60 cm long lined with wet blotting-paper, or in recent models with moist ceramic, and closed at each end by a thick glass plate. A collimated beam of light is produced by the lamp and lens at the upper end of the tube, and its transmission through the chamber measured by means of a selenium photo-cell in the base.

To perform a count, the tube is flushed with the air to be measured and then brought to a selected overpressure by means of the pump. About half a minute is allowed for thermal equilibrium and vapour saturation to be attained, and in this time the lamp current is adjusted to give a standard photocell response. The overpressure is then released suddenly by means of the exhaust-tap, and a fog is formed by condensation on nuclei active at this supersaturation. The fractional change in photocell response is a measure of the fog density, and is termed the extinction.

Unlike the Aitken counter this is not an absolute instrument, and requires to be calibrated. Such a calibration has been performed by Nolan and Pollak (1946) who established one count by comparison with an Aitken counter, and obtained a

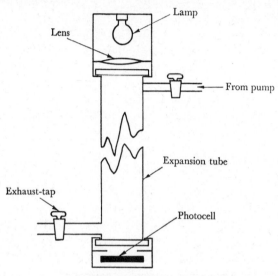

Fig. 4.5. Schematic diagram of a photo-electric nucleus counter.

complete calibration curve by successive dilution of a standard aerosol stored in a large gasometer.

The principal advantage which the photo-electric counter has over the Aitken counter is that one reading, taking less than a minute to perform, gives a reliable count of nucleus concentration. It is also very easy to alter the saturation ratio at which the count is made by simply changing the overpressure. Pollak (1957) gives the counting-range of such an instrument as 100–440,000 nuclei per cm.³

Numerous improvements to this basic design have been made by Pollak and his co-workers (Pollak and Murphy, 1953; Pollak and O'Connor, 1955), and a modification employing expansion into another chamber at reduced pressure has been discussed by Rich (1955).

The thermal gradient cloud chamber

We now consider two counters in which the supersaturation is produced in an entirely different manner from the expansion method used by the counters above. Both chambers were originally designed to produce a high continuous supersaturation in clean air for the study of the ionization tracks left by cosmic ray particles, and their application to cloud physics has been very recent.

The first counter relies upon a thermal gradient to produce supersaturation, and was originally discussed by Langsdorf (1936) whose chamber used methyl alcohol as the condensable

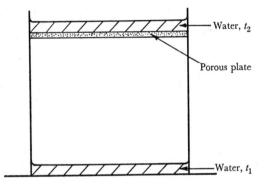

Fig. 4.6. A thermal gradient cloud chamber.

vapour. Relating our discussion now to water-vapour, the principle of operation can be seen from fig. 4.6. The experimental chamber consists of a cylinder, in the bottom of which is water at a temperature t_1. The top of the cylinder is closed with a porous plate which is kept wet with water at a higher temperature t_2.

Water-vapour is distributed through the chamber by mixing and diffusion, and provided there is no heat loss to the side walls the water-vapour pressure in a given air-parcel will be linearly related to its temperature. Since, however, the saturation vapour-pressure of water is not linearly related to its temperature, this mixing and diffusion leads to a supersaturated condition as shown in fig. 4.7. The vapour in the chamber must at all points have values of p and t lying on the broken line

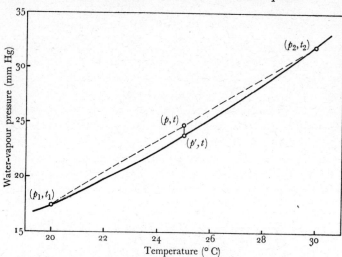

Fig. 4.7. The principle of the thermal gradient cloud chamber (after Wieland, 1956).

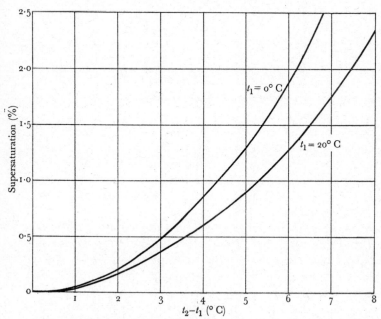

Fig. 4.8. Supersaturation produced in a thermal gradient chamber (after Wieland, 1956).

Experimental Methods

shown. The equilibrium vapour-pressure of water at temperature t is, however, p', so that the supersaturation is $(p-p')/p'$. This has a maximum near the temperature $(t_2-t_1)/2$, and if there is some stirring or convection within the chamber, all particles may be expected to experience this maximum supersaturation at some time. Fig. 4.8 shows this supersaturation as a function of the temperature difference between the top and the bottom of the chamber.

A counter making use of this principle has an advantage over an expansion counter since its supersaturation is constant, at least until appreciable droplet growth has occurred. It is, however, less convenient to use than the expansion chamber, and for this reason it is most useful in the range of supersaturations from 0·1 % to a few per cent, which range is not easily covered by an expansion chamber.

Wieland (1956) has described a counter of this type in which twenty small cylinders form the test space, and a hot top having a linear temperature gradient from one end to the other produces a range of supersaturations so that a nucleus spectrum can be obtained.

There is, however, one serious drawback to this type of counter. This is due to the effect of the side walls. If these are at any point cooler than the vapour which comes in contact with them, then quite large supersaturations may be produced and upset the measurement. The chemical gradient cloud chamber which we shall describe next overcomes this difficulty.

The chemical gradient cloud-chamber

In 1936 Vollrath (1936), following up an observation on the fuming of concentrated acids, constructed a cloud chamber similar in appearance to that of fig. 4.6, but with water at the top and concentrated HCl in the bottom of the enclosure, both being at the same temperature. The same components have been employed in a somewhat different manner by Holl and Muhleisen (1955) in the construction of a nucleus counter, and more recently Twomey (1959 a, b) has used a chamber with dilute HCl for the study of nuclei at low supersaturations.

The process whereby supersaturation is produced in this chamber is somewhat similar to that involved in the thermal gradient type. It relies upon the non-linearity of vapour-pressure curves, this time as a function of composition, rather than of temperature. In this case it is necessary to consider two diffusing components, since water-vapour diffuses downwards while HCl gas diffuses upwards. In a given air-parcel the pressure of water-vapour is a linear function of the concentration of HCl, but the equilibrium water-vapour pressure over HCl solution is not a linear function of concentration, because the van't Hoff factor is not a constant. In general this can lead either to supersaturation or to undersaturation, depending upon the components involved, but for the case of HCl and water supersaturation exists. We shall not go into this matter in any further detail; a complete discussion has been given by Twomey (1959a) in his paper describing this cloud chamber. The maximum supersaturation achieved depends only upon the concentration of the acid solution employed, and may therefore be easily and accurately determined. Supersaturation ranging from a few hundredths of a percent to over 100 % can thus be simply produced, and because of mixing within the chamber all particles experience the maximum supersaturation in the chamber at some time. Fig. 4.9 shows maximum supersaturation as a function of HCl concentration over the range of interest. Again the expansion chamber is more convenient for measurements at high supersaturations, and this chamber finds most useful employment in measurements at low supersaturations.

In Twomey's version of this counter a set of glass beakers is used with a range of acid concentrations. The droplet density is evaluated by an ultramicroscope method, the fog being illuminated by a narrow, well-defined beam from a mercury arc, and photographed at right angles to the direction of illumination. The counter is thus absolute in the same way as is the Aitken counter, and requires no calibration.

Two objections might be made to this method, both based upon the fact that the condensing-phase is not pure water. In the first place, as we made clear in our discussion of nucleation theory, the nucleation rate depends strongly on the surface-tension. In the concentrations used, however, there is negligible change in this quantity for the system HCl–water. This consideration does, nevertheless, place a restriction on the components which can be used in such a chamber. A second, and perhaps more serious objection is the possibility of chemical reaction altering the efficiency of nuclei. The concentrations of HCl vapour involved are, however, very small, and it is most unlikely that any chemical reactions will occur through the vapour attacking the nuclei. In the case of solution droplets, it is assumed that HCl vapour dissolves in the droplet and this is unlikely to cause any chemical reaction with most naturally

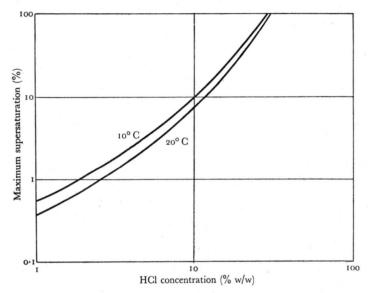

Fig. 4.9. Supersaturation produced in a chemical gradient chamber using dilute HCl (after Twomey, 1959a).

occurring droplets. In the case of insoluble nuclei, if a reaction does occur once a droplet has formed it is of no importance, since nucleation has already taken place. Twomey (1959b) reports a comparison of his counter with a similar chamber employing expansion to give the supersaturation, and the excellent agreement between the two sets of results even at high acid concentrations implies that side-effects of this nature are unimportant.

THE SAMPLING OF AEROSOL PARTICLES

As we have already remarked, one of the main difficulties in the detailed examination of aerosol particles is the collection of a representative sample. Large particles can obviously be collected by simply allowing them to settle in still air on to an appropriate sampling-plate. This is satisfactory for particles larger than a micron or so in diameter, but below this size other methods must be used. We shall discuss some of these in this section.

Thermal precipitation

It has long been a matter of common observation that hot objects in dusty atmospheres are surrounded by a thin dust-free space. This phenomenon was investigated as long ago as 1883 by Aitken (1883-4), but the exact mechanism involved is still a matter for discussion (Rosenblatt and La Mer, 1946; Saxton and Ranz, 1952). Briefly, any particle in a temperature gradient in a gas is subject to a force which is in such a direction as to drive the particle towards regions of lower temperatures. The origin of this force depends in detail upon the momentum transfer of gas molecules striking the particle, and we shall not discuss it here. The velocity with which the particle moves is proportional to the temperature gradient, and increases with decreasing particle diameter provided the particle diameter is greater than the mean free path of the gas molecules. The force also depends upon the material of the particles, being greatest for particles with low thermal conductivity.

A very simple form of thermal precipitator can be made by drawing the air to be sampled between two parallel plates, one of which is heated and the other cooled. Particles tend to be precipitated on to the cool plate, and, if the plates are close enough together and of sufficient length, virtually all the suspended particles can be removed from the air. There is a rough sizing effect along the length of the plate which is an advantage from the point of view of subsequent examination. Precipitators of this type have often been used to remove the dust particles and produce clean air, but only a few workers, among them Watson (1936), have used the method to precipitate samples for examination. Watson's precipitator, in which the source of heat was an electrically heated wire placed between two glass coverslips, effectively deposited particles from 0·01 μ to about 10 μ in diameter.

Impaction

Simple settling of particles is only an efficient collection method for particles more than a few microns in diameter. An obvious method of increasing the sedimentation rate is by use of a centrifuge, and this approach was followed by Sawyer and Walton (1950) who designed a conical centrifuge, which, as well as hastening the precipitation, sorted the particles by size along the deposition area. This instrument was suitable for particles down to 0·5 μ dia.

A rather simpler method, which we shall now discuss, involves blowing the air containing the suspension against some form of obstacle. The larger particles have enough momentum to deviate from the flow lines of the air and strike the obstacle, where they may be retained by a suitably treated surface. This is termed the impaction method. A closely related procedure involves moving a small obstacle through still air, and since the collection process is the same we shall treat the two methods together.

The efficiency of collection is found to depend on the particle-size, the velocity of the airstream and the geometry of the obstacle, in such a way that to collect small particles we require a fast airstream and a small obstacle (that is, a high surface-curvature). Detailed particle-trajectories have been calculated by Langmuir and Blodgett (1946), but it is not worth while to consider these in detail here. More recently Ranz and Wong (1952) have extended these results and compared them with experiment. Fig. 4.10 shows the collection efficiency of a system consisting of an infinitely long, round or rectangular jet of diameter or width d_j impinging on a flat plate of infinite extent, while fig. 4.11 shows the collection efficiency of various obstacles of diameter or width d_j placed in an airstream of infinite extent. The curves in all cases are plotted in terms of a dimensionless parameter ψ defined by

$$\sqrt{\psi} = (\gamma \rho_p u_0 / 18 \eta d_j)^{\frac{1}{2}} d_p, \qquad (4.15)$$

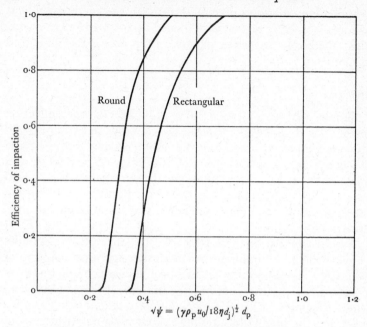

Fig. 4.10. Theoretical impaction efficiencies of aerosol jets (Ranz and Wong, 1952).

where d_j defines the obstacle or jet as we have noted, d_p is the diameter of the aerosol particle and ρ its density, u_0 is the undisturbed airstream velocity and η the viscosity of air, all in c.g.s. units. γ is a constant defined by

$$\gamma = 1 + (2l/d_p)[1 \cdot 23 + 0 \cdot 41 \exp(-0 \cdot 44 d_p/l)], \quad (4.16)$$

where l is the mean free path of gas molecules.

The impaction efficiency approaches unity for large particles and falls off for small particles, but it is not worth while to give numerical values because of the large variety of experimental arrangements which can be used. Obstacles range from cylinders many inches in diameter used by Langmuir (1944) to collect cloud droplets, through 1 mm rods and slides of Woodcock and Gifford (1949) to the $0 \cdot 1\,\mu$ spider threads of Dessens (1946, 1949). The narrower cylinders require smaller airspeeds to collect fine particles, but their area is so small that a long collection time is required to obtain a good sample.

The Sampling of Aerosol Particles

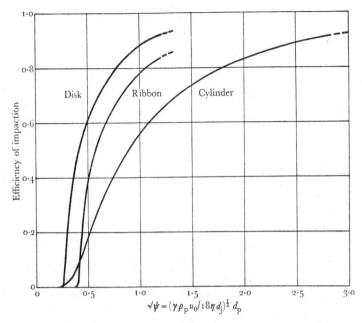

Fig. 4.11. Theoretical impaction efficiencies of an aerosol on obstacles of various shapes (after Ranz and Wong, 1952).

A point that emerges from this discussion is that particles collected by impaction are by no means an unbiased sample of the aerosol, and it is always necessary to apply a weighting factor in favour of small particles to correct the measured size-distribution.

Several forms of impactor have been designed and used at various times. In the Zeiss konimeter (Junge, 1952) air passes through a small aperture and impinges on a flat plate which is coated with vaseline or lacquer to retain small particles or droplets. In a two-stage instrument of this type used by Junge (1953) and shown in fig. 4.12 the first opening is relatively large so that only large particles are collected on the first impaction plate. After leaving this stage the air passes through a smaller opening and a second plate collects the finer particles. This is a considerable convenience for subsequent microscopic examination. A rather similar instrument having four graded jets in series is the cascade impactor described by May (1945).

The air-flow in these instruments is conveniently provided by a small suction pump acting on the last stage.

We shall discuss the sampling of aerosols by small cylindrical obstacles in connexion with particular investigations later. In this case use is often made of natural wind or the speed of passage of an aircraft to collect the sample, and the obstacle is simply mounted in a suitable way and exposed to the air.

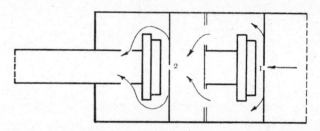

Fig. 4.12. The two-stage konimeter. Large particles are collected in the first stage and small particles in the second (Junge, 1953).

EXPERIMENTAL RESULTS

The results to be discussed in this section bear on three main questions—the geographical distribution of nuclei, their principal sources and their size-distribution. Early work concentrated almost exclusively on the first two questions, and an immense number of measurements have been made, mostly using Aitken or Scholz counters. Only comparatively recently has interest shifted to the larger nuclei, and today much work is concerned with their size and properties. An excellent survey of the earlier work has been given by Landsberg (1938), and much of the work on larger nuclei up to 1952 has been summarized by Junge (1952, 1953). We shall treat the material under three headings, considering first the older work on Aitken nuclei, then discussing measurements on the large particles in continental aerosols, and finally the more recent work on nuclei over the sea. In the last section of this chapter we shall consider that part of the nucleus spectrum which is of particular importance in the formation of clouds.

Aitken nuclei

In our discussion of the Aitken nucleus counter we mentioned that the supersaturation reached in the expansion, though not well determined, was certainly greater than 100 % and perhaps more than 200 %. This being the case, an Aitken counter activates almost all the nuclei present in the aerosol, only small ions and particles with extremely high contact angle being excepted. An Aitken count is, therefore, essentially a measure of the total nucleus concentration of the atmosphere.

In natural condensation, as we shall see later, the expansion rates involved are very slow, so that condensation on the more active nuclei removes excess water-vapour almost as quickly as it becomes available, and the supersaturation is never likely to exceed a few percent. Aitken nucleus counts, therefore, have little relevance to cloud formation, and we shall discuss them only briefly.

Table 4.2. *Aitken nucleus count in nuclei/cm^3 for different types of localities (Landsberg, 1938)*

Locality	Places	Observations	Average	Average maximum	Average minimum	Absolute maximum	Absolute minimum
City	28	2500	147000	397000	49100	4000000	3500
Town	15	4700	34300	114000	5900	400000	620
Country inland	25	3500	9500	66500	1050	336000	180
Country seashore	21	2700	9500	33400	1560	150000	0
Mountain							
500–1000 m	13	870	6000	36000	1390	155000	30
1000–2000 m	16	1000	2130	9830	450	37000	0
> 2000 m	25	190	950	5300	160	27000	6
Islands	7	480	9200	43600	460	109000	80
Ocean	21	600	940	4680	840	39800	2

Table 4.2, compiled by Landsberg (1938) from the results of many thousands of observations, gives some idea of the Aitken counts in various environments. The range between extreme values is immense—a factor of 10^6—and there is even a range of 10^2 in average values for different localities. High counts are associated with human habitation and industrial activity, and the count falls off, though not to zero, in regions remote from such centres.

These observations are consistent with our conclusion that

combustion produces very large numbers of small nuclei, and indeed direct measurements in rooms lit with gas flames and near other burning substances show this to be the case (Coste and Wright, 1935). On the other hand, natural dust, soil dust and sand swept up by high winds have been found to have relatively little influence on the count (Boylan, 1926), an effect explained by the fact that this dust consists of a relatively very small number of large particles which are numerically quite insignificant beside the small nuclei.

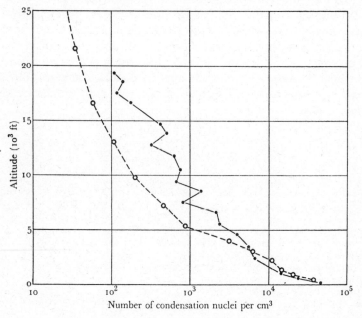

Fig. 4.13. Mean curves for concentration of condensation nuclei at various altitudes according to Wigand O--O and Weickmann ●——● (after Weickmann, 1957b).

Falling counts on mountain slopes indicate the earth's surface to be the principal source of nuclei, and these few measurements have been supplemented by those taken by Wigand (1919) in balloon ascents and more recently by Weickmann (1957b) from an aeroplane, which show the same decrease in count with increasing altitude. These later data are shown in fig. 4.13. Weickmann's curve is an average of curves taken on very many

Experimental Results

days and indicates the general decrease of count with height. Individual curves deviate greatly from this behaviour and show marked changes above or below inversions, clouds, etc. These general observations support the theory that the majority of these nuclei originate at the earth's surface and are borne aloft by convection. Their number is rapidly depleted by coagulation and the scouring action of precipitation processes which act to maintain an equilibrium distribution. Nuclei formed high in the atmosphere are clearly numerically of little significance in the Aitken count.

The size-range of the main constituents of the Aitken count has been determined in two ways. It is found that a large number of these small particles bear a single electronic charge, and if this is assumed to hold for the majority of particles their sizes can be determined from Stokes' law by measuring their mobility in an electric field. Israel and Schulz (1932) in this way found the majority of particles to have radii between about $1 \cdot 6 \times 10^{-6}$ cm and $1 \cdot 5 \times 10^{-5}$ cm, the distribution varying considerably from place to place, smaller particles predominating in cities. A more direct method has been used by Linke (1943) and Hosler (1950) who formed droplets on the particles by expansion in an Aitken counter, collected the droplets on the graticule, evaporated them, and observed the residues under an electron microscope. They found a size-range from 5×10^{-7} to 2×10^{-5} cm.

All this evidence shows that the majority of Aitken nuclei are rather small and hence require considerable supersaturations to become active in condensation. The further observation that the number of droplets in cloud ranges from about 30 to 1000 per cm^3 whilst Aitken counts range typically from 10^3 to 10^5 per cm^3 shows that it must be only the most active of the nuclei which participate in cloud formation. In the next sections we shall, therefore, concentrate on these larger nuclei.

Large continental nuclei

Junge (1952, 1953, 1955) has made large numbers of measurements of the concentration of atmospheric particles in the size-range greater than $0 \cdot 2\,\mu$, using the two-stage konimeter we discussed above. These measurements were made both in

Europe and eastern America and are therefore probably representative of continental regions with considerable population density. He found that in the range 0·1–10 μ of particle radius the number distribution is given approximately by

$$\frac{dn(r)}{d\log r} \propto \frac{1}{r^3}, \qquad (4.17)$$

where $n(r)$ is the number of particles per cm³ having radius greater than r. An alternate expression of this distribution law is

$$\frac{dm(r)}{d\log r} = \text{constant}, \qquad (4.18)$$

where $m(r)$ is the total mass per cm³ of particles having radii greater than r. Above 10 μ and below 0·1 μ radius the number of particles is much less than predicted by (4.17). Since particles in this size-range are produced primarily by a dispersion mechanism which gives rise to a log-normal size-distribution, Junge suggests that the observed distribution may be produced by superposition of many such log-normal distributions having different parameters. Since the law (4.17) is by no means universally followed this appears to be an adequate explanation.

When the size-data on Aitken nuclei is combined with that on large nuclei it is possible to construct a typical size-distribution model for a continental aerosol. Junge (1952) has done this, and his diagram is shown in fig. 4.14 which is largely self-explanatory. The two peaks below 3×10^{-7} cm are due to loose aggregations of air and water molecules on ions, and are not, properly speaking, foreign particles. The maritime component we shall discuss in a later section.

Few direct determinations of the chemical composition of large (0·1–1 μ) and giant (> 1 μ) particles have been made, but several independent methods have given consistent results. Junge examined samples collected in Europe and America with his two-stage konimeter by means of chemical analysis and in some cases by electron diffraction. He concluded that the large particles are composed predominantly of ammonium sulphate, whilst the giant particles contain large amounts of sodium chloride, and may therefore be originally of maritime origin. Similar conclusions were reached by Yamamoto and Ohtake (1953) on the basis of the appearance of particles under the

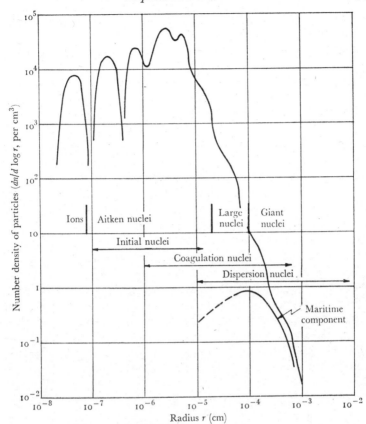

Fig. 4.14. Average size-distribution of nuclei in continental air-masses (after Junge, 1952).

electron microscope. They found a maximum for sea-salt nuclei at about $3\,\mu$ dia. and for 'combustion products' (ammonium sulphate or sulphite) at about $0.3\,\mu$ dia., though both components existed over the range 0.1–$10\,\mu$. This is also evidence supporting the theory that the $1/r^3$ distribution is a superposition of differing log-normal distributions.

Twomey (1954, 1955) and Fournier d'Albe (1957) have sampled continental aerosols for particles of sea-salt greater than $1\,\mu$ dia. and found that in dry conditions such nuclei can be carried hundreds of miles inland from the sea over which they originated, without great decrease in their concentration.

Precipitation, however, rapidly removes them from the air-mass, which is to be expected since they are very efficient condensation nuclei. Twomey also found a few particles of other hygroscopic substances, probably ammonium sulphate or sodium carbonate in his samples, and, particularly during summer months, mixed nuclei consisting of sea-salt and insoluble soil-particles were common.

Several analyses of precipitation products have confirmed these findings, though they cannot, of course, give any information on relative nuclear sizes. Köhler (1925a, 1926) did much pioneer work on the chloride content of rime deposited from supercooled clouds, whilst recently Houghton (1955) and Gorham (1958) have analysed rainwater samples. Houghton, working in America, determined anions only, and found large amounts of chloride, particularly in air with a recent trajectory over sea, together with sulphate which appeared always to be present. The acidity of some of the samples led him to believe that the sulphate might be present often as sulphuric acid, though under some conditions it was apparently combined as a salt. Similar general conclusions were arrived at by Gorham for water samples collected in the Lake District in England. Houghton's analysis also revealed the presence of considerable amounts of insoluble material, presumably derived from the soil.

Summarizing the experimental evidence, it appears that continental aerosols have three main components in the size-range above $0 \cdot 1\ \mu$. The first is sea-salt, which is the predominant constituent of nuclei greater than $1\ \mu$ in diameter. The second is a sulphate component, which may be present either as sulphuric acid, or combined as a salt, perhaps ammonium sulphate, and which predominates in nuclei between $0 \cdot 1$ and $1\ \mu$ in diameter. This component may change from place to place and may also contain other hygroscopic materials. The third component consists of insoluble particles probably derived from the soil, and its concentration depends upon the degree of desiccation of the soil, as well as average ground wind. The relative importance of these three components will clearly depend upon details of the history of the air-mass.

Sea-salt nuclei

We have already discussed ways in which small droplets can be produced from a liquid surface by the action of wind and the bursting of bubbles. In the paragraphs above we have also seen that sea-salt particles form an appreciable component of aerosols over the continents, and we shall now examine the distribution of sea-salt nuclei over the oceans themselves.

The major work in this field was performed by Woodcock and Gifford (1949) and by Woodcock (1950, 1952, 1953) who collected the salt particles in the form of hemispherical droplets on hydrophobic slides exposed from an aircraft. The results of these measurements are qualitatively just what one would expect. Over the range of measurement the particle-count decreases with increasing particle-radius in a manner which resembles the tail of a log-normal distribution, as shown in fig. 4.15. As the wind-speed increases so the concentration of nuclei of each size increases, but in addition the relative abundance of large particles increases.

The particle concentration falls off with increasing height in an approximately exponential manner with little or no size-effect. The form of decrease with height is a little difficult to explain in detail, but Junge (1957b) suggests that it follows if the concentration of nuclei in the air-mass is increasing, rather than in a steady state. This is quite reasonable since most of the nuclei are removed from parts of the air-mass from time to time by precipitation.

Nuclei produced from sea-salt are predominant in truly maritime air since the nuclei of continental origin have been removed by precipitation. In continental air sea-salt nuclei are numerically insignificant beside the host of smaller and less active nuclei produced by other processes. Because they are large, however, they can be readily distinguished and identified, and Twomey's measurements (Twomey, 1955) show that over the continent in dry maritime air-masses, the size-distribution is essentially the same as that determined over the sea by Woodcock. As we shall see later these large nuclei play a specially important part in the processes of cloud formation and in the development of precipitation.

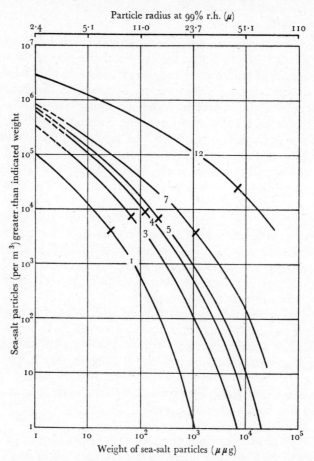

Fig. 4.15. Effects of varying wind-force, shown as a parameter, on the size-distribution of large sea-salt particles near cloud-base level. Short transverse lines on each curve are median values (Woodcock, 1953).

NUCLEI AT LOW SUPERSATURATIONS

From the rather lengthy discussion of atmospheric condensation nuclei which we have just completed, one point clearly emerges. This is that the concentration of condensation nuclei normally present in the atmosphere is so much greater than the concentration of water droplets found in clouds that only a very small fraction of the nuclei can be involved in their formation. The normal range of droplet concentrations in clouds and fogs is from a few tens to at most a few thousands of droplets per cm³, so that it is upon the few thousand most active nuclei that we should concentrate our attention.

Most investigations of large nuclei have been in terms of their size, since this is conveniently measured. Size alone, however, does not provide an adequate index of the efficiency of a

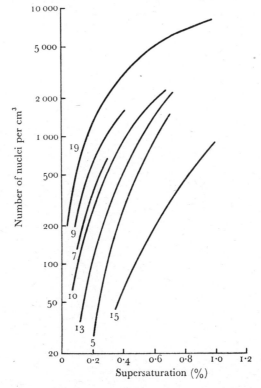

Fig. 4.16. Nucleus concentrations in typical air-masses. Curves 5, 13 and 15 are for Föhn conditions and curve 19 for a 'mixed' air-mass (Wieland, 1956).

nucleating particle, and though it is possible to relate size to activity when chemical composition is known, it is much more convenient to have a direct determination of nucleus spectrum as a function of supersaturation.

Though this had been realized by many people, the first investigation of the condensation nucleus spectrum at low supersaturations was not made until 1956 when Wieland (1956) used his version of the temperature gradient cloud chamber which we discussed earlier in this chapter to investigate aerosols between 0 and 1 % supersaturation.

Fig. 4.16 shows some of Wieland's curves for air-masses of various origins observed in Switzerland. Curves 5, 13 and 15 are for Föhn air-masses which have risen high over mountains

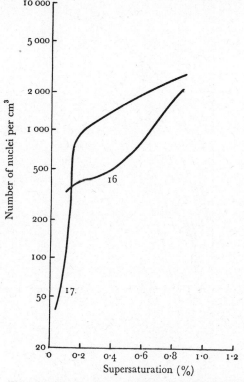

Fig. 4.17 Nucleus concentrations of modified air-masses.
17 = rain shower; 16 = wind discontinuity (Wieland, 1956).

and been cleared of nuclei by precipitation before descending to the region in which measurements were made. The air-mass 19 on the other hand is described as 'mixed' and has presumably had a long continental trajectory. The curves in fig. 4.17 show the effects of a rain shower (curve 17) and of a wind change (curve 16) on these distributions.

Nuclei at Low Supersaturations

It is clear from these curves that sufficient nuclei can be activated for the formation of an average cloud at supersaturations of a few tenths of a percent, and this estimate is borne out by the fact that the rain shower described in curve 17 of fig. 4.17 appears to have depleted the air-mass of nuclei

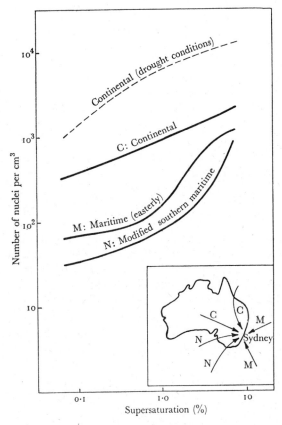

Fig. 4.18. Nucleus content of typical air-masses (Twomey, 1959*b*).

active at less than about 0·2 % supersaturation. This is, however, only suggestive since the exact recent history of the air-mass is not known.

More recently measurements have been made in Australia by Twomey (1959*b*) using the chemical gradient cloud chambers described earlier in this chapter. These measurements were made at Sydney which is on the coast, and it was possible to describe air-masses as continental or maritime with considerable exactness, only a few air trajectories leading to an intermediate type which was termed 'modified maritime'. Twomey's results for these situations are shown in fig. 4.18. Comparing these results with those of Wieland shown in fig. 4.16 it is seen that the general agreement is reasonably good, but Wieland's curves rise more rapidly than do Twomey's. This may be a real effect or may be due to one of the shortcomings of the thermal gradient chamber which we discussed previously.

Continental air-masses are found to be much richer in efficient condensation nuclei than are maritime air-masses, and this is in general agreement with the larger droplet concentrations usually found in continental clouds. We shall discuss this in more detail in the next chapter.

These observations are particularly interesting as they point to the land as the major source of condensation nuclei active at low supersaturations. This is in agreement with the general findings on large nuclei which we have already discussed, and we have mentioned several mechanisms by which hygroscopic particles may be produced or multiplied over land.

From this discussion it will be evident that, though the general features of the origin and distribution of condensation nuclei in the atmosphere are reasonably well established, we are still very far from a complete understanding of the details of the processes involved. The general features which have been established are, however, sufficient for us to proceed with a discussion of the way clouds form upon these nuclei in the atmosphere, and this we shall do in chapter 6.

CHAPTER 5

THE MICROSTRUCTURE OF NON-FREEZING CLOUDS

INTRODUCTION

THE observational study of clouds before about 1940 was confined largely to measurements made from the ground and qualitative observations from aircraft, and for that reason it proved very difficult to obtain a detailed picture of the microstructure of clouds other than fogs. Since that time, however, measurements of cloud properties from aircraft have become common procedure, and a large amount of information has been collected about the structure of clouds of various types. It will be our purpose in this chapter to review briefly the types of measurements which can be made in clouds, the instrumental techniques used, and the results obtained. After we have presented this information in as coherent a form as possible, we shall examine, in chapter 6, the detailed mechanism of cloud growth.

We have already discussed, in chapter 1, some aspects of the structure of clouds, especially those relating to air motion and water content. Our present chapter will therefore concentrate on three microphysical quantities—the concentration of cloud droplets, their size-distribution, and the liquid water content of the clouds they comprise. We shall also have something to say about the stability of different types of clouds, though the question of the form of precipitation will be postponed, for the most part, to chapter 7. It should perhaps be pointed out explicitly that we are restricting our attention, for the present, to clouds consisting entirely of water droplets. In clouds containing ice-crystals additional phenomena appear, as we shall see in a later chapter.

EXPERIMENTAL METHODS

Cloud droplet samplers

The most obvious method of examining the droplet population of a cloud is to catch all the droplets in a given volume by some precipitation mechanism, and examine them microscopically. Most methods for determining droplet concentration and size-distribution do essentially just this, and vary only in the means by which they attempt to overcome the difficulties involved.

These difficulties are of two types. First, the sampling technique is usually not equally efficient for droplets of all sizes, and some means of correction for this must be found. Secondly, once collected, the sample must be protected from changes caused by coalescence or evaporation of the droplets until photographic records or counts have been made.

Though electrostatic precipitation has been used to collect droplet samples (Pauthenier and Brun, 1940), most collection methods rely upon inertial impaction of droplets on to a suitably prepared slide held in a fast stream of air. We have already discussed this method of collecting aerosol particles and seen that it suffers from the disadvantage that the collection efficiency decreases for small droplets, and very small droplets are not collected at all. This variation in collection efficiency can be corrected for, using the calculations of Langmuir and Blodgett (1946) to which we referred in chapter 4. This procedure is satisfactory provided the concentration of droplets with a low or zero collection efficiency (typically those less than a few microns in diameter) is not important.

Three different methods have been developed to overcome coagulation, shattering and evaporation of the droplet sample on the slide. In the first of these, used by many workers (Fuchs and Petrjanoff, 1937; Diem, 1942, 1948; Mazur, 1943; Weickmann and aufm Kampe, 1953; Levin and Starostina, 1953, among others), the slide is covered with a layer of oil, the object being that impacted droplets will become immersed in the oil, which will then prevent evaporation and coalescence. Whilst the method works fairly well the droplets often tend to remain incompletely submerged in the oil with consequent

PLATE I

Imprints made by cloud droplets in the carbon film on a glass-rod collector (Squires).

Experimental Methods

possibility of evaporation and coalescence. For this reason it is necessary to examine the slides as soon as possible after exposure, and Diem (1942), for example, used a standard delay of only 15 sec.

The second method aims at keeping a replica of the cloud droplets, rather than the droplets themselves, and for this purpose the slide is covered with a thin film of magnesium oxide powder, obtained by burning a magnesium ribbon near the slide. The film is soft and droplets leave clear holes which are related to their sizes. This method was developed by Mazur (1943) and May (1947, 1950) and has since been widely used. It has a considerable advantage over the oil method in that evaporation effects do not occur, and overlapping impressions are easily recognized as such so that coalescence does not constitute a problem. Further, the impressions are relatively permanent and can be examined at leisure. There are, however, other difficulties associated with preparation of the film, and changes of film-texture with burning conditions of the magnesium and with exposure to air reduce its accuracy and reproducibility. An alternative method using a thin carbon film deposited from burning hydrocarbon gas or liquid (May, 1945) has been reported as giving much better results (Squires, 1958c). A typical droplet sample taken by this method is shown in Plate I.

The third method uses a collector slide covered with water-soluble dye (May, 1945; Okita, 1958a) and has the advantages associated with the replica method. Droplets produce small clear spots surrounded by rings of more intense colour which are easily measured. All these methods, except for the MgO film which is rather thick, respond to all droplet sizes normally collected by the slide, and so do not cause further trouble once a calibration has been performed.

For all these impaction techniques a short, accurately known exposure time is required so that droplet concentrations can be calculated. In typical clouds and at aircraft speeds an exposure time of about 0·01 sec gives a sample of satisfactory density, and several instruments have been devised to expose one or more slides for times of this order. May (1945) developed a single shutter with exposure time of $\frac{1}{25}$ to $\frac{1}{100}$ sec, but more convenient gun-like devices have been developed exposing several slides in succession so that measurements can be made during cloud traverses. Such instruments have been described by Diem (1942), Squires and Gillespie (1952), Brown and Willett (1955) and Owens (1957). These typically use glass rods instead of slides and expose them either by a reciprocating action, or by firing them across the airstream into a collecting tube.

Measurements made with cloud droplet samplers such as these give, in principle, complete information on droplet concentration, size-distribution and cloud water content. We shall next discuss some less direct means of obtaining the same information, and then describe some instruments designed to measure liquid water content.

Rotating cylinder collectors

Since the collection efficiency for droplets of different sizes of a cylinder exposed to a moving cloud varies with the diameter of the cylinder, the rate of water collection by cylinders of different sizes is related to the droplet size-distribution of the cloud. From Langmuir's calculations it is, in fact, possible to work back from such water-collection data to find an approximation to the original droplet distribution. Several instruments using this principle have been designed and used, of which the best known is that installed on Mount Washington in the United States (Schaefer, 1945). This consisted of six coaxially mounted cylinders each 10 cm long and respectively $\frac{1}{16}$ in., $\frac{1}{4}$ in., $\frac{3}{8}$ in., 1 in., 2 in. and 3 in. dia., rotating at 12 rev/min. The cylinders were of cotton material or of porous alumina, and after a few minutes exposure to the cloud they were demounted and weighed to determine the amounts of water collected. From tables the droplet spectrum of the cloud and its liquid water content could then be determined (Mount Washington Observatory, 1946; Brun et al. 1955). As distinct from the droplet sampling methods, this approach yields an average distribution over several hundred metres of cloud instead of a nearly instantaneous reading over a distance of the order of a metre.

Optical methods

Several indirect optical methods have been used to find information about cloud parameters. Köhler (1925a) determined drop sizes by measuring the angular diameters of the first and higher order diffraction rings of a corona, and made an estimate of the width of the size-distribution from the diffuseness of the rings.

A more refined method has been described by Oura and Hori (1953), who made measurements on the energy distribution in the diffraction pattern produced by passing a parallel beam of red light through the cloud and deduced approximate size-distributions.

Various types of optical transmission and polarization measurements can also be made to yield information on drop size-distribution and liquid water content. Driving, Mironov, Morozov and Khvostiknov (1943) made measurements of transmission as a function of wavelength in the visible spectrum, whilst Keily (1954) has done similar work in the infra red, measurements of this type giving both drop spectrum and water content. Alternately, on the basis of a measured or assumed drop size-distribution, liquid water content can be found approximately from a single transmission measurement (aufm Kampe, 1950; aufm Kampe and Weickmann, 1952). Further discussion has been given by Atlas and Bartnoff (1953) and by Fritz (1954).

These optical methods usually give values averaged over paths of 10 m or so, and are relatively quick in response. On the other hand, the indirect means of measurement involved make some of the results obtained rather suspect, when, as sometimes happens, they differ greatly from those obtained more directly.

Liquid water meters

Though liquid water content can be determined from measurements of droplet concentration and size-distribution, it is useful to have an independent instrument to measure this quantity in a continuous manner, and several such devices have been constructed.

Vonnegut (1949a) has described an instrument in which cloud droplets are collected by impaction on a porous sintered metal surface already saturated with water and connecting to a water-filled system. Capillary forces draw the collected droplets into the pores of the surface, and displace a corresponding liquid volume in a graduated capillary tube connected to the system. This is a direct-reading instrument, but does not have a very fast response time, nor is it readily adaptable for continuous recording.

Warner and Newnham (1952) have developed an instrument with fast response time and automatic recording which is well suited to making measurements during aircraft traverses of warm clouds. In this instrument, a paper tape is driven steadily past a slit exposed to the impact of cloud droplets. The electrical resistance of the paper tape is found to vary reproducibly with its water content, so that a measurement of the transverse resistance of the tape after exposure gives a measure of the amount of impacted water. The electrical signal is recorded continuously and gives a record of the cloud water content with only a small time-lag, and with a resolution of the order of a few tenths of a second. This allows the fine structure of the water content within a cloud to be examined fairly readily.

Both the capillary collector and the tape method depend upon impaction for the collection of cloud water and hence discriminate against small droplets. This is not usually serious since most of the cloud water content resides in droplets greater than 10 μ dia. in typical clouds, and for these the collection efficiency is high. The overall accuracy of the tape method is estimated at $\pm 20\%$ which, while it leaves something to be desired, is adequate for most purposes.

Another type of instrument having a relatively fast response contains a hot wire upon which droplets impinge (Neel and Steinmetz, 1952). Evaporation of the droplets cools the wire and the amount of cooling is measured by a thermocouple or resistance thermometer. Comparative tests (Barnett, 1958) show this type of instrument to agree closely with the paper-tape method, with which its performance is comparable. A similar comparison (Warner and Squires, 1958) between the paper-tape instrument and the cloud droplet sampler of Squires and Gillespie also gave satisfactory agreement and confirmed that both methods were probably reliable to within $\pm 20\%$.

Other instruments have been described in which liquid water is collected on a rotating cylinder or disk following the same principle as in the Mount Washington apparatus. Variations discussed include cooling the disk so that the droplets freeze

Experimental Methods

and measuring the rate of build up of the ice deposit (Day, 1955). This type of instrument, of course, involves rather long averaging times and is not suitable for detailed study of cloud structure.

FACTORS DETERMINING MICROSTRUCTURE

Before we go on to consider the results of measurements of the properties of clouds, it will be helpful to recognize those factors which determine the microstruture, so that rational distinctions and classifications can be made.

We have already discussed some of these points in chapter 1 where we considered the effects of cloud-base temperature, up-draught velocity, turbulence and cloud depth upon the general behaviour of a cloud. These factors lead to the general classification of clouds, familiar to meteorologists, and based largely upon their external appearance. To these we must add a general specification of the type of air-mass in which the clouds form. This is less obvious, and its importance has only recently been recognized. A classification is rather difficult, but we can distinguish broadly between maritime air-masses which have spent many days or weeks over the ocean, and continental air-masses which have spent similar long times over the interior of continents. We should note, however, that maritime air-masses can travel appreciable distances inland without losing their characteristic nucleus content, and similarly continental air may be found many hundreds of miles out to sea.

Few measurements of cloud properties are available in the literature giving data for all of these contributing variables, but we can nevertheless examine the data broadly on the basis of whether it applies to layer or convective clouds and whether the air-mass is predominantly continental or maritime. For the remainder of this chapter we shall examine the experimental data which has been collected and see what broad pattern emerges.

LIQUID WATER CONTENT

Since we have already discussed the results of observations of liquid water content in chapter 1, we shall mention them only briefly here. One of the main points brought out by measurements with instruments having fast response times is that liquid water content varies very rapidly from place to place within a cloud. The maximum values of water content are almost always found to be considerably less than the adiabatic value, and of course mean values are lower still. A value of about 0·5 g/m³ is probably typical for liquid water content averaged throughout a normal cumulus cloud, the peak value within the cloud being perhaps twice this quantity.

DROPLET CONCENTRATION

Like the liquid water content, the droplet concentration in clouds undergoes rapid spatial variations, and almost clear patches are often found within comparatively dense clouds. Since sampling methods examine only a minute amount of cloud it is therefore usually necessary to have a large number of measurements to obtain a representative droplet count for the cloud as a whole. The various magazine-type samplers which we have already discussed are therefore very useful and convenient for these measurements.

Because the importance of the origin and nucleus content of the air-mass involved in cloud formation was not recognized until quite recently, most earlier observations of cloud droplet concentration varied widely for a given type of cloud for no apparent reason. This variation we now attribute to the air-mass itself and when this is taken into account, as we shall see below, a coherent pattern emerges.

Recently Squires and Warner (1957; also Squires, 1956, 1958a, b) have made an extensive series of measurements on clouds which can be unambiguously classed as of maritime, transitional or continental origin, and of slow or fast lift rate. The maritime clouds were measured near the island of Hawaii and off the south and east coasts of Australia, no systematic geographical differences being found. The continental clouds were all measured over inland Australia.

The results of measurements of droplet concentrations are shown in fig. 5.1. and in table 5.1. For the three types of cloud originating in the same air-mass, the distinction between layer-clouds and convective clouds is clear. Hawaiian orographic cloud has an updraught of only 15–25 cm/sec and the resulting median droplet concentration is only 10 cm^{-3}. The dark stratus behaves similarly. Maritime cumuli, on the other hand, have updraughts of several metres per second and the resulting median droplet concentration was found to be 45 cm^{-3}. This general

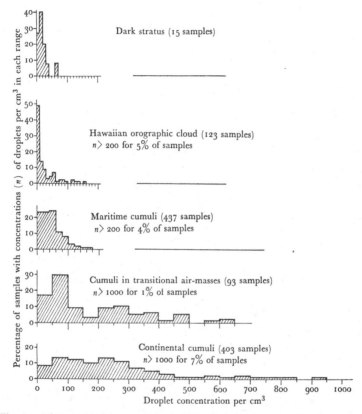

Fig. 5.1. Histograms of the percentages of samples taken in each of five cloud types for which the droplet concentrations fell in the ranges indicated by the graduations on the horizontal axes. The scale of concentrations is common to the five histograms (Squires, 1958a).

tendency for layer-clouds to have lower droplet concentrations than convective clouds has also been shown by the results of many other measurements under less clearly defined conditions (for example, Diem, 1948).

The distinction between maritime and continental clouds of

the same type is even more striking. Whilst Squires found a median droplet concentration of 45 cm^{-3} for maritime cumuli, the corresponding figure for continental cumuli was 228 cm^{-3}, with clouds in transitional air-masses having an intermediate concentration as expected. Very similar results were found by Battan and Reitan (1957) who found an average droplet concentration of 55 cm^{-3} in cumuli over the ocean near Puerto Rico, and 200 cm^{-3} in clouds over the central United States. Similar direct comparisons are not available for layer-clouds, but comparison of figures given by Squires for maritime clouds and Diem for clouds of unknown origin suggest the distinction may be equally striking in this case when additional data has been collected.

Table 5.1. *Data on the frequency of values of the droplet concentration per cm^3 in three types of cloud (after Squires, 1958b)*

	Hawaiian orographic cloud	Maritime cumuli	Continental cumuli
First quartile droplet concentration	4	22	119
Median droplet concentration	10	45	228
Third quartile droplet concentration	30	70	310
Maximum droplet concentration observed	370	470	2800
Total number of observations	123	438	403

A certain amount of work has also been done on the variation of droplet concentration from place to place within a cloud. Apart from the rapid fluctuations on which we have already commented, there is a steady decrease in droplet concentration with height above cloud base (Zaitsev, 1950; Weickmann and aufm Kampe, 1953; Squires, 1958a, c). This decrease has been reported as ranging from about a factor of 2 to a factor of 5 or more, and is clearly related to the dilution mechanism by which the liquid water content is reduced in the upper parts of clouds. Zaitsev also concluded, on the basis of rather few measurements, that there was a decrease in droplet concentration towards the sides of clouds, but Squires (1958c) found that the

Droplet Concentration

droplet concentration, like the liquid water content, does not, on the average, begin to fall off until within about 100 m of the edge of the cloud.

DROPLET SIZE-DISTRIBUTION

From what we have found out about liquid water contents and droplet concentrations we can immediately say something about drop sizes in different cloud types. We have noticed that, averaged over a large volume, the liquid water content in convective clouds is roughly half a gramme per cubic metre, so that it immediately follows that clouds with large droplet concentrations consist of small droplets, whilst those with small droplet concentrations contain many large droplets. Layer-clouds should thus have larger droplets than more active convective clouds, and maritime clouds larger than continental clouds; this generalization is borne out by measurements, as we shall see directly.

Since the shape of a droplet size-distribution is of importance, as well as the average size of the droplets, we must first make a few remarks on this subject. Two distributions are in common use. When droplet numbers are being considered a plot of $n(r)$ against r, where $n(r)dr$ is the number of droplets in range dr about r, is often used. An alternate distribution is that of $r^3 n(r)$ against r, which is useful since $\frac{4}{3}\pi r^3 n(r)\, dr$ is the contribution to the liquid water content of droplets within dr of r. The mean, median or mode of either one of these distributions may be used to specify a measure of droplet size, but since the shape of the spectrum varies widely it is really necessary to specify the complete spectrum.

Because the sampling of cloud droplets is a rather difficult procedure when all factors are considered, and the exact technique used in sampling and taking appropriate averages may have a considerable influence on the final result, comparisons between clouds of different types are best made between measurements by the same observer. For this reason we shall give details of some fairly complete sets of results and then comment more briefly upon less extensive measurements.

Since Squires (1958 a) is almost the only worker to make explicit distinction between maritime and continental air-masses, we shall present his results first. This is done in graphical form in fig. 5.2. The dark stratus, orographic cloud and trade-wind cumuli all originate in essentially the same maritime airmass. Going from orographic cloud and dark stratus to trade-wind cumulus, there is a progressive increase in droplet concentration, decrease in average drop size, and narrowing of the droplet spectrum, all with very little change in liquid water content. This illustrates the effect of cloud type when the nucleus population remains constant.

Continental cumuli almost appear as an extension of this series. The droplet concentration is higher, the droplet size smaller, and the distribution narrower still. Since the convective activity and water content are quite similar to those of maritime clouds, the distinction is presumably due to the nucleus content of the two air-masses involved, as we shall see later.

Squires' observations on layer-clouds are supported by measurements on sea fog by Houghton and Radford (1938) who found a median volume diameter of 46 μ, droplet concentration of only about 3 cm^{-3}, and a liquid water content of 0·13 g/m^3.

The distinction between maritime and continental cumuli is supported by the data of Battan and Reitan (1957) referred to previously which shows a rather broad spectrum for maritime cumuli, with median diameter about 30 μ, while continental cumuli had narrower spectra and median droplet diameters near 10 μ. These figures agree very closely with Squires' results shown in fig. 5.2.

Other observations of droplet spectra have not specified the origin of the air-mass in which the clouds have formed, and for this reason are more difficult to assess. Diem has made a series of measurements in clouds of different types and his results are summarized in fig. 5.3. We see that stratus and nimbo-stratus clouds show rather broad spectra and low droplet concentrations compared with strato-cumulus or fair-weather cumulus. Alto-stratus has a higher droplet concentration than might be expected, but this cannot be evaluated in the absence of further details. Perhaps the most notable deviation is the case of cumulus congestus, which shows a rather low droplet concentration, a broad spectrum and a large mean droplet diameter. This cloud type was also found to have a relatively high water content.

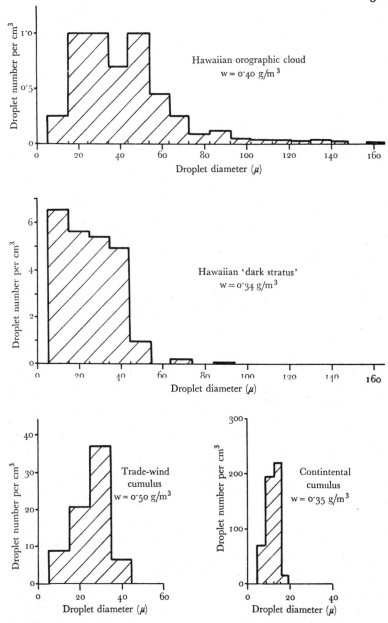

Fig. 5.2. Droplet spectra in clouds of various types, with liquid water contents w as shown. Cumulus samples are taken 2000 ft above cloud base, orographic and dark stratus values are average. Note change in ordinate scale from figure to figure (after Squires, 1958a).

This distinction between small and large cumuli has been noted by other workers to a more or less marked extent. Weickmann and aufm Kampe (1953) found an average liquid water content of 1·0 g/m³ and droplet concentration of 300 cm⁻³ in fair-weather cumuli, whereas in the upper part of

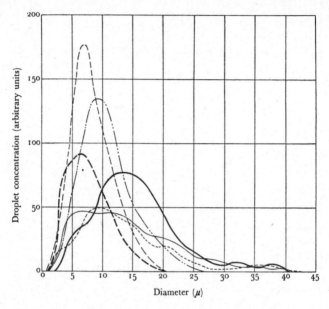

Fig. 5.3. The mean droplet size-distributions for various cloud types (Diem, 1948). Fair-weather cumulus, - - -; cumulus congestus, ———; strato-cumulus, - - -; alto-stratus, — · · —; nimbo-stratus, ———; stratus, - - - -.

a cumulus congestus average values were 3·9 g/m³ and 64 cm⁻³ respectively. Though the mode droplet diameters were the same in both cases (about 10 μ), the droplet spectrum was much broader in the case of cumulus congestus, this accounting for the higher liquid water content.

Results showing a similar trend have been reported by Durbin (1956), who considered groups of cumuli with thicknesses in the ranges 750–2500 ft and 3850–7000 ft, and by Battan and Reitan (1957). The differences observed were, however, very much less pronounced than found by Weickmann and aufm Kampe. Some of these measurements are compared in fig. 5.4.

A partial explanation of this effect is found in the spatial variation of microstructure within cumuli, which has been investigated by Zaitsev (1950) and by Squires (1958a). Both found that average droplet diameter and width of size-distribution increased with height above cloud base, whilst the

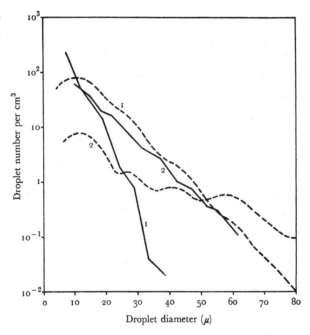

Fig. 5.4. Various measurements of drop size-distributions in small and large cumuli, mostly of continental origin. Weickmann and aufm Kampe ---; Battan and Reitan ——. 1 = fair-weather cumulus; 2 = cumulus congestus.

droplet concentration fell. This is illustrated in figs. 5.5 and 5.6. Small cumuli do not show the development characteristic of the tops of large cumuli, and the spectra tend to be narrow. Large cumuli have more large drops, and the effect is accentuated if samples are taken near cloud-top level, as was done by Weickmann and aufm Kampe.

We should perhaps mention in passing that some optical measurements in clouds on mountain-tops (Driving et al. 1943; Keily, 1954; see also Fritz, 1954) suggest the presence of very large numbers of droplets less than 1 μ dia. Direct measure-

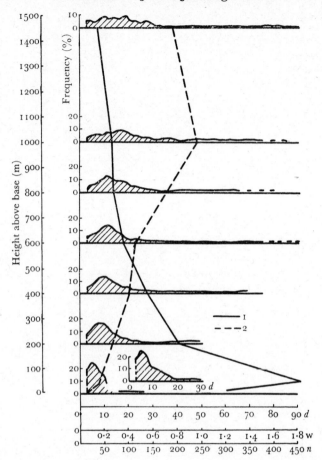

Fig. 5.5. Average drop size-distribution, droplet concentration and liquid water content according to height in cumulus congestus. d, droplet diameter in microns; n (curve 1), droplet concentration/cm^3; w (curve 2), liquid water content in g/m^3 (from Zaitsev, 1950).

ment of droplets as small as this is very difficult, and as yet no confirmation has been obtained by other methods. Whilst mountain-top conditions are not typical of clouds in the free air, the existence of large numbers of small drops seems most unlikely even in this case, in view of the demonstrated validity of theoretical treatments which predict no such effect. We shall discuss these treatments in the next chapter.

Droplet Size-distribution

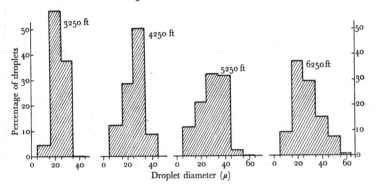

Fig. 5.6. The mean droplet spectrum at four levels in a trade-wind cumulus off the east coast of Hawaii. Cloud base was at 2250 ft and cloud-top at 9500 ft. Droplet concentrations (n) per cm³ and liquid water content (w) g per m³ are: 3250 ft: $n = 74$, $w = 0.50$; 4250 ft: $n = 51$, $w = 0.53$; 5250 ft: $n = 27$, $w = 0.50$; 6250 ft: $n = 30$, $w = 0.42$ (Squires, 1958a).

Distribution shapes

From Diem's results, shown in fig. 5.3 and those of Squires in fig. 5.2 it appears that cloud drop size-distributions are generally bell-shaped, more or less asymmetrical, and with a tail extending towards large radii. This is the general shape of the log-normal distribution which is characteristic of many natural processes (Levin, 1954b).

Best (1951) has examined the earlier experimental data of Mazur, Diem and others, and found that in many cases the drop size-distribution, on a volume basis, can be fitted by the formula

$$1 - F = \exp[(-x/a)^n], \qquad (5.1)$$

where F is the fraction of liquid water contained in droplets less than x in diameter, and a and n are constants. These parameters vary from cloud to cloud, but the average value of n is about 3·3. This is again an asymmetric bell-shaped curve and fails to represent well those distributions having a second hump in the liquid water distribution at large droplet sizes.

THE STABILITY OF CLOUDS

The general stability-pattern of clouds has been known for a long time, but has excited little comment. Explicitly putting aside, for the present, consideration of any clouds which may

Microstructure of Non-freezing Clouds

contain ice-crystals in their upper parts, it is well known that clouds of considerable vertical development are more likely to rain than clouds of smaller depth, so that the depth of a cloud type before precipitation commences may be taken as a rough measure of the stability of its microstructure.

The most important observational distinction between the stability of clouds of different types is that between clouds of

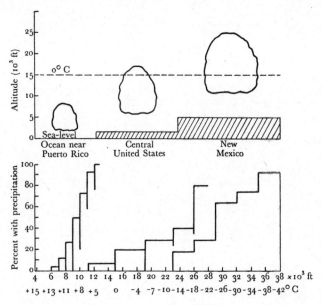

Fig. 5.7. Precipitation likelihood as a function of cloud-top height and cloud-top temperature for cumulus clouds in the Caribbean Sea area, Central United States and New Mexico. The clouds pictured in the upper part of the figure are of the size, and at the location, of those clouds in the three regions which have a 20% precipitation probability (Battan and Braham, 1956).

continental and maritime origin. Even in general appearance these clouds are different. Continental clouds tend to have flat bases and to grow to considerable heights in compact structures with sharply defined edges; maritime clouds, on the other hand, tend to have ragged edges, and a less compact structure. This general pattern and the observation that maritime clouds rain more easily than do continental clouds was remarked upon by Houghton (1950), and has recently been considered explicitly

The Stability of Clouds

by Squires and Twomey (Squires, 1956, 1958a; Squires and Twomey, 1958).

This stability-pattern has received detailed support by studies of the Cloud Physics Group at the University of Chicago (Byers and Hall, 1955; Battan and Braham, 1956; Braham, Reynolds and Harrell, 1951) who studied clouds in different localities by radar methods and took a census of the cloud heights for which precipitation elements became detectable. These studies were carried out over the Caribbean Sea, over continental United States, and over New Mexico and form a sequence from maritime to extreme continental conditions. The results are summarized graphically in fig. 5.7 and show excellently the instability of maritime clouds, and the great stability of clouds originating in extreme continental air-masses.

The immediate conclusion to be drawn from these observations and our previous discussions of microstructure is that instability in clouds is associated with large average cloud droplet size and broad droplet spectra. A similar conclusion is reached if we examine fig. 5.3 giving Diem's data for various cloud types. Those clouds normally associated with rain have large droplet sizes and broad spectra in general, though here matters are complicated by other differences in cloud structure. We shall examine reasons for this association in the next chapter.

CHAPTER 6

THEORY OF THE DEVELOPMENT OF NON-FREEZING CLOUDS

THE GROWTH OF CLOUD DROPLETS

WE have already discussed, in chapter 3, the way in which a liquid droplet nucleates on a foreign particle in a supersaturated vapour, and in chapter 4 we have examined the population of such aerosol particles in the free atmosphere. A study of the formation of a cloud involves two further lines of thought. In the first place we must discuss the way in which the size of an individual droplet increases by condensation from a supersaturated environment, and secondly we must consider the interaction between the droplets growing upon all the members of the nucleus population.

Individual droplets

Consider a solution droplet of radius r at rest in a supersaturated environment in which the concentration of vapour molecules at a great distance from the droplet is n_0. This vapour diffuses towards the droplet and condenses upon it, and from the requirements of continuity of flow, if the growth-rate of the droplet is small, the concentration of vapour molecules at any point R approximately satisfies the equation

$$\nabla^2 n(\mathrm{R}) = 0. \tag{6.1}$$

When $\mathrm{R} = \infty$, n must have the value n_0, and near the surface of the drop it must equal n_r, the equilibrium vapour concentration over the droplet surface.† The appropriate solution of (6.1) is then

$$n(\mathrm{R}) = n_0 + (n_r - n_0)r/\mathrm{R}. \tag{6.2}$$

† This boundary condition is not strictly correct when condensation or evaporation is occurring at an appreciable rate. A correct expression involves consideration of the kinetics of vapour molecules near the droplet and of the accommodation coefficient of its surface. A general discussion has been given by Schrage (1953) and a more simple treatment of the present case by Rooth (1957). The end-result

If the diffusion coefficient of vapour molecules is D, then the rate at which the mass of the droplet is increasing is

$$\frac{dM}{dt} = 4\pi r^2 m D \left(\frac{dn}{dR}\right)_r, \qquad (6.3)$$

where m is the mass of a molecule. Substituting from (6.2) we find

$$\frac{dM}{dt} = 4\pi r D m (n_0 - n_r), \qquad (6.4)$$

so that the rate of increase of the volume of a droplet is proportional to its radius, rather than to the surface-area as might be expected without a close examination.

Expressing M in terms of r and the n's in terms of partial pressures we have

$$r\frac{dr}{dt} = D \frac{\rho_V}{\rho_L} \frac{(p - p'_r)}{p}, \qquad (6.5)$$

where p is the partial pressure of vapour at a large distance from the droplet and p'_r that in equilibrium with the surface of the droplet. ρ_L is the density of liquid water, and ρ_V the density of water-vapour in the environment.

Derivation of this equation involves two approximations. In the first place we have assumed that the system is essentially in a steady state. This is not true since the radius of the droplet is increasing, but for small supersaturations the growth-rate is sufficiently small that the approximation is valid. Secondly, we have assumed that the diffusion coefficient is a constant and that the diffusion equation applies to regions comparable with the mean free path of a vapour molecule. Howell (1949) quotes work by Langmuir (1944) which shows that in fact the diffusion constant has its normal value for diffusion to droplets larger than a few microns in radius, but for very small droplets the effective diffusion coefficient is less than this. However, Squires (1952a) has shown that droplets grow so quickly through this stage that no appreciable error is made by using the normal uncorrected diffusion coefficient, so we shall not consider this additional complication here.

is an effective reduction of the diffusion coefficient for the growth of small droplets. It does not seem likely that this will be of importance unless the accommodation coefficient for a liquid water surface turns out to be very much less than unity.

Development of Non-freezing Clouds

We have made one other important assumption, namely that we have assumed the droplet to be at rest in its environment. This is physically impossible in the atmosphere, and in practice the droplet will always be falling with its terminal velocity relative to the immediately surrounding air. The effect of this motion is to bring new parcels of air continually into the vicinity of the droplet, and this ventilation effect increases the rate of supply of moisture. The magnitude of the effect depends upon the characteristics of the air-flow past the droplet and is best expressed in terms of the non-dimensional parameters familiar in aerodynamics, the Reynolds number Re, the Prandtl number Pr, and its analogue Pr' for the flow of vapour rather than heat.† Treatments of this problem have been given by Frössling (1938), by Kramers (1946) and by Kinzer and Gunn (1951) and a review has been given by Richardson (1953). Most of these authors have been concerned with the related problem of evaporation from drops falling through unsaturated air, but calculations have been made specifically for condensation by Squires (1952 a). Inclusion of the ventilation effect modifies (6.5) to the form

$$r\frac{dr}{dt} = D\frac{\rho_V}{\rho_L}\frac{(p-p'_r)}{p} f(\text{Re, Pr'}). \qquad (6.6)$$

The function $f(\text{Re, Pr'})$ has been calculated by Squires for values of interest in our present discussion, and the results are given in table 6.1.

Table 6.1. *Values of the ventilation factor* $f(\text{Re, Pr'})$ *in (6.6) for* 10° C, 800 mb *(Squires, 1952 a)*

$r(\mu)$	10	30	50	70	100
Re	0·014	0·37	1·5	3·2	8·0
$f(\text{Re, Pr'})$	1·03	1·14	1·28	1·41	1·66

It can be seen that for droplets less than about 10 μ in radius the ventilation effect may be neglected, and even for droplets

† The Reynolds number is defined by $\text{Re} = 2ru/\nu$, where r is the droplet radius, u the velocity of the drop relative to the air, and ν the kinematic viscosity of air (viscosity/density). The Prandtl number is $\text{Pr} = \nu/\kappa'$, where κ' is the thermal diffusivity of air (conductivity/specific heat per unit volume). The analogue of Pr for the flow of vapour is $\text{Pr'} = \nu/D$, where D is the coefficient of diffusion of water-vapour in air.

The Growth of Cloud Droplets

as large as 50 μ in radius it affects the result only to the extent of about 30 %.

Our next problem is to evaluate the vapour-pressure p'_r in equilibrium with the droplet during its growth. This involves three factors—the curvature of the droplet surface, the presence of solute, and the heating of the droplet by release of latent heat of condensation. Let us first evaluate the temperature rise caused by this latent heat, and then combine it with the other effects.

The heat-flow equations are almost identical with those we have already considered for vapour-diffusion. We make similar simplifications by neglecting the slight increase of heat energy stored in the drop as its radius increases and neglect also the deviation of the effective thermal conductivity of air from its normal value for very small particles. Equations (6.1) and (6.2) then apply with T substituted for n, and analogously to (6.4) we have†

$$\frac{dM}{dt} = \frac{4\pi r \kappa}{L}(T_r - T_0), \qquad (6.7)$$

where κ is the thermal conductivity of air, L the latent heat of vaporization of water, T_r the temperature of the drop surface and T_0 that of the environment. As before we should consider the effects of ventilation and this leads finally to an equation analogous to (6.6), namely,

$$r\frac{dr}{dt} = \frac{\kappa}{L\rho_L}(T_r - T_0)f(\text{Re}, \text{Pr}). \qquad (6.8)$$

Now the effect of temperature on saturated vapour-pressure is given by the Clausius-Clapeyron equation (Glasstone, 1947, p. 227) which, if the volume occupied by a liquid molecule can be neglected compared with that of a molecule in the vapour, can be written

$$\frac{dp}{dT} = \frac{L}{TV},$$

where L is the latent heat of vaporization per gram, and V is the volume occupied by one gram of the saturated vapour. This

---

† Note that throughout the derivation all heat quantities are expressed in c.g.s. mechanical units.

equation can be integrated immediately if we make the approximation that the vapour itself obeys the ideal gas laws, and gives

$$\ln \frac{p(T_r)}{p(T_0)} = \frac{LM_0}{R}\left(\frac{1}{T_0}-\frac{1}{T_r}\right), \tag{6.9}$$

where M_0 is the gram molecular weight of water and R is the gas constant. For small temperature differences this equation can be written

$$\frac{p(T_r)-p(T_0)}{p(T_0)} \approx \frac{LM_0}{RT^2}(T_r-T_0) \tag{6.10}$$

and this is sufficiently accurate for our present purposes.†

Now in chapter 3 we considered the effects of surface-curvature and solute-content on vapour-pressure, and (3.42) shows that

$$(p'_r/p_\infty) \approx 1 + \frac{a}{r} - \frac{b}{r^3}, \tag{6.11}$$

where a relates to the effects of surface-tension and b to the presence of solute. If we write the supersaturation of the environment relative to a plane surface of pure water as

$$S = \frac{p-p_\infty}{p_\infty} \tag{6.12}$$

then (6.10), (6.11) and (6.12) can be combined to give the supersaturation relative to the growing droplet as

$$\frac{p-p'_r(T_r)}{p} \approx S - \frac{a}{r} + \frac{b}{r^3} - \frac{LM_0}{RT^2}(T_r-T_0). \tag{6.13}$$

The intermediate temperatures and vapour-pressures can now be eliminated between (6.6), (6.8) and (6.13), by simple algebra, to give the growth equation

$$r\frac{dr}{dt} = G\left(S-\frac{a}{r}+\frac{b}{r^3}\right)f(\text{Re, Pr}'), \tag{6.14}$$

where
$$G = \frac{D\rho_V}{\rho_L}\left[1+\frac{DL^2\rho_V M_0 f(\text{Re, Pr}')}{RT^2\kappa f(\text{Re, Pr})}\right]^{-1}, \tag{6.15}$$

and κ is the thermal conductivity of air.

† This equation is often used to express supersaturations in terms of differences in dew-point temperature. Thus a supersaturation of $1°$C implies a supersaturation of about $6\frac{1}{2}\%$ at $10°$ C, this relation varying somewhat with temperature.

The Growth of Cloud Droplets

This equation is quite general and applies to the evaporation of large raindrops as well as to the growth or evaporation of cloud droplets. In the latter case table 6.1 shows it is a reasonable approximation to set both the ventilation factors f equal to unity.

Droplet populations

Equation (6.14) describes the way in which a single droplet grows in a supersaturated environment. We are interested, however, in the case of a large population of droplets growing in a rising current of air to form a cloud, and there will clearly be some interaction between the growing droplets. Langmuir (1944) has shown that with droplets at rest with respect to the air, the separation of droplets in normal clouds is so great compared with their diameters that any individual interactions between droplets is negligible. When droplet motion is considered this conclusion is also true, since a typical growing cloud may contain drops of say 7 μ radius, separated from each other by about 1000 μ on the average, and falling relative to the air about 5000 μ/sec. Every growing droplet therefore experiences essentially the same environment, and the droplets influence each other only through their combined influence on this common environment.

The general outline of the condensation process in a steadily rising current of air is very straightforward. As the air rises it expands approximately adiabatically and the saturation ratio increases. Once the saturation point has been passed, condensation begins to occur on the suspended nuclei, the most efficient being activated first. The supersaturation continues to rise and more and more nuclei are activated and begin to grow to droplets; the rate of increase of supersaturation is, however, now rather slower because the growing droplets are removing the excess vapour from the air. Since large droplets remove water-vapour more quicky than do small ones (6.4) the excess vapour is soon being removed from the air as quickly as it is being made available by expansion, and thereafter the supersaturation falls towards zero.

A more detailed discussion clearly requires consideration of the lift rate and of details of the nucleus population, along with use of the growth equations we have developed, and to this we now turn. Satisfactory discussions of this process are all comparatively recent. Kraus and Smith (1949) have given a treatment for several concentrations of nuclei considered to become active at zero supersaturation. However, their treatment is based on (6.5) and neglects the effects of latent heat and ventilation. This is not a very serious omission since it affects only the magnitude of G in (6.14), but this of course changes the final numerical results. Howell (1949) has made very detailed calculations for various updraughts and model nucleus spectra using (6.14) but with the omission of the ventilation factor, which as we have seen is of little importance. Tsuji (1950) has considered the effect of ventilation, as also has Squires (1952a), and recently Twomey (1959b) has made calculations with actual observed nucleus spectra.

Consider a parcel of air which is rising vertically. In the absence of any condensation the supersaturation will increase according to a law of the form

$$\frac{dS}{dt} = Q_1 \frac{dh}{dt}, \qquad (6.16)$$

where h is vertical height and Q_1 is substantially constant. If, however, condensation is occurring, then this will tend to relieve the supersaturation, and the amount by which the supersaturation is lowered will be proportional to the mass ω of liquid condensed per cm³. Combining this effect with (6.16) we have

$$\frac{dS}{dt} = Q_1 \frac{dh}{dt} - Q_2 \frac{d\omega}{dt}. \qquad (6.17)$$

The expressions Q_1 and Q_2 have been evaluated by Squires (1952a) and, converting his results to the system of units in use here, we have

$$Q_1 = \frac{L M_0 g}{R T^2 c_p} - \frac{M_a g}{R T} \qquad (6.18)$$

and

$$Q_2 = \frac{L^2 M_0}{T P M_a c_p} + \frac{R T}{p M_0}. \qquad (6.19)$$

In these equations P is the total atmospheric pressure, M_a the average molecular weight of air and c_p its specific heat at constant pressure; p is the partial pressure of water-vapour and

The Growth of Cloud Droplets

M_0 its molecular weight, and g is the acceleration due to gravity.

In (6.17), dh/dt is specified by the updraught in the cloud and $d\omega/dt$ can be found from the equations of nucleation and growth as a function of S and of the nucleus spectrum. From (6.14) we have

$$\frac{d\omega}{dt} = \sum_r 4\pi r^2 \frac{dr}{dt} \rho_L$$

$$= \sum_r 4\pi r \rho_L Gf \left(S - \frac{a}{r} + \frac{b}{r^3}\right). \qquad (6.20)$$

In this form (6.17) can be integrated numerically for an assumed nucleus population, and this is essentially the method adopted by all the workers to whom we have referred.

The results of such treatments are most clearly shown by the figures computed by Howell (1949). Fig. 6.1 shows the assumed nucleus spectra for two of these calculations. Spectrum A contains a rather small number of nuclei and the rate of lift used in the calculation (60·4 cm/sec) is fairly high. The computed droplet distribution and saturation curves are shown in fig. 6.2.

From this figure it is clear that the supersaturation behaves in just the manner we have described. For times less than about 10 sec condensation is negligible and the supersaturation rises linearly with time. As condensation begins the supersaturation increases more slowly and reaches a maximum of 0·35 % after 25 sec, after which it slowly decreases.

The drop size history shows other interesting features. Below their critical supersaturations drop sizes remain essentially in equilibrium, with very little time-lag. Once nucleation occurs, however, droplet growth is rapid, and because small droplets increase in radius much more quickly than do large droplets, the size-distribution soon becomes remarkably uniform. As the supersaturation falls the haze droplets which have not been activated as nuclei shrink back to smaller sizes.

The nucleus spectrum marked B in fig. 6.1 is more typical of atmospheric conditions, and for the calculation Howell uses a lift rate of 30·2 cm/sec, which is also moderate. The resulting curves are shown in fig. 6.3 which is very similar to fig. 6.2. The peak supersaturation is this time about 0·072 %,

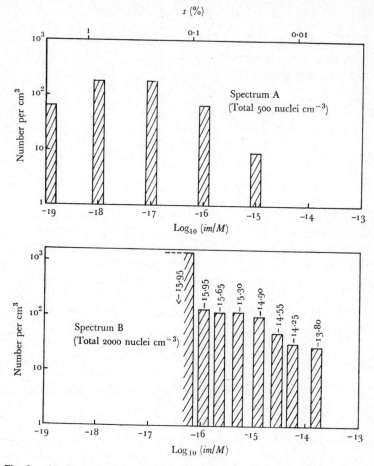

Fig. 6.1. Nucleus spectra used in Howell's calculations. im/M is nuclear size in gram equivalents; s is the critical supersaturation, per cent, at which the various nuclei become active. Shaded blocks are used pictorially only; in the calculation nuclei of discrete sizes are used.

reached after about 70 sec. Again, the droplet spectrum tends to narrow as condensation proceeds, but it is rather broader in this case.

An important point which arises from these calculations is that the supersaturation in a growing cloud is very small, typically about a tenth of a percent, and it can surpass 1 % only

The Growth of Cloud Droplets

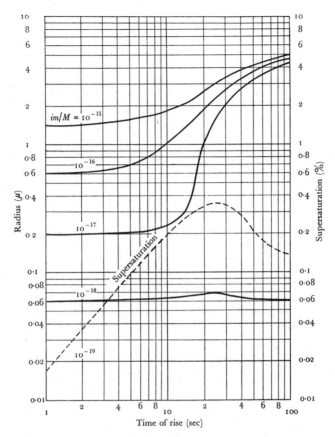

Fig. 6.2. Drop growth curves for spectrum A at a lift rate of 60·4 cm/sec. The course of the supersaturation is also shown (Howell, 1949).

under extreme conditions. Howell's calculations, of course, were made for model nucleus distributions not directly related to any actual measurements. His conclusions about the behaviour of the supersaturation and about its maximum value are, however, borne out by Twomey's (1959b) calculations for actual observed nucleus spectra, as we shall see later.

Since, as we have remarked, almost all the nuclei which become activated grow to cloud droplets, a determination of the maximum supersaturation attained together with a knowledge of the nucleus spectrum would allow one to calculate the

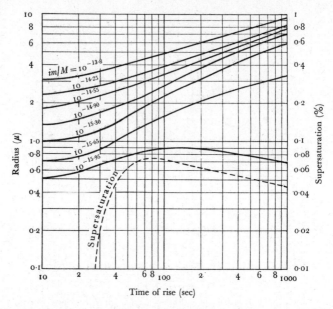

Fig. 6.3. Drop growth curves for spectrum B at a lift rate of 30·2 cm/sec. The course of the supersaturation is also shown (Howell, 1949).

resulting droplet concentration. Approximate calculations of this nature have been performed by Squires (1952a) and by Twomey (1959b).

Following Twomey, we observe that once the drops have grown to an appreciable size, it is a reasonable approximation to neglect the effects of the solute and of surface-curvature, so that the growth law (6.14) for an individual drop becomes

$$r\frac{dr}{dt} \approx GfS \tag{6.21}$$

or
$$r^2 \approx 2Gf \int_\tau^t S\, dt, \tag{6.22}$$

where τ is the time at which the particular nucleus becomes activated. The approximate form of (6.20) is similarly

$$\frac{d\omega}{dt} = 4\pi\rho_L Gf S \Sigma\, r. \tag{6.23}$$

The Growth of Cloud Droplets

Now if we suppose that the number of nuclei with critical supersaturations between s and $s+\Delta s$ is $n'(s)\Delta s$, then (6.17), (6.22) and (6.23) can be combined to give

$$\frac{dS}{dt} = Q_1 \frac{dh}{dt} - Q_3 S \int_0^S n'(s) \left[\int_\tau^t S\, dt\right]^{\frac{1}{2}} ds, \qquad (6.24)$$

where
$$Q_3 = 2\pi\rho_L Q_2 (2Gf)^{\frac{3}{2}} \qquad (6.25)$$

and $S = s$ at $t = \tau$.

This equation cannot be solved directly, but a first approximation can be used for the S function in the integrand (for example, $dS/dt = Q_1 dh/dt$) and the equation used to find the approximate behaviour of S with time. Proceeding in this way Twomey calculated upper and lower bounds to S_{max} and derived explicit expressions for the maximum supersaturation and the concentration of cloud droplets. If the number of condensation nuclei active at supersaturation s is given by

$$n(s) = cs^k \qquad (6.26)$$

which represents a good approximation to experiment, with k usually in the range 0·2–0·5, then

$$n'(s) = kcs^{k-1} \qquad (6.27)$$

and Twomey finds the number of cloud droplets to be[†]

$$n \approx c^{2/k+2} \left[\frac{6 \cdot 8 \times 10^{-6} u^{\frac{3}{2}}}{kB(\frac{3}{2}, \frac{1}{2}k)}\right]^{k/(k+2)}, \qquad (6.28)$$

where u is the uplift velocity in cm/sec and $B(\frac{3}{2}, \frac{1}{2}k)$ is a beta-function. This relation is stated to have an accuracy of better than 15% in most cases. An expression essentially equivalent to this, though expressed in a different form, has been derived by Squires (1958b) by an entirely different argument.

To illustrate the behaviour of the supersaturation as a function of time in typical maritime and continental clouds, we have integrated (6.24) numerically for various updraught velocities, and for Twomey's observed average maritime and continental condensation nucleus spectra ($c = 310$, $k = \frac{1}{3}$ and

---

[†] Converting from supersaturation expressed in degrees to the fractional supersaturations we have been using.

$c = 6000$, $k = \frac{2}{5}$ respectively). The results of these calculations are shown graphically in fig. 6.4. The behaviour is clearly very similar to that of the models calculated by Howell and shown in figs. 6.2 and 6.3. Points worthy of special notice are the systematic differences in behaviour for different uplift velocities u, and the distinction between maritime and continental clouds. Continental clouds reach peak supersaturations after times of 2–15 sec, while corresponding times for maritime clouds are

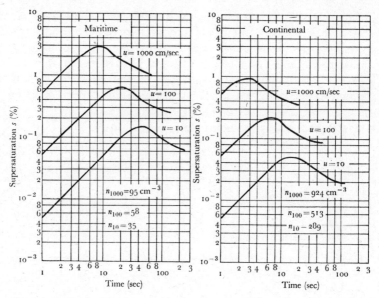

Fig. 6.4. Variation of supersaturation with time in clouds forming in an updraught velocity u in typical maritime and continental air-masses. The quantities n_V are the droplet concentrations generated.

10–50 sec. The peak supersaturation reached in continental clouds is only about a third as great as that achieved in maritime clouds under the same updraught conditions. We shall comment on the droplet concentrations generated in the next section.

Comparison with experiment

Since Twomey (1959a) has made measurements of the condensation nucleus spectrum at low supersaturations in maritime and continental air-masses, it is of great interest to use these

The Growth of Cloud Droplets

values in the theory we have outlined above, and to compare the calculated results with experiment. This comparison has been made by Twomey (1959b). For layer-clouds he assumed an updraught velocity of 10 cm/sec and for cumuli performed calculations for velocities of 1 and 10 m/sec. The results of the calculation are shown in table 6.2 together with the experimental results of Squires (1958b) and of Battan and Reitan (1957). It is immediately clear that the general agreement is very good, and that the distinction between maritime and continental clouds on the one hand and between layer and convective clouds on the other is borne out by the theory. We may therefore feel confident that the theory is fairly reliable, and takes account of at least the major factors influencing the formation of clouds.

Table 6.2. *Calculated and observed droplet concentrations per cm^3 in clouds of various types*

Cloud type	Calculated (Twomey, 1959b)	Observed mean (Battan and Reitan, 1957)	Observed median (Squires, 1958b)	Observed maximum (Squires, 1958b)
Maritime layer	35–60	—	10	370
Maritime cumuli	57–90	55	45	470
Continental cumuli	500	200	288	2800

When we come to compare the droplet spectra predicted by the theory with observed distributions, however, the agreement is far from good. Fig. 6.2 shows that as condensation proceeds the droplet spectrum narrows, and the same tendency can be seen in fig. 6.3. This behaviour is described by (6.14), which shows that, to a first approximation, the rate of increase of droplet radius is inversely proportional to the radius, so that small droplets grow in radius more quickly than their larger companions.

To illustrate this, fig. 6.5, due to East (1957), shows the effect of continued condensation, in the absence of other factors, upon a typical droplet size-distribution. The curve for $\alpha = 1$ g/kg liquid water content is relatively broad, but by the time a mixing-ratio of 10 g/kg has been reached the distribution is very sharply peaked.

It is thus evident that simple condensation alone is unable to explain the observed wide droplet size spectra found in many clouds. In the following sections we shall examine some of the other processes occurring within the cloud and see to what extent they may tend to broaden the distribution.

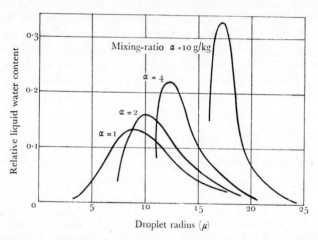

Fig. 6.5. Modification of drop size-distribution by condensation in the absence of coalescence; the liquid water mixing-ratio increases, but the curves are normalized to constant area (East, 1957).

BROAD DROPLET SPECTRA

Several suggestions have been made of ways in which broad droplet spectra could be produced by variants of the uniform condensation mechanism. Howell, Wexler and Braun (1949) considered a model in which different air-parcels within the cloud rise at different speeds and hence develop different droplet populations. At a higher level in the cloud these parcels become mixed and produce a broad droplet distribution. A mechanism of this type has been further examined by Squires (1952 b). The critical part of the differentiation process occurs very near to the condensation level since, from figs. 6.2 –6.4, it is within the first 30 m or so of ascent above this level that the droplet concentration is determined. Once above this level the uplift speed is immaterial provided only that it is not high enough to activate new nuclei. Mixing between air-parcels, of course, must not take place until most of the liquid water has been condensed, otherwise the original differentiation will be lost.

Whilst processes of this type undoubtedly take place within clouds, and these authors have shown that with suitably chosen velocities they can account approximately for at least some drop distributions, it remains to be seen whether this is the dominant process.

A rather different mechanism has been suggested by Best (1952a) and by Mason (1952b) for layer-clouds. These clouds have relatively long lives, of the order of several hours, and differences in updraught velocities are not likely to be large. The boundaries of the cloud are regarded as sinks for droplets which immediately evaporate in the dry air outside. The life of a droplet within the cloud varies from droplet to droplet, with the result that a small fraction of the droplets remain within the cloud for a very long time, and grow to large sizes.

Another process which may be capable of modifying the droplet spectrum is the collision and coalescence of droplets with one another. This topic will be discussed in detail in later sections of the present chapter, and for the moment we shall content ourselves with a few general remarks.

Recent careful calculations have shown that collisions between droplets can only occur when one of the pair is greater than about 18 μ in radius. Coalescence processes are thus only of importance in modifying the spectrum of clouds having rather large average droplet size.

Before these accurate calculations became available, the effect of coalescence upon droplet spectra was investigated by Best (1952a), Das (1956), East (1957) and Elton, Mason and Picknett (1958), using less accurate collision efficiencies calculated by Langmuir. When an error of a factor 10^4 in Das's treatment is corrected these discussions suggest that appreciable modifications of droplet spectra can occur in times of the order of 10 min through the action of coalescence alone provided that the average droplet radius is greater than about 15 μ. Accurate treatments using the new collision efficiency values are not yet available, but the result is likely to remain substantially the same, except that radii perhaps 20 % greater may be required for corresponding effects.

This discussion suggests that mixing may be the agent primarily responsible for spectrum broadening in continental clouds, while coalescence may assume considerable importance in clouds of pronouncedly maritime character. Study of the subject is, however, far from exhausted.

COALESCENCE OF WATER DROPLETS

We have mentioned briefly in the previous section that the collision and coalescence of water droplets within a cloud may be important in determining the population of large drops present, though it appears to be of less importance than condensation as far as most of the small droplets are concerned. We shall now consider this coalescence process in some detail as preparation for a discussion of the stability of clouds and of the development of precipitation within them.

In chapter 4 we examined the coagulation of smoke particles because of their random motion of thermal agitation, neglecting any effects due to fall of the particles under gravity. If we apply the same analysis to the collision of water droplets, using (4.12), we can estimate the importance of this Brownian motion in the coalescence process. In a typical cloud there may be perhaps 300 droplets cm^{-3}, all roughly $10\,\mu$ dia. Equation (4.12) then states that the collision frequency is only $0 \cdot 1\ cm^{-3}\ hr^{-1}$ or $30\ m^{-3}\ sec^{-1}$. This represents a completely negligible proportion of the cloud droplets during the life of an ordinary cloud, so that second collisions with these larger droplets virtually never occur. Collision due to Brownian motion can thus be neglected as a mechanism for the production of appreciable numbers of large droplets.

With Brownian motion neglected and the effect of gravity considered, the problem now becomes an aerodynamic one. The cloud consists of droplets having a range of sizes, all falling with different terminal velocities. The larger droplets overtake the smaller ones and may collide with them, though the collision is complicated by the flow of air around each of the droplets. Coalescence itself is another stage of the process, since it is by no means obvious that two droplets will automatically coalesce if they collide.

Theoretical treatments

An initial treatment of the collision problem was given by Langmuir (1948) as an adjunct to the work on the collection efficiency of impaction obstacles, which we have already discussed. Since he was interested primarily in the collection of small cloud droplets by relatively large drops falling quickly, he made the approximation of considering the small drops to some extent as mass points. This involves only a small correction when the collected droplet is very small compared with the collector drop, but of course it introduces serious errors when the two drops become comparable in size.

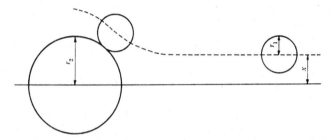

Fig. 6.6. Critical trajectory for a droplet of radius r_1 which is just collected by a larger drop of radius r_2. Collection efficiency is defined as $E = x^2/r_2^2$.

We shall not consider the calculation in any detail since it is complicated, and involves final solution using a differential analyser. Langmuir calculated two extreme cases corresponding to viscous flow at low Reynolds numbers and aerodynamic flow at large Reynolds numbers and interpolated between these to obtain appropriate values for the droplets considered. Referring to fig. 6.6, we have drawn the limiting trajectory for a droplet of radius r_1 which is just collected by a larger droplet of radius r_2. Langmuir defined the collection efficiency E by

$$E = x^2/r_2^2 \qquad (6.29)$$

so that E is the fraction of those droplets, whose centres lie within the volume swept out by the falling drop, which are ultimately collected by it. He calculated a table of values of E as functions of r_2 and r_1, and these results are shown in graphical form in fig. 6.7. The values of E are everywhere less than 100 % and decrease as the size of the collected droplets decreases. The reason for this is illustrated in fig. 6.6 where the trajectory, relative to the collector drop, of a smaller droplet which is just collected is seen to be appreciably curved because of the air-flow pattern around the large drop. The deviation of the path

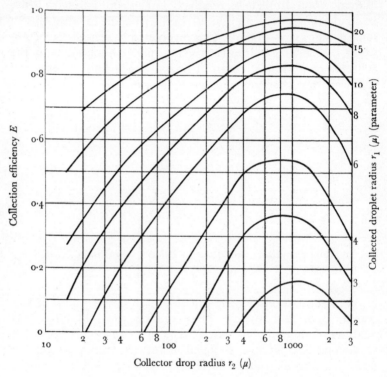

Fig. 6.7. Collection efficiency E for a drop of radius r_2 falling through a cloud of droplets of radius r_1 (after Langmuir, 1948).

of the small droplet from a straight line depends upon its inertia. The larger droplets are not greatly deflected, and the collision efficiency is high. Smaller droplets tend increasingly to follow the lines of air-flow and must lie close to the axis of motion of the large drop if they are to be collected by it. If the cloud droplets are very small they will not be collected whatever their initial trajectory.

Langmuir's solution to the collision problem represents, however, only a first approximation since interpolation methods were used and the effect of the droplet on the motion of the larger drop was neglected. The results are therefore least correct in the region in which the size of the droplet is appreciable compared with that of the drop.

More recently Pearcey and Hill (1957) have given a treatment using Oseen's approximation for flow around a sphere and taking into account the effect of the droplet motion on that of the drop by superposing the individual flow-patterns. Their solution employed a large digital computer and within the framework of the assumptions the results are accurate, rather than being interpolations between extreme cases as was the case with Langmuir's results. The calculations were directed explicitly at examining collisions between droplets nearly equal in size, a situation to which Langmuir's assumptions are clearly inapplicable.

Fig. 6.8 shows the results of these calculations in graphical form. They are expressed in terms of a linear collision cross-section σ, defined with reference to fig. 6.6 by

$$\sigma = x/r_2 \qquad (6.30)$$

and related to Langmuir's collection efficiency E by

$$E = \sigma^2. \qquad (6.31)$$

Whilst the general trend of the results is similar to that found by Langmuir, there is an important difference in that collection efficiencies much greater than unity are found for large droplets of approximately equal diameters. The physical reason for this is brought out by the computed trajectories. When a large droplet is falling, the Reynolds number is high and the droplet has a pronounced wake in which the air tends to be dragged along behind the droplet. The motion of a larger drop overtaking the droplet is influenced by the wake and it tends to be deflected towards the trajectory of the droplet. This effect is greatest when the relative motion of drop and droplet is small, that is when their sizes are nearly equal, because the deflecting force due to the wake has a longer time in which to act.

Although the calculations of Pearcey and Hill represent a great advance on those of Langmuir they are still an approximation, and are least accurate in the case of small droplets of comparable sizes. Thus while their computed collection efficiencies are probably accurate for droplets several tens of microns in radius, their reliability is questionable in the small cloud-droplet region in which we are primarily interested.

Very recently, Hocking (1958, 1959) has examined the flow around droplets at low Reynolds numbers where the Stokes' approximation is valid for the flow-pattern, and found an exact

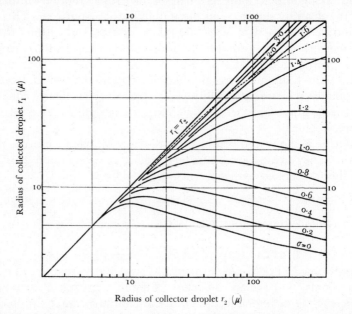

Fig. 6.8. Range of droplet sizes capable of collision. The figure shows contours of constant linear cross-section σ as a function of both droplet radii. Collision cannot occur for droplets which fall below the $\sigma = 0$ curve, and cases above the straight line $r_1 = r_2$ have no meaning. The dotted line represents geometrical cross-sections (Pearcey and Hill, 1957).

solution to the equations. In its range of validity (for droplets less than about 30 μ in radius) this solution enables trajectories to be calculated, again with the aid of a large digital computer, and collection efficiencies determined. The results are shown in fig. 6.9 where E is given as a function of the ratio r_1/r_2.

The most important point which emerges from this calculation is the fact that when the collecting drop has a radius of

18 μ or less, its collection efficiency is zero for all smaller droplets. This radius is much greater than the cut-off value of about 6 μ found by Pearcey and Hill, and this has important consequences as we shall see later.

Study of the computed trajectories for small droplets just escaping collision shows that the droplets roll around each other with a spacing between their surfaces amounting to only a few per cent of the radius of the smaller droplet. This suggests that any mechanism which might cause distortion of the droplet

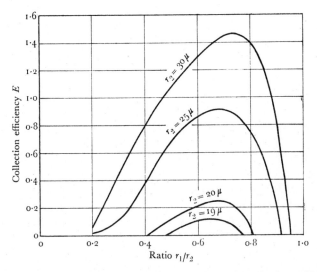

Fig. 6.9. The collision efficiency of small drops. A drop of radius r_2 μ collides with droplets of radius r_1 μ lying in a horizontal circle of area E times the cross-section of the drop. The values of E are shown for various drop radii as functions of the ratio r_1/r_2 (Hocking, 1959).

surfaces at their closest approach might be effective in increasing the collection efficiency for such droplets from zero to some finite value.

Because the regions of overlap between these three calculations are small it is difficult to fit them together into any coherent approximation valid for all drop sizes. It is, however, useful to compare the cut-off values found for collection efficiency by the three approaches. This has been done in table 6.3, the lower rows being generally the more accurate.

From this table it is clear that for coalescence to begin in a cloud it is necessary that there be present initially some droplets greater than about 19 μ in radius. As a drop grows, however, its collection efficiency increases and ultimately it becomes capable of colliding with virtually any droplet in the cloud. We shall return to consider this process at a later stage.

Table 6.3. *Critical radius r^* of a falling drop below which no collection of droplets of radius r can occur. Radii are in microns (1, Langmuir, 1948; 2, Pearcey and Hill, 1957; 3, Hocking, 1959)*

r	1.5	2	3	4	5	6	7	10	15	20
r^*_1	600	350	140	58	31	20	14	—	—	—
r^*_2	—	—	450	100	46	25	15	10	15	20
r^*_3	—	—	—	—	25	22	20	19	19	23

Experimental evidence

Two distinct types of experiment can be performed to investigate the collision-coalescence process. In experiments of the first type actual cloud droplets are studied and the results are immediately applicable to natural clouds. These experiments, however, are rather difficult to perform because of the very small size of cloud droplets and the rather high terminal velocity of larger drops. The results are, therefore, usually of a statistical nature and ingenious experimental arrangements are required if single collisions are to be studied.

The second type of experiment aims at overcoming these difficulties by the use of a modelling method whereby cloud droplets are simulated by spheres of macroscopic size moving through a liquid. As we shall see later, there are fundamental faults in this method which render its results of only qualitative interest.

Considering first the direct experiments, we find that their number is rather small. Findeisen (1932) observed that when a fog was aged in an enclosed space the drop size increased, and he interpreted this as evidence that collision between cloud droplets usually resulted in coalescence. The same result could, however, be produced in an inhomogeneous fog by the growth of large droplets at the expense of the smaller ones because of vapour-pressure differences. Several microscopic examinations of the behaviour of very small fog droplets have also been reported (Dady, 1947; Swinbank, 1947) in which it was stated that, though many collisions were observed between droplets of radius about 2 μ, no coalescences ever occurred. This result is consistent with Hocking's calculations.

An investigation of the collection efficiency of large drops, 3 mm in diameter, falling through a cloud of smaller droplets ($r = 6$–$100\,\mu$) was later carried out by Gunn and Hitschfeld (1951), who weighed the drops before and after passage through a column of cloud 3 m in height. Calculations based on Langmuir's theory and the measured drop size-distribution in the cloud were in excellent agreement with the experimental results, confirming, in a limited region, not only Langmuir's calculation, but also his assumption that coalescence always occurs on collision.

More recently Telford, Thorndike and Bowen (1955) have made a careful study of the coalescence of water droplets of approximately equal size. This region is, of course, of prime importance in studies of the early stages of cloud instability. A uniform cloud of droplets approximately 150 μ in diameter and having a diameter spread of only about 6 % was produced by a spinning disk. These droplets were fed into a vertical wind-tunnel the speed of which was adjusted so that all droplets moved slowly upwards through the field of view. Any larger droplets formed by coalescence of two cloud droplets then moved down through the field of view, and were thus easily recognized. The collisions were studied photographically, using a film moving at right angles to the direction of motion of the droplets, and a modulated light-source.

The results of these experiments indicated an unexpectedly high collection efficiency, $E = 12\cdot6 \pm 3\cdot4$, and this was explained qualitatively on the basis of capture of the upper droplet by the wake of the lower. We have already discussed the quantitative development of this theory by Pearcey and Hill (1957), and their treatment shows that an E value of $12\cdot6$ ($\sigma = 3\cdot6$) for droplets of radius 80 μ can be accounted for if the difference in diameters is about $6\frac{1}{2}$ %, which is in reasonably good accord with the experimental situation. This high value of E and the agreement with experiment once again suggest that the assumption that droplets always coalesce upon collision is substantially valid.

In a recent experiment Kinzer and Cobb (1958) have traced the growth of single droplets as they fall through a dense fog, in this way obtaining collection efficiencies for a wide range of collector drop sizes. They produced a dense and moderately uniform fog, containing 1500–3000 droplets/cm^3 with radii 5·5–8 μ, in a vertical wind-tunnel, and observed a single droplet through a long-focus microscope. By varying the wind-speed they were able to keep this drop stationary in the field, and hence to follow its growth by measuring the power supplied to the driving-fan, appropriate calibration being made.

They found collection efficiencies approaching the geometrical value for droplets about 8 μ in radius, that is, comparable in size with the cloud droplets. The collection efficiency then fell steadily to about 0·2 for droplets 20–40 μ in radius, rose to nearly unity for 200 μ drops and then fell steadily again to small values for droplets greater than 1000 μ.

Whilst the behaviour of collecting droplets larger than 20 μ radius is roughly in accord with the theory, the high collection efficiency for small droplets is anomalous. It is, however, in this small-size domain that the experimental procedure becomes most difficult, and extremely close humidity and temperature control is necessary if size change by condensation or evaporation is to be avoided. It is not clear to what extent the apparent anomaly may be due to these causes.

Turning now to model experiments, we first point out their basic faults. The modelling principle chosen is to match the Reynolds numbers of spheres falling through a liquid to equal those of cloud droplets falling in air. The flow-patterns are then identical and the interactions of different spheres can be observed. The fallacy is, however, that the model only holds for uniform unaccelerated motion, and therefore breaks down at the very point it is desired to investigate (Pearcey and Hill, 1956). This failure is due to the fact that in the air most of the kinetic energy of motion is associated with the droplet, whilst in a liquid environment, because of its greater density, much of the kinetic energy is carried by the medium surrounding the drops. The models are, however, useful in demonstrating qualitatively some of the features of droplet collision.

The model constructed by Sartor (1954) consisted of water drops, a few millimetres in diameter, falling through mineral oil. He observed many of the effects to which we have referred, the pushing aside of small droplets and the effect of droplet wake for example, but despite many collisions there were no coalescences. The reason for this failure to observe coalescence is fairly clearly the existence of the strongly oriented structure of the oil-water interface, and there is unlikely to be any similar effect with water droplets in air.

A somewhat similar model investigation has been carried out by Schotland and Kaplin (1956), the model in this case consisting of steel balls falling through a concentrated sugar solution. The results are again only of semi-quantitative interest, but evaluation is made difficult by the fact that several conflicting versions have been published (see also Schotland, 1957a, b, c). Assuming the most recent version (Schotland, 1957c) to be correct, the collision efficiencies follow the pattern calculated by Pearcey and Hill with the exception that overall values were higher and there was apparently no cut-off for small droplets.

These experimental results then, confirm in general terms the calculations discussed above, and we can proceed to use these with confidence of their approximate validity in the size-ranges to which they apply.

Electrical effects

Though we have not mentioned it explicitly, all our discussion so far has assumed that no electric charges are present on the droplets involved, and that no strong electric fields are applied. It is quite obvious on general grounds that if the droplets involved in a collision bear unlike charges, the collision efficiency will be increased, while like charges will decrease the chances of collision. It is also obvious that, because water is a dielectric, a strong electric field will induce dipoles upon the droplets which will normally be attractive, and hence increase the probability of collision. The question is whether these effects are significant for the charges and fields which may ordinarily exist in clouds.

Very little theoretical consideration has been given to the collision efficiencies of charged droplets, which is not surprising when the difficulties of the neutral case are recalled. A few calculations have, however, been made for the case of a charged drop falling through a cloud of smaller neutral droplets (Pauthenier and Cochet, 1950; Pauthenier and Loutfoullah, 1950; Cochet, 1951) for values of droplet radius for which

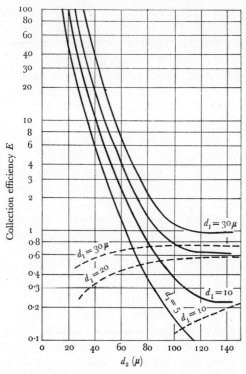

Fig. 6.10. Collection efficiency for a drop of diameter d_2, bearing a charge of 4×10^{-4} e.s.u., falling through a cloud of droplets of diameter d_1. Broken curves are Langmuir's calculations for uncharged drops (Cochet, 1951).

Stokes' law is valid. The electrostatic interaction is in this case between the charge on the large drop and the dipole induced on the smaller droplet. Fig. 6.10 shows the results of Cochet's calculation for a drop bearing a charge of 4×10^{-4} e.s.u or approximately 8×10^5 electronic charges. It can be seen that the collision cross-section is greatly increased above Langmuir's values for an uncharged drop, for drop radii less than about $60\,\mu$. As we shall see later, however, the charge assumed on the drop was at least three orders of magnitude larger than those typically found on cloud droplets, so the calculation greatly overestimates the importance of the effect in natural clouds.

Levin (1954a) carried out a similar calculation for single droplets bearing much smaller charges and found curves similar to those of fig. 6.10. These calculations suggest 30 μ dia. as the upper limit of appreciable effect, and, since the assumed charge (0·5 × 10^{-7} d_2 e.s.u., where d_2 is in microns) is at least an order of magnitude larger than the experimental values, we must expect effects to be confined to droplets of still smaller size.

We may use this calculation to estimate the importance of other interactions. Dipole-dipole interactions, as for example between neutral droplets in an external field, are normally weaker than the interaction considered and so can usually be neglected as far as collision efficiency is concerned. This may not apply for very small drops, however, and it is possible that coalescence processes upon collision may be affected.

Interactions between pairs of charged droplets, on the other hand, are much stronger than the charged-uncharged reaction, and it is quite possible that naturally occurring cloud-droplet charges may be large enough to affect the collision-coalescence process (Levin, 1954a).

Several of the experiments which we have already discussed included some investigation of electrical effects. Gunn and Hitschfeld (1951), for example, found no effect when they charged their 1·6 mm radius collector-drop to ±0·2 e.s.u. and concluded that such charging was therefore unimportant as far as the later stages of growth of a raindrop are concerned. It is difficult to estimate any extrapolation of fig. 6.10 with which to compare this result, so that nothing can be deduced about the agreement with theory.

Kinzer and Cobb (1958) similarly investigated the effect of small charges on the accretion rate of a drop falling through a cloud of droplets. In this case both the collector-drop and the cloud particles were charged. Little effect was found on the collection efficiency for drops greater than about 8 μ radius, though there was some tendency for droplets to coalesce spontaneously when they came within 2 or 3 μ of each other.

Telford *et al.* (1955), on the other hand, found very marked effects in clouds of uniform small charged droplets. The experimental arrangement was such that all drops were charged and it could be arranged that either all droplets had the same sign, or else that equal numbers of oppositely charged droplets were formed. When all drops are charged the interactions are, of course, very strong. It was found that unlike charges of the order of 6×10^{-4} e.s.u. per 130 μ-diameter drop increased the coagulation rate by a factor of from 2 to 20, depending upon the droplet concentration. On the other hand, when all droplets were charged with the same polarity, coalescence appeared to be completely inhibited. It seems very likely that most of this effect is caused by changes in the collision efficiency, but it is possible that the coalescence process could also be altered. We shall discuss this again later.

Finally, we recall that Sartor (1954), in his model experiments using water droplets in mineral oil, observed many collisions but no coalescence between drops. However, when he applied an electric field vertically in the tank, coalescences began to occur. The field had negligible effect on the droplet trajectories for fields less than 200 V/cm or so, but the coalescence efficiency on collision rose from zero to 100 % for an applied field of about 240 V/cm, the variation being approximately linear. Sartor made no calculation of the relation of these field strengths in his model to those in a cloud situation, and once again, of course, the model is not really valid for quantitative discussions.

Before the importance of electrical effects can be evaluated, we must have some data on the electrical fields and droplet charges which may exist in clouds.

Most measurements of droplet charge involve measuring the total charge per unit volume of cloud, and this is, of course, unsatisfactory if comparable numbers of positively and negatively charged droplets are present. Measurements of this kind made by Webb and Gunn (1955) on thirty-five different clouds indicated that the net charge per unit volume on the droplets was very small and did not exceed one electronic charge per drop in most cases. This, of course, does not rule out the possibility of large charges accurately balanced between positive and negative drops.

Twomey (1956), on the other hand, developed a technique for determining the charge on individual droplets by observing their motion in an electric field. He found that about 50 % of cloud droplets bore measurable charges and that in clouds consisting entirely of water, 80 % of the charged droplets were positive. The appearance of larger numbers of negatively charged droplets seemed to be related to the presence of ice-crystals in the cloud. The charge on individual droplets was roughly proportional to the square of their radii, and the charge on a droplet 20 μ dia. was typically about 4×10^{-7} e.s.u., or 1000 electronic units, thus

$$Q \approx 3 \times 10^{-9} r^2, \qquad (6.32)$$

where Q is in e.s.u. and r in microns.

Phillips and Kinzer (1958) have recently performed an experiment similar to that of Twomey, but with quite different results. In fair-weather conditions in cloud on a mountain peak, they found approximately equal numbers of positively and negatively charged droplets, and the mean absolute charge was only about 6 electronic units for droplets of 8 μ dia. Thundercloud droplets, on the other hand, showed much greater electrification, typically a few hundred electron charges, and droplets might either be predominantly of one sign, or else equally mixed, in different clouds.

Most theories of electrification processes have been directed towards explanation of the violent effects observed in thunderstorms, and relate largely to charge separation caused by raindrops or ice-particles falling at fairly high speeds. In many of these theories the ice-phase plays an important part, and this appears also to be the case in practice. For further details of some of these processes the reader is referred to the review given by Mason (1957, ch. IX); we shall here consider those processes which may be active in clouds composed entirely of water droplets.

Wilson (1929) considered a falling droplet to be polarized by the atmospheric electric field, so that when falling through a concentration of ions of both signs, the lower half of the droplet was a more efficient collector of oppositely charged ions than was the upper half. This theory was later worked out in detail by Whipple and Chalmers (1944), who showed that if the drop is falling relative to ions of both signs, the ultimate charge collected is $-0.515 Xr^2$, where r is the drop radius and X the static field in e.s.u. If ions of only one sign are present the result is $\pm 3 Xr^2$, whilst if the drop is moving slowly relative to a mixed ion population, the charge is zero.

Ions of both signs are usually present in the atmosphere, so that a charge will develop only upon droplets falling more quickly than the downward-moving ions. A typical value for the electric field is $X = 150$ V/m (Gish, 1951) and ionic mobilities are about 1 cm^2/Vsec (Wait and Parkinson, 1951), so that only droplets with terminal velocities considerably greater than 1·5 cm/sec can be charged in this way. This limit corresponds to droplets about 25 μ in diameter, so that the mechanism cannot cause charging of small cloud droplets. If the air shows a marked excess of ions of one sign, charging can take place and the charge on a drop of radius r cm should be typically about $1·5 r^2$ e.s.u., which is of the order of the results found by Twomey.

Gunn also has proposed several electrification mechanisms for water clouds. One of these (Gunn, 1956) is a modification of Wilson's process, in which relative motion of droplets and ions is to some extent neglected, and the droplet considered as a sphere immersed in a fluid of positive and negative ions of different mobilities (λ_+, λ_-) and polarized by an external field X. The ultimate charge is now $3 Xr^2 f(\lambda_+, \lambda_-)$, where the function f is typically of order 0·1. The results are thus very similar to those discussed above.

In two further theories Gunn has considered ionic charging in the absence of appreciable electric field effects. In the first discussion (Gunn, 1954, 1955 a) charging is considered to result from random collisions with atmospheric ions. This process results in a charge distribution symmetrical about zero charge, and with mean absolute charge of about 10^{-8} e.s.u. on a 20 μ diameter drop. The mean charge varies approximately as the square root of droplet radius. These results are at least in qualitative agreement with the experiments of Webb and Gunn and of Phillips and Kinzer which we have discussed above.

Gunn (1955 a, b) then goes on to show that by association of droplets bearing such small charges, a different charge distribution can be built up which is again symmetrical about the origin, but varies now as r^2, and yields a charge of about 5×10^{-7} e.s.u. on a droplet 20 μ in diameter. This agrees with Twomey's results as to charge magnitude, but not as to uniformity of sign.

This brief discussion shows that many problems remain to be solved relating to droplet electrification, even if we restrict ourselves to non-freezing clouds. The results do, however, allow us to draw some conclusions about the importance of electrical effects in cloud coalescence processes. We need only one more piece of information, and that is the value of the electric field within clouds. This is fortunately reasonably easy to measure, and average fields near the earth's surface are usually 100–300 V/m in normal weather. When thunderstorms are present the situation is quite different, and fields as high as 1000 V/cm have been measured within thunderclouds (Gunn, 1951).

Returning now to coalescence: we noted that several experiments showed that the collection efficiency of large drops was unaltered by the presence of moderate charges. On small droplets, however, it was possible to produce considerable effect. Examining the results of Telford *et al.* (1955), however, we see that the drop diameter which they used was 130 μ, and the charge on such a drop is almost certainly less than 10^{-4} e.s.u. Their results then suggest that a charge of this magnitude would alter the collection efficiency by less than 10 % of its original value. Gunn (1955a) has carried out some calculations which similarly suggest that electrical effects are probably 'just noticeable' for average cloud drops, and hence not a major factor in the development of clouds.

Similar comments apply to the effects of static fields. These are normally more than an order of magnitude smaller than those found by Sartor (1954) to change drop trajectories and should therefore have only a small effect. An exception occurs, of course, in the case of thunderstorms, where much larger fields are found. The evidence suggests, however, that ice-crystal processes producing precipitation are the cause of these fields, so that, whilst electrical effects may be important in the later development of precipitation, they are not a major influence in its initiation.

Coalescence

We have said very little, as yet, about the mechanism of coalescence itself. We recognized that the aggregation of water droplets involves two stages, collision and coalescence, but most of the experiments we discussed were unable to separate the two processes. We shall only find it possible to make a few qualitative remarks about this as yet largely unexplored field.

When two droplets collide with each other, several things must happen before they can coalesce. The air between the two droplets must be expelled so that their surfaces come into intimate contact, the surface-structure must be destroyed over the area of contact so that the drops can actually join, and they must then deform to a single sphere. The first two steps in this process represent the principal barriers to coalescence.

The ease with which two drops come into close contact on collision depends on their sizes and relative velocities. Very small droplets are unable to penetrate the air-layers flowing round other droplets and no collisions occur, as shown by the calculations of Langmuir, of Pearcey and Hill and of Hocking. Droplets from about ten to a few hundreds of microns in diameter, on the other hand, are able to collide, and appear to come into close contact with no difficulty. For larger drops, of the order of a millimetre in diameter, a new phenomenon occurs. The drops deform on contact, an air-layer is trapped between two nearly flat surfaces and takes some time to flow out. This phenomenon appears to account, at least in part, for the skating of droplets of about this diameter over flat water surfaces.

Once close contact has been achieved, the surface separating the drops must be destroyed. Polar liquids like water have a surface in which the outer few layers of molecules are oriented with their dipoles parallel, to form an electrical double layer, and this structure must be broken down to the normal random one over the area of contact before the drops can coalesce. This barrier seems to be very small in the case of pure water drops, but the presence of monolayers on the droplet surfaces, particularly when these are of long-chain organic compounds, may make destruction of the surface-structure extremely difficult. An effect of this sort clearly took place in the experiments of Sartor using water drops in oil.

There is little doubt that strong electric forces may have considerable influence on these processes. In the case of an air-film between large deformed drops, a field renders the configuration unstable so that coalescence very quickly takes place, as observed in the familiar fountain experiment. This deformation effect is not, however, appreciable in droplets of cloud-droplet size. Charges or image forces will tend to draw the drops together, and strong fields may influence the stability of the surface-structure considerably. As before, the question is to what extent these effects are important in clouds.

The experiments of Gunn and Hitschfeld and of Telford, Thorndike and Bowen show that the coalescence efficiency must approach unity over a considerable range of droplet sizes; droplets of 10 μ radius are successfully collected by a 1 mm drop, while droplets 70 μ in radius mutually coalesce with high efficiency. It is possible that coalescence barriers become appreciable for very small droplets, but for these the collision efficiency is already so small that they make negligible contribution to any coalescence process. Since most cloud droplets reach radii of 6 μ or more by condensation, we are probably justified in assuming coalescence always to occur upon collision, but more work is needed on this subject.

Conclusions

In the preceding sections we have discussed what is known of the collision and coalescence of small water droplets. Whilst it is clear that much work remains to be done, a fairly good general understanding of the processes and orders of magnitude involved is now available. It may be useful to summarize very briefly the main features.

Considering first only uncharged droplets, calculations and experiments indicate that the collision efficiency is low if both of the droplets involved are less than about 18 μ in radius. The collision efficiency increases with increasing droplet size, as a general rule, except for droplets greater than about 1 mm radius for which collection efficiency for small drops falls somewhat. Droplets greater than about 50 μ in radius and of nearly equal size have large collision efficiencies because of interaction of the upper drop with the wake of the lower.

Coalescence appears to take place readily upon collision, at any rate for droplets from about ten to a few hundred microns in radius. When both droplets lie outside this range the position is not clear, but since most processes of interest in cloud physics take place within this range of sizes it is a good approximation to assume a coalescence efficiency upon collision of unity.

As far as electrical effects go, the evidence appears to show that normal atmospheric electric fields and droplet charges have little effect on the stability of the droplet population, at any rate for droplets larger than about 10 μ radius, or, if Gunn's charges are typical, even for much smaller droplets.

The electric fields and droplet charges developed during thunderstorms, however, would seem to be large enough to affect appreciably the stability of the cloud microstructure.

THE STABILITY OF CLOUD MICROSTRUCTURES

The coalescence process provides a mechanism by which the size of certain droplets in a cloud, initially somewhat larger than average, can increase fairly rapidly so that precipitation can occur. Whilst we shall defer consideration of details of this precipitation mechanism to the next chapter, we shall here briefly examine the role which microstructure plays in the general stability of a cloud.

A consideration of collision efficiency, such as we have made in the previous section, suggests that clouds containing large droplets may be more unstable than clouds containing an equal amount of water in the form of small droplets, but we must look at this a little more closely to reach any definite conclusions.

In a cloud the physically relevant parameter is not the collection efficiency E itself, but rather a quantity giving the number of coalescences per centimetre of droplet fall, or perhaps the number of coalescences per cm^3 per sec. The study of such discrete coalescences on a statistical basis is unfortunately a difficult problem, and we shall be content to consider a smoothed-out growth process in which liquid is accreted continuously at a rate appropriate to the liquid water content and collection efficiency involved.

During the early stages of droplet growth by coalescence all droplets are rather small and of comparable sizes, so that

Hocking's values of collection efficiency are those which we must study. From a glance at fig. 6.9 it is immediately clear that if a cloud contains no droplets larger than 18 μ in radius then collision will not occur and the cloud will be stable against coalescence. This applies in general terms to continental clouds whose droplet spectra extend typically from 2 to 10 μ radius. For drops larger than 18 μ radius, however, the collection efficiency increases sharply, and if there are appreciable numbers of these drops then coalescence will be important. This will be the case with maritime clouds whose droplets typically range from 3 to 22 μ in radius.

To put this more quantitatively we must consider the size-distribution of cloud droplets in greater detail, and in this we shall follow the discussion given by Twomey (1959c). Twomey observes that in convective clouds, independent of their origin, the droplet spectrum in the early stages, before coalescence becomes appreciable, follows a general pattern in which the droplet radii are distributed approximately normally, and the ratio of standard deviation to mean has a value close to 0·2. (This is approximately the same as having a set of narrow log-normal distributions all having the same standard deviation.) Making use of this generalization together with the experimental fact that the mean liquid water content of such clouds rarely differs greatly from 1 g/m^3 it becomes possible to compute the behaviour of typical droplet populations with confidence that the model chosen represents a fairly typical cloud microstructure.

On this basis and with a value of standard deviation over mean radius of 0·15, assumed to be typical of conditions before coalescence begins, Twomey calculated the proportion of liquid water collected by drops of various sizes falling through droplet populations typical of maritime and continental clouds. These curves are reproduced in fig. 6.11. Shown also in this figure is the rate of increase in the radius of the collector-drop with distance fallen.

With the aid of this figure we can now examine two possible types of instability—that resulting from the growth of the largest drops in the normal droplet spectrum, and that due to the growth of a small number of very large drops essentially extraneous to the normal spectrum.

Development of Non-freezing Clouds

In the maritime cloud about two-thirds of the droplets are bigger than the critical 18 μ radius above which collisions can occur and from fig. 6.11 it is clear that the larger droplets are able to collect a considerable portion of the liquid water lying in their path of fall. A cloud of this type may therefore be expected to be unstable, to suffer modification of its droplet

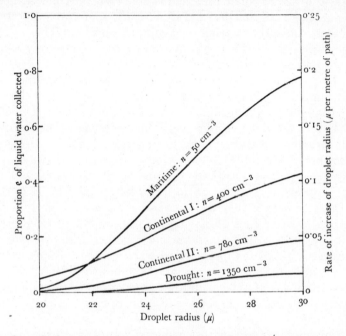

Fig. 6.11. Coalescence in a heterogeneous cloud containing 1 g/m³ liquid water, with normal distribution of radii and standard deviation equal to 15% of the mean radius. The dimensionless ordinate e gives the proportion of the liquid water in the volume swept by a falling droplet which is actually collected by the droplet; the right-hand ordinate gives the rate of growth of the collector-drop in μ/m of path (Twomey, 1959c).

distribution by coalescence, and probably to produce precipitation.

The cloud marked 'continental I' on the other hand, having 400 droplets per cm³, has only about 1 in 10^{13} of its main droplet population bigger than 18 μ radius. We may therefore expect the droplet spectrum to remain stable with respect to coalescence.

Under almost all conditions, however, one finds small

numbers of giant salt-nuclei in the air-mass from which the cloud is derived, and these may produce drops in the 20–30 μ size-range which do not fit into the general spectrum. From fig. 6.11 such a 30 μ droplet in a maritime cloud is able to intercept 0·8 of the liquid water within its path and grows at an initial rate of 0·2 μ in radius per metre of path, or about 1·2 μ per min, this rate increasing with time. The same drop introduced into a cloud of continental-I type grows at only half this rate, whilst in a drought-time cloud the growth-rate is less than one tenth of its value in the maritime cloud. Clouds having large numbers of droplets are therefore relatively stable to the introduction of a few large drops, and any precipitation developed in this way may take as much as an order of magnitude longer than similar growth in a maritime cloud. In the next chapter we shall examine this precipitation growth in greater detail.

CHAPTER 7

RAIN FROM NON-FREEZING CLOUDS

INTRODUCTION

THE fact that rain can fall from clouds whose temperatures are everywhere warmer than freezing is now sufficiently well known to require no documentation. Precipitation ranging from fine drizzle to heavy storms of large drops has been observed at different times, the type of precipitation depending in detail upon the clouds involved.

In this chapter we shall discuss in a quantitative manner the way in which precipitation develops in a cloud consisting entirely of water droplets. An arbitrary division is usually made between cloud droplets and precipitation elements at a radius of about 100 μ. This is mostly a matter of convenience, but has a physical basis in that the terminal velocity of drops 100 μ in radius is nearly a metre per second, so that they are able to fall against the sort of updraughts normally met in clouds.

After a consideration of the way in which the precipitation depends on the structure of the cloud in which it is being generated, we shall conclude the chapter with a brief discussion of the modification of the raindrops during their passage through the relatively dry air between the cloud base and the ground.

THE FORMATION OF RAIN

The question of the formation of rain in clouds consisting entirely of water droplets was investigated in 1939 by Findeisen (1939). He first established that raindrops could not form in a reasonable time by condensation alone, and then went on to consider the effect of coalescence on the basis of the assumption that falling drops collected all smaller droplets in their path. He calculated that in a typical cloud a droplet initially of 30 μ radius would grow to precipitation element size (100 μ) after a fall of only 150 m, and reach the size of a large raindrop after 1200 m. He rejected this calculated value, however,

The Formation of Rain

as 'contrary to experience' and assumed some mechanism must operate to prevent the continued growth of large drops by coalescence.

Langmuir examined this theory again in 1948 (Langmuir, 1948), and calculated more accurate collision cross-sections for falling droplets, as we have discussed in detail in the preceding chapter. Though his collection efficiencies were much less than the geometrical values, his calculation again showed that rain could be produced in clouds of moderate size with typical updraughts, provided some rather large drops ($r \sim 40\,\mu$) were available as nuclei for raindrops. He further suggested that a chain reaction might be initiated by the break-up of very large raindrops, yielding fragments suitable for repeated growth to shatter-size by coalescence.

Bowen (1950) later made use of Langmuir's collection efficiencies to calculate in detail the effect of variations in water content and updraught velocity on the production of raindrops, and showed that the calculated behaviour was borne out by radar observations of the development of coalescence rain.

About the same time Houghton (1950) examined precipitation mechanisms and concluded that whilst most mid-latitude rain appears to originate with ice-crystals which later grow by accretion, large drops may be produced in deep tropical clouds by the action of coalescence alone, without an ice-phase being present. Later still Ludlam (1951c) carried out an analysis very similar to that of Bowen for convective clouds, and Mason (1952b) discussed the development of drizzle in layer-clouds. We shall discuss these investigations in more detail after we have developed some quantitative expressions for the growth of drops by coalescence.

The growth equations

Consider a drop of radius r_2 falling with terminal velocity u_2 through a uniform cloud of smaller droplets of radius r_1 and terminal velocity u_1. Then if the efficiency of collection of the small droplets by the drop is E and the liquid water content of the cloud is w (in g/cm³), the rate of increase in mass of the large drop is
$$\frac{dm}{dt} = \pi r_2^2 (u_2 - u_1) wE, \tag{7.1}$$

where we have replaced the discrete collection process by a continuous growth. From this equation

$$\frac{dr_2}{dt} = \mathrm{w}(u_2 - u_1)E/4\rho_\mathrm{L}, \qquad (7.2)$$

where ρ_L is the density of liquid water. To be completely accurate we should also include a term taking account of the increase in radius caused by condensation from the slightly supersaturated environment within the cloud, but for the moment we shall omit this complication.

Suppose now that in the cloud there is an updraught of velocity u, then the true upward velocity of a large drop is

$$\frac{dh}{dt} = u - u_2, \qquad (7.3)$$

where h is height above a fixed level. Then eliminating t between (7.2) and (7.3)

$$\frac{dr_2}{dh} = \frac{\mathrm{w}(u_2 - u_1)E}{4\rho_\mathrm{L}(u - u_2)}. \qquad (7.4)$$

This can be integrated to yield

$$\int \mathrm{w}\,dh = 4\rho_\mathrm{L} \int \frac{(u - u_2)dr_2}{(u_2 - u_1)E}, \qquad (7.5)$$

which gives a relation between height and drop-size as a function of the conditions of updraught and water content in the cloud, and of the initial drop sizes.

Langmuir (1948) has considered this equation for the simple case where w, u and r_1 are constant within the cloud, and u_1 is negligible compared with u_2. Equation (7.5) can then be written

$$h - h_0 = \frac{4\rho_\mathrm{L}}{\mathrm{w}}\left[u\int\frac{dr_2}{u_2 E} - \int\frac{dr_2}{E}\right], \qquad (7.5)$$

where the integrals extend from the r_2 value at h_0 to that at h. Langmuir tabulated numerical values of these two integrals based upon his calculated collision efficiencies, and these have been used by most other workers.

The general behaviour of growing drops can be seen from (7.6). A drop somewhat larger than average cloud droplet size is initially carried upwards in the updraught, since $u_1 < u$ and

The Formation of Rain

the first integral is larger than the second. As the drop grows u_2 increases until it exceeds u, at which stage the second integral increases more quickly than does the first and the drop falls, growing all the time. The course of this development was considered in detail by Bowen (1950) with the inclusion of effects due to condensation. Langmuir (1948) pointed out the possibility of a chain reaction. If when $h = h_0$ in (7.6) the growing drop has a sufficient radius to shatter, providing more than one drop of the size from which it grew at the initial level h_0, then the process is regenerative. The critical radius for shatter of water drops falling at terminal velocity is, however, about 3 mm (Lenard, 1904; Blanchard, 1950) and we shall see later that only under conditions of very high liquid water content and high updraught velocity could drops of this size be produced in a cloud. The scarcity of drops of even 2 mm radius in observed raindrop size-spectra shows that such conditions must be extremely rare. We shall therefore confine our attention, for the most part, to the more normal growth of precipitation within a cloud.

In this section we shall discuss the growth of raindrops within a cloud under the combined influence of condensation and coalescence. A complete solution of this problem is closely related to the modification of a nearly uniform droplet spectrum by coalescence, and, as we remarked before, no really satisfactory treatment has yet been given. It is, however, possible to elucidate many features of the process by study of a simplified model such as that of Bowen (1950). We shall examine the behaviour of such a model now, and later see in what ways it should be modified in order to give a better representation of phenomena in a real cloud.

We assume the model cloud to consist of uniform droplets of radius r_1 rising in a uniform updraught of velocity u, and trace the history of a somewhat larger drop of radius r_2 formed, for example, by coalescence of two of the cloud droplets. This drop grows by coalescence with cloud droplets according to (7.2), and at the same time its radius increases by condensation according to (6.21), which we rewrite

$$r_2 \frac{dr_2}{dt} \approx GfS, \qquad (7.7)$$

where S is the fractional supersaturation within the cloud and Gf has the approximate value $1 \cdot 0 \times 10^{-6}$ at $10°$ C and 800 mb pressure.

The vertical motion of drops and droplets in the cloud is expressed by the relation

$$\frac{dh}{dt} = u - u_1, \qquad (7.8)$$

where h is height above cloud base and u_1 the terminal velocity of the droplet considered. For small cloud droplets u_1 is given to sufficient accuracy by Stokes' law which yields

$$u_1 = \frac{2g\rho r_1^2}{9\eta}, \qquad (7.9)$$

where g is the acceleration due to gravity, ρ is the density of water and η the viscosity of air. For droplets growing only by condensation, combination of (7.7)–(7.9) gives

$$\frac{dh}{dt} \approx u - \frac{4g\rho GfS}{9\eta} t. \qquad (7.10)$$

This represents a parabolic trajectory, but the downward acceleration is so small that the cloud droplets essentially follow the cloud updraught as shown for a particular case in fig. 7.1 in which S was assumed to have a constant value of $0 \cdot 1 \%$.

Initial double drops, however, as well as growing by condensation, can coalesce with the smaller cloud droplets. Since the relative velocities and collision efficiencies are initially small, coalescence is comparable in efficiency with condensation. As the drop grows larger, however, its rate of collection of small drops increases and its trajectory can be calculated from (7.2), (7.7) and (7.8) with a more general expression for the terminal velocity replacing (7.9). Bowen has performed this calculation with the additional simplifications of neglecting the terminal velocity of the cloud droplets and assuming the liquid water content of the cloud to be uniform. The results for a particular case are shown in fig. 7.1. Initially the double drops are carried without lag by the updraught, but after 20–30 min, in this case, they begin to fall behind. Growth then proceeds more rapidly and after a further similar period of time the drops become large enough to fall against the updraught. The final growth

The Formation of Rain

stage is rapid, and only about a quarter of the total growth time is taken up by the fall of the large drops to cloud-base level again, where they emerge as raindrops.

Although this model provides only a very simplified picture of the development of raindrops, it can be used to obtain at least a semi-quantitative idea of the effects of various parameters upon the process. The first, and most important, parameter is the updraught velocity u. This effects the growth of raindrops

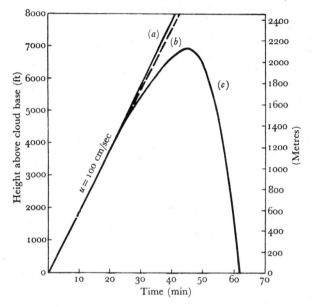

Fig. 7.1. The motion of (a) the air, (b) cloud droplets, and (c) droplets which have grown by coalescence, in a cloud in which the mean upward air-velocity is 100 cm/sec, average droplet diameter 20 μ and cloud water content 1 g/m³ (Bowen, 1950).

both directly by its influence on their motion, and indirectly through its effect on the supersaturation within the cloud. Bowen took as an approximation

$$S \approx 10^{-5}u, \qquad (7.11)$$

which is seen from fig. 6.4 to be of the right order of magnitude for typical cases. With this relation, and the initial assumptions we have already discussed, he calculated the curves shown in

Rain from Non-freezing Clouds

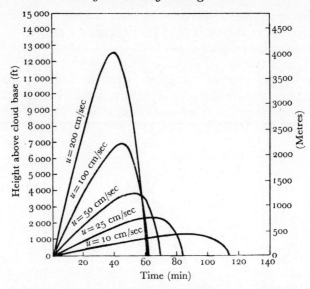

Fig. 7.2. Trajectories of drops which grow by coalescence in clouds having a range of vertical air-velocities from 10 to 200 cm/sec (Bowen, 1950).

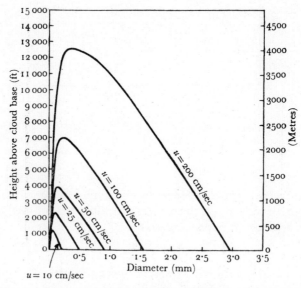

Fig. 7.3. Change in drop diameter with height for a range of vertical air-velocities (Bowen, 1950).

figs. 7.2, 7.3 and 7.4. The first of these figures shows the trajectories of growing droplets as functions of time, while the second relates droplet diameter to height above cloud base. Fig. 7.4 shows the results of similar calculations for clouds with different liquid water contents.

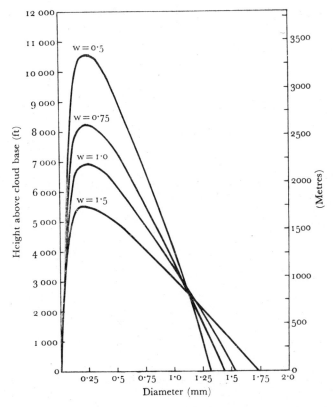

Fig. 7.4. Change in drop diameter with height for different values of cloud water content, w, in g/m^3 (Bowen, 1950).

From these figures three conclusions emerge, apart from the general one that the process appears capable of producing large raindrops in reasonable lengths of time. The first of these is that clouds with strong updraughts, if they rain at all, should produce rain after a shorter period of time than clouds with weaker updraughts. The second conclusion is supplementary and requires that clouds with vigorous updraughts must have considerable vertical development if they are to produce rain, whilst less vigorous clouds require less depth. Finally, the raindrops which fall from a large, vigorous cloud should be considerably larger in size than those falling from a cloud with lower updraught and less vertical development.

Ludlam (1951c) has made somewhat similar calculations for drops of several arbitrary sizes introduced into the base of a

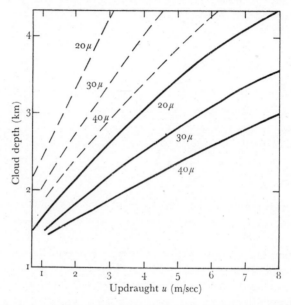

Fig. 7.5. Variation of minimum cloud depth for shower development as a function of updraught u, for initial droplet radii of 20–40 μ in adiabatic clouds with base temperatures of 5° C (broken lines) and 20° C (Ludlam, 1951c).

growing cloud. He, however, neglected condensation and used an average value for the collision efficiency E instead of Langmuir's tables. His results are somewhat different from Bowen's, since in his cloud model the water content has the adiabatic value instead of the lower uniform value used by Bowen. The same general trends are, however, evident, as can be seen from fig. 7.5. Here Ludlam has plotted the cloud heights necessary in order that the growing-drop should have attained a radius of 150 μ at the cloud summit. A drop of this radius can fall against updraughts of 1 m/sec, which is probably the sort of value encountered near the summit. Such a drop can also survive falls of a few hundred metres through air at 90% r.h. and so has a good likelihood of surviving in turbulent conditions near to the cloud-top.

Mason's calculation (Mason, 1952 b) was explicitly concerned with stratus rather than cumulus clouds, so that much smaller updraught velocities were considered. The assumptions and equations were the same as those of Bowen except that the growth of a drop condensing on a hygroscopic nucleus of given size among a population of droplets $6\,\mu$ in diameter was calculated. Since a calculation of this kind gave drops too small to reach the ground unless improbably high liquid water contents were assumed, Mason suggested a turbulent mechanism which allowed a small fraction of the growing drops to spend long times within the cloud and grow to larger sizes. This mechanism is reasonable in such layer-clouds, but is not, of course, required to explain precipitation from cumulus clouds.

Modification of the model

Several modifications are required in the model we have just discussed to make it correspond more nearly with conditions in a real cloud. We shall consider these in turn.

Two alternative water content models have been proposed—one in which the water content has a constant value throughout the cloud, and one in which it follows an adiabatic law. The measurements which we discussed in chapter 1 show that, despite small-scale fluctuations, the former model is more realistic, and we shall retain it. Condensation also appears to be an appreciable factor in the early stages of drop growth, so it must be retained in an improved model.

The principal shortcoming of the models we have discussed lies in their treatment of the smaller cloud droplets. In a more realistic model not only should their fall velocity u_1 be taken into account, in the early stages of growth at any rate, but also the fact that coalescence of the larger drop with them is a discrete rather than a continuous process. We should not expect a model showing these improved features to differ radically from its simpler counterpart, but it is certainly possible, as we shall see later, for the time-scale of some of the processes to be changed considerably.

Another, and indeed a crucial assumption of the model requires clarification. This is the size and origin of the larger drops which are ultimately to grow to raindrops. Bowen, for the sake of simplicity, assumed them to form near the base of the cloud by chance coalescence of two cloud droplets. Ludlam, on the other hand, considered them to originate as large sea-spray droplets borne up to cloud-base level before having a chance to evaporate.

The number of such larger drops required is not large. An average cloud may contain about half a gramme of liquid water per cubic metre, and an average raindrop is about a millimetre in diameter. If all the cloud is converted to raindrops of this size we thus require only about 10 large drops per litre, and since such complete conversion is impossible only two or three large drops per litre are required to ensure essentially complete precipitation from the cloud. If any sort of chain reaction occurs then much lower initial numbers of large droplets may be sufficient.

The measurements of Woodcock (1953) over the sea and of Twomey (1955) over land, which we have already discussed in chapter 4, suggest that sea-salt nuclei having mass greater than 10^{-10} g are nearly always present in the atmosphere in concentrations of the order of 1 per litre. These nuclei are likely to contribute droplets with radii about 20 μ at cloud base, and so constitute a fairly general source of larger than average cloud droplets.

At the same time cloud-droplet spectra usually approximate a log-normal distribution, as we have discussed in chapter 5, and there are always significant numbers of larger than average droplets. Part of this broadening of the droplet distribution is due to mixing of air-parcels having different histories within the cloud, and part to coalescence. Whichever agency predominates, the important thing is that a fairly wide droplet spectrum exists throughout the cloud.

Combining these two generating mechanisms, it is clear that we must consider not only large droplets introduced into the base of the cloud, but also larger than average droplets formed at higher levels within the cloud itself.

The Formation of Rain

A considerable step towards a more realistic model has been made by Telford (1955) who replaced the continuous-growth accretion process by one involving discrete capture of smaller droplets, and at the same time took into account the terminal velocities of all the droplets involved. The calculation then becomes very formidable, and we shall simply indicate some of the results obtained.

Consider a cloud of uniform droplets containing a relatively small number of larger droplets. Whereas earlier models typified by (7.2) picture these droplets as growing continuously by coalescence, Telford took account of the fact that coalescences are discrete events whose times of occurrence are statistically distributed. Thus, whilst the average growth-rate of drops by coalescence is approximately that given by the assumption of continuous growth, small numbers of drops grow at rates much greater or much less than this average value.

Considering those few drops which have made a coalescence collision after a rather short time, they are now in a more favourable position than their fellows to make a further collision, because of their larger size. These second collisions are similarly statistically distributed giving a further widening of the spectrum, and so on. Telford's detailed calculations show that the shape of the resulting droplet spectrum is substantially determined by the first twenty captures. After this the statistical fluctuations are unimportant, and the continuous growth equations are sufficiently accurate.

The importance of this statistical mechanism, apart from supplying a basis for the general evolution of broad droplet spectra, is that it shows that a small but significant proportion of the original droplets grows very much faster than the average rate, and may produce quite large drops after relatively short growth times. Since cloud droplets occur in concentrations of the order of 100 cm^{-3} whilst raindrops are less than 1 per litre, it is the development of the largest drop in 10^5 or 10^6 which is important in the production of rain, and this growth may be very much faster than the average growth rate.

To illustrate this, fig. 7.6 shows the results of a calculation of the growth of a group of drops 12·6 μ in radius in a cloud of 10 μ drops, after growth for 4000 sec. For simplicity in this calculation Telford assumed $E = 1$ for all drops, but this does not radically affect the results. It can be seen that, whilst the radius predicted by the continuous growth equation is only 0·53 mm, 10% of drops have radii greater than 2 mm and a significant number have radii approaching 4 mm. The effect is, therefore, of great importance in the development of rain, and reduces by a considerable factor the time required for large drops to develop.

Since on any theory the first ten coalescences take up most of the growth time it is interesting to compare times for these.

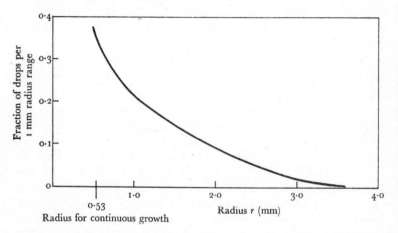

Fig. 7.6. Number of drops having radii between r and $r+\delta r$ after 12·6 μ radius drops have grown for 4000 sec in cloud of 10 μ radius drops at concentration 1 g/m³. Collection efficiency assumed unity (Telford, 1955).

Telford considers a cloud containing about 200 droplets/cm³, 10 μ in radius, together with 12 cm⁻³ of droplets 12·6 μ in radius, formed by coalescence of pairs of the smaller droplets. Assuming $E = 1$, the time required for 100 drops/m³ to experience ten collisions is found to be about 5 min, whilst the time for such growth to occur on the basis of continuous accretion is no less than 33 min. A similar ratio of times would be expected if size-dependent collision efficiencies were to be used.

Because of its computational complexity this theory has not yet been applied to a calculation of the development of cloud-droplet spectra using a realistic initial spectrum and Hocking's collision efficiencies. Whilst then a complete treatment must include these statistical effects, we must for the present be content with less sophisticated calculations.

The other modifications required to Bowen's original treatment relate to collision efficiencies. In the original calculations the terminal velocities of the cloud droplets were neglected, and the growth equation (7.2) used in the approximate form

$$\frac{dr_2}{dt} = \frac{wu_2 E}{4\rho_L}, \qquad (7.12)$$

Langmuir's values being used for E. This equation is rather seriously in error in the early stages of drop growth. A more nearly correct equation is

$$\frac{dr_2}{dt} = \Sigma E \frac{(u_2 - u_1)}{4\rho_L} \Delta w, \qquad (7.13)$$

where Δw is the contribution to the liquid water content made by droplets of terminal velocity u_1, and the summation is over all groups of droplets smaller than the collector drop. The quantity e calculated for clouds of various types by Twomey and given in fig. 6.11 may be used to aid in this calculation through the relationship

$$e = \Sigma \frac{E(u_2 - u_1) \Delta w}{u_2 w}. \qquad (7.14)$$

The principal effect of these changes is in the initial stages of growth, and in fact Langmuir's values for E must be used for large drops. Since, however, most of the growth time is spent in growth through the first few tens of microns this may have a large effect on the overall rate of precipitation development. In the next section we consider this in more detail.

Maritime and continental clouds

We have already commented several times upon some of the similarities and differences between cumuli formed in maritime and continental air-masses. A feature of particular interest is the fact that maritime cumuli rain much more easily than do their continental counterparts, and we have related this distinction to the difference in microstructure, and in particular to droplet size in the two types. In this section we shall examine more quantitatively the precipitation process for clouds of each type.

At the top of table 7.1 are shown the physical parameters of a typical maritime and a typical continental cumulus, derived from the work discussed in chapters 5 and 6. Updraught and liquid water content are the same for each cloud, but the maritime cloud consists of large drops and the continental cloud of small drops. Since to a good approximation the ratio of standard deviation to mean in the distribution of droplet radii is about 0·15 in the early stages of most clouds, we can estimate the content of drops of a given size, and the table shows the radius exceeded by about one drop in every litre of cloud. These are the drops which will grow fastest, and their concentration is sufficient to give moderate precipitation if they grow to raindrop size. The next section of the table shows the history of these drops.

Table 7.1. *Development of precipitation in typical clouds*

		Maritime	Continental
Updraught		100 cm/sec	100 cm/sec
Liquid water content		1 g/m^3	1 g/m^3
Supersaturation		0·2 %	0·1 %
Droplet concentration		50 cm^{-3}	200 cm^{-3}
Average droplet radius		17 μ	11 μ
Droplet radius for 1/litre		30 μ	19 μ
1/litre	Final radius	0·55 mm	0·9 mm
	Time	40 min	120 min
	Height	1000 m	3800 m
30 μ drop	Final radius	0·55 mm	0·55 mm
	Time	40 min	85 min
	Height	1000 m	2400 m
40 μ drop	Final radius	0·45 mm	0·4 mm
	Time	30 min	65 min
	Height	600 m	1400 m

In the maritime cloud one drop per litre exceeds 30 μ in radius. These drops rise with the air, growing by condensation and coalescence, and are big enough to fall against the updraught after rising 1000 m. On emerging from cloud base they are a little over a millimetre in diameter and so constitute average raindrops.

In the continental cloud one drop per litre exceeds 19 μ in radius, and this is just large enough for collisions to occur. Because the collection efficiency is very low in the early stages the growing drops are borne high into the cloud and a long

The Formation of Rain

time is required for them to grow to raindrops. Because of their high trajectories, however, the drops which do form are rather large when they reach cloud base.

For any cloud more continental in character than the one we have considered the large droplets will still be below the coalescence limit of 18 μ and precipitation will not form by such a homogeneous process.

Under almost all conditions, however, there are present in the cloud, in concentrations of about one per litre, specially large droplets which have formed on giant sea-salt nuclei. These droplets occur in both maritime and continental clouds, and their radii are usually several tens of microns. In the lower parts of table 7.1 we trace the growth of such droplets in our two typical clouds. Droplets initially 30 μ and 40 μ respectively in radius entering at cloud-base level produce raindrops from both clouds, but the times required and cloud depths necessary are about twice as great for the continental cloud as for the maritime cloud.

Whilst these calculations are not rigorous they suggest that precipitation can develop by a coalescence mechanism in maritime clouds in times of the order of 40 min provided that the cloud depth is greater than about 1000 m. In moderately continental clouds, on the other hand, times of 1–2 hr and cloud depths of 3000 m or so may be required. Clouds of extreme continental character would require still longer times and greater depths for precipitation to develop by this mechanism, and in fact the ice-crystal process is generally more important.

These results are in good general agreement with the observations of Battan and Braham (1956) which we discussed in chapter 5. They found that maritime clouds required depths of 2000 m, continental clouds 4000 m and desert clouds 5000 m for a precipitation probability of 20 %. This suggests that our treatment is essentially correct, though it is recognized that many rough approximations have been made in the calculations.

SEA-SALT NUCLEI AND RAIN

Interest in the relationship between cloud droplets, raindrops and salt-particles dates back to the early measurements of salinities by Köhler which we discussed in chapter 3, and is still maintained at the present time. Simpson (1941) showed that it is extremely unlikely that each droplet in a cloud forms on a nucleus of sea-salt, but suggested that the considerable number of much larger salt-particles normally present in the air might be of significance in the behaviour of clouds. This idea was later taken up by Woodcock and his co-workers, and much of the present interest in the subject is due to their studies and measurements.

Woodcock (1952) has put forward a hypothesis relating to the part played by giant sea-salt nuclei in the growth of precipitation. This hypothesis demands a one-to-one correspondence between the raindrops produced by a cloud and the large salt-particles originally contained in it. It is further assumed that the correspondence is such that the largest raindrops grow on the largest salt-particles. The growth mechanism envisaged is similar to that which we have already discussed in detail—an initial growth by condensation followed by the accretion of cloud droplets of a uniform low salinity.

Two features of this hypothesis are in immediate apparent conflict with our previous discussion of the growth of precipitation in convective clouds. The salt-particle concentration is assumed to be sufficiently high to completely outweigh the effects of the statistical growth mechanism, and the largest salt-particles are supposed to produce the largest raindrops, whereas our treatment suggests that they produce the smallest drops.

Further examination of Woodcock's work, however, shows that the hypothesis has been applied almost exclusively to orographic cloud, and there may be important differences in details of the rain-producing mechanism between clouds of this type and those of convective structure. We shall therefore consider the hypothesis explicitly in relation to orographic clouds.

Application to orographic clouds

Since most of Woodcock's experimental results relating to his salt-nucleus hypothesis have been gathered from the orographic cloud on the island of Hawaii, we must first look briefly at the physical characteristics of this cloud to see in what ways it may differ from the convective clouds discussed previously. Fig. 7.7,

Sea-salt Nuclei and Rain

taken from a paper by Wexler (1954), shows the general form of the orographic cloud. The cloud-top is limited by an inversion and cloud-depth decreases steadily in the direction towards which the wind is blowing.

Fig. 7.8 shows an idealized form of this cloud and gives the trajectories of giant salt-nuclei of two sizes as they grow to

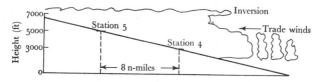

Fig. 7.7. Schematic diagram of cloud forms in orographic rain on the island of Hawaii (Wexler, 1954).

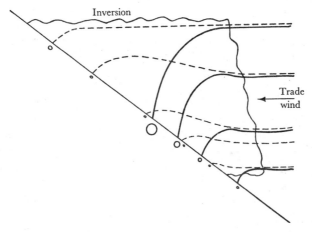

Fig. 7.8. Schematic diagram of the growth trajectories of giant salt-nuclei in orographic cloud. Solid curves are for large initial nuclei and broken curves for smaller nuclei. Resultant raindrop sizes are shown near the foot of each trajectory.

raindrops. It can be seen that the situation is quite different from that existing in a convective cloud; instead of entering from the base in an updraught, the nuclei enter the cloud at all levels on the windward side. In this figure the solid lines represent the trajectories of very large salt-drops which require only a short time to grow to raindrop size and are therefore deposited near the foot of the slope. The size of the raindrops formed depends upon the depth of cloud through which the drops fall. Relative sizes are indicated near the foot of each trajectory.

The broken lines in the figure show the paths of drops growing upon smaller salt-particles. These take longer to grow to precipitation size and are carried further up the slope by the wind. Since the depth of cloud here is less than at the foot of the slope, the resulting raindrops are smaller. At any given point, then, the largest raindrops have formed upon the largest salt-particles, and there is a monotonic relation between drop size and initial particle size. The relation is not strictly one-to-one with the salt-particles in the windward air-mass, however, because of the sorting of different particle-sizes as we proceed up the slope. A more detailed discussion of a model of this type has been given by Komabayasi (1957).

This part of the salt-nucleus hypothesis then appears to be a valid approximation for orographic clouds of the type found over Hawaii. It is not, however, valid in more general cases. It now remains to see whether there are sufficiently many salt-particles present of sufficient size to cause precipitation before it is produced by a random coalescence mechanism.

Keith and Arons (1953) have made calculations of the growth of sea-salt particles in the unsaturated atmosphere below cloud base, and have shown that under typical conditions salt-particles greater than about 3000 $\mu\mu$g will have attained radii greater than 30 μ at cloud-base level. Our discussion earlier in this chapter suggests that only droplets greater than about 30 μ in radius can produce raindrops more quickly than does the random coalescence process. It is, therefore, the concentration of salt-nuclei greater than about 3000 $\mu\mu$g which is of importance in evaluating the salt-nucleus hypothesis.

Extensive measurements of the concentration of large salt-nuclei in maritime environments have been made by Woodcock, as we discussed in chapter 4. Referring to fig. 4.15 we see that the number of nuclei greater than 3000 $\mu\mu$g in mass exceeds 0·1 per litre only for winds of force greater than 5 on the scale used by Woodcock. Since this is about the minimum number of nuclei required to produce rain of the observed intensities, it is evident that moderately high winds are required to supply these particles. Whilst agreement is somewhat marginal at this point, our knowledge of the processes involved is not sufficiently accurate for us to conclude that insufficient numbers of nuclei are normally present. The hypothesis, therefore, appears justified for the island of Hawaii under conditions of moderately strong wind.

Salinity measurements

No direct tests of the salt-nucleus hypothesis have as yet been devised, and the principal indirect evidence in its favour is based on the salinity measurements made by Turner (1955) and by Woodcock and Blanchard (1955).

Turner used a modified form of the raindrop spectrograph described by Bowen and Davidson (1951) to separate the raindrops collected into six or seven size-ranges, the salinity of each of which was measured. He found a relatively high salinity for small raindrops, which decreased to a minimum for drops about half a millimetre in diameter, and then rose again for larger drops. Daily mean salinities at cloud base ranged from 0·34 to 8·5 parts per million of NaCl.

Woodcock and Blanchard made separate measurements of raindrop spectrum, nucleus spectrum and mean salinity, but no direct salinity measurements for different drop sizes. On the basis of this data and the relation implied by the salt-nucleus hypothesis, they then calculated curves for salinity as a function of drop size. These curves show salinity minima for drop diameters near 1 mm and are in good general agreement with Turner's measurements.

One further observation supporting the hypothesis, again for Hawaiian orographic cloud, has been made by Squires and Twomey (1957). They made measurements of the concentration of giant sea-salt nuclei at the windward edge of the cloud, and also at the lee-side in the saddle of the mountains. The measurements indicated a considerable depletion of the larger sea-salt particles during passage of the air through the cloud, and the inference is that these nuclei were removed by precipitation.

All these experiments, though not in themselves conclusive, give fairly strong support to the salt-nucleus hypothesis as far as rain from Hawaiian orographic cloud is concerned.

Application to convective clouds

When we examine the salt-nucleus hypothesis in relation to convective clouds we find some immediate discrepancies. In convective clouds the height of the base is fixed and most of the moist air containing nuclei enters the cloud through its base. Any large droplets present will therefore follow trajectories similar to those of fig. 7.3, the largest nuclei resulting in the smallest raindrops.

Twomey (1955) has measured the concentration of giant sea-salt nuclei over land, and found that values as high as those over the sea are often obtained. Rain, however, appears to be a potent agent for the removal of these nuclei, and their concentration is usually very low in air-masses in which continued widespread cloud formation has taken place.

Because of these variations it is difficult to estimate the importance of salt-nuclei in rain formation. When they are present in sufficient numbers in a cloud of moderate drop size they may give rise to a considerable portion of any rain which falls. As the number of salt-nuclei increases, however, there is usually a decrease in the average size of the cloud droplets (Squires and Twomey, 1958), presumably because the air-mass has been dry for a considerable time and many smaller continental nuclei have been added. Such clouds will be more stable, despite the increased number of salt-nuclei, because of the greatly reduced collection efficiency of the large drops for the smaller cloud droplets. The cloud may then grow to considerable heights without precipitation forming, and the ice-crystal mechanism will become an additional competing factor.

Conclusions

Before going on to discuss some of the characteristics of rain from non-freezing clouds, it may be as well to summarize the conclusions reached in this chapter concerning its initiation.

The greatest numerical uncertainties in the coalescence process relate to its earliest stages where all the droplets are relatively small and uniform in size. The marked distinctions in behaviour between maritime and continental clouds can be explained adequately on the basis of their different average drop

Sea-salt Nuclei and Rain

size and the sharp variation in collision efficiency with size for droplets around $18\,\mu$ radius. The times predicted for the development of precipitation seem to be in reasonably good agreement with observation, though the margin of uncertainty in the calculations is still rather large.

The salt-nucleus hypothesis, when shorn of its restrictive assumption relating raindrop and nucleus sizes, provides a mechanism for by-passing this initial stage of difficult growth. The concentration of salt-nuclei normally present is just about sufficient to account for observed rain intensities, but this may not be the case in some situations.

The distinction between the stabilities of maritime and continental clouds appears to depend primarily upon the difference in size of the average droplets within the cloud. A complete absence of giant salt-nuclei may enhance the stability of a cloud to some extent, but a large excess of such nuclei, being usually accompanied by an increase in the number of smaller nuclei over the continent, does not lead to greatly decreased stability.

The role of electrical effects in coalescence processes is still uncertain. Electric fields and charges normally present in clouds are probably insufficient to greatly disturb the collision rate, but in clouds of marginal stability may have an appreciable influence.

In thunderclouds, on the other hand, the immense fields generated during precipitation initiated by ice-crystals may well be intense enough to greatly increase the collision efficiency of water droplets in lower parts of the cloud. Coalescence rain may then augment the ice-initiated precipitation to lead to the downpours often associated with this type of cloud.

Whilst understanding of all these processes is rapidly increasing, cloud stability still represents one of the most important fields for further fundamental work in the whole subject of cloud physics.

THE CHARACTERISTICS OF RAIN

Two properties of rain can be readily observed from the ground—its intensity and its size-distribution—and these are clearly of fundamental importance. Sampling from aircraft enables some idea of vertical structure to be obtained, and hence a study can be made of the modification undergone by the raindrops in their descent from cloud base to ground level. Similarly, radar techniques enable a broad picture of a precipitating cloud to be obtained which often gives valuable information about the origin of the precipitation.

In the following section we shall examine some of the instrumental methods commonly used in the study of rain, and then go on to discuss the characteristics of rain and the way in which they are modified between the cloud base and the ground.

Experimental methods

Conventional methods for examining the characteristics of rain usually involve collecting a raindrop sample in such a way that information can be obtained about the drop size-distribution as well as the total amount of precipitation. The simplest measurement method was introduced by Wiesner (1895) and, with minor modifications, has since been used by many workers (Lenard, 1904; Defant, 1905; Niederdorfer, 1932). A piece of filter-paper with its lower surface dusted with a water-soluble dye (rhodamine, eosin, methylene blue or bromocresol green) is exposed to the rain on a suitable frame. The drops moisten the paper, and on drying leave circular stains whose diameters are simply related to those of the raindrops. Though the method is agreeably simple, evaluation of drop size-distribution is very tedious. Niederdorfer (1932) has described some of the other complications which arise from non-uniformity of paper-stock, and from velocity-dependent and splashing effects with large drops.

The splashing of large drops has been largely eliminated in a modification of this method used by Blanchard (1949), who used as measuring surfaces fine screens of brass or nylon mesh coated with soot or powdered sugar. On passing through the screen raindrops then leave easily measured clear circular spots.

In another technique the raindrops fall into finely divided flour, in which they form beads of dough. After treatment the size of the beads can be related to the size of the original drops. The method is not an attractive one at first sight, but it was used

The Characteristics of Rain

successfully by Bentley (1904) and in the hands of Laws and Parsons (1943) it gave excellent quantitative results.

A very different method, originally used by Schindelhauer (1925), counts the number of raindrops falling by the electrical pulses which they generate upon striking a microphone diaphragm. A development of this principle described by Cooper (1951) measures the pulse amplitude as well and so yields the size-distribution. With this instrument carried aloft on a meteorological balloon the vertical variations of the raindrop spectrum can be examined.

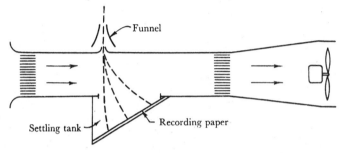

Fig. 7.9. The raindrop spectrograph (Bowen and Davidson, 1951).

A simple instrument which overcomes most of the disadvantages of the filter-paper method has been described by Bowen and Davidson (1951) and is illustrated in fig. 7.9. In principle the device is a form of mass spectrograph in which falling raindrops are deflected by a horizontal air-current. The drops fall through the funnel, which is carefully designed to eliminate the effects of drops which strike its sides, and are deflected by the airstream produced by the electric fan and honeycomb ducts. The distance through which the drops are deflected is approximately inversely proportional to their mass, so that if the sample contains a distribution of drop sizes they are spread out along the bottom of the tunnel according to size. There they impinge on sensitized filter-paper, which is moved slowly at right angles to the airstream, giving a continuous record of variations of the spectrum with time. The only disadvantage of this instrument is the requirement that the rain to be measured be falling vertically when it enters the funnel.

Several optical methods have been described, of which the

most effective is that of Mason and Ramanadham (1953). The raindrops are allowed to fall through a narrow beam of light, and the light scattered by them is focused on to the slit of a photomultiplier tube. The height of the light-pulse depends upon the size of the scattering-drop, and electronic sorting of the resulting voltage pulses allows counts of the drops in several size-ranges to be made automatically. This instrument has the advantage of being independent of the direction of fall of the rain, and of positively eliminating any effects due to mechanical definition of the sampling volume.

Radar techniques

The principles of radar are sufficiently well known that no review of them is necessary here. It is sufficient to note that concentrations of drops or other small particles scatter centimetre radio-waves to a sufficient extent to give a detectable radar echo, so that by application of conventional radar techniques the spatial distribution of the scattering particles can be found.

If the scattering particles are spheres of diameter d and if their concentration is n per unit volume, then the radar echo is proportional (Ryde, 1957) to the quantity:

$$\frac{1}{\lambda^4} \Sigma n d^6, \qquad (7.15)$$

where λ is the wavelength of the radiation used ($\lambda \gg d$) and the summation is over all the size-groups present. Since in the performance of a system the attenuation in the intervening path must be considered and this increases with decreasing wavelength, there is a limit to the shortness of wavelength desirable for the detection of particles of a given size. Wavelengths used lie typically in the range 1–10 cm and allow the detection of significant densities of particles in the size-range typical of raindrops. Aerials are usually arranged to sweep either horizontally, which gives rise to a plan of the rainfall distribution results, or vertically, which yields a section through the cloud mass. This latter method is more generally useful in cloud physics studies, whilst the former is often used in synoptic meteorology.

PLATE II

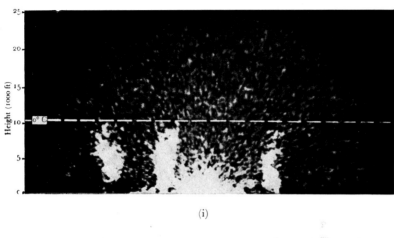

(i)

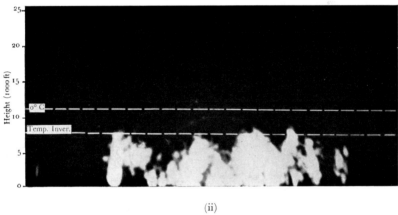

(ii)

(i) Radar record of non-freezing showers extending to the 0° C isotherm; (ii) non-freezing showers limited by a temperature inversion (Day, 1953).

The Characteristics of Rain

Much work has been done on the detailed characteristics of radar echoes. General reviews have been given by Ligda (1951) and Wexler (1951), whilst the effects of range and attenuation in distorting the apparent echo-pattern has been discussed by Atlas and Banks (1951). If some form of standard drop size-distribution can be assumed, then measurement of echo-intensity gives a direct measure of liquid water content (Atlas and Bartnoff, 1953). Atlas (1954) has shown that there appears to be a general correlation between rainfall intensity and the parameters of the drop size-distribution and has used this information in the estimation of cloud parameters from radar echo-intensities.

Measurements of Doppler shift in the radar echo-frequency give information about drop motion and hence about turbulence and winds, while Bartnoff and Atlas (1951) have shown that a study of the audio-frequency components of the Doppler shift together with attenuation measurements can be used to obtain information about the size-distribution of the falling drops.

When melting snow-crystals as well as water-drops are involved in the precipitation it is no longer a good approximation to consider the scattering-particles as spheres. The scattering of microwaves by ellipsoidal particles has been considered by Labrum (1952) and by Atlas, Kerker and Hitschfeld (1953) who show that deviations from sphericity are of great importance in the explanation of some of the features of the radar echoes observed from clouds containing ice-crystals. We shall consider this further in a later chapter.

Radar observations

Since radar techniques give an overall view of the distribution of precipitation elements within a cloud, we shall review some of the general results which have been obtained by these means before going on to discuss more specific measurements.

From the point of view of cloud physics the radar presentation giving most useful information is that giving echo-strength as a function of position in a vertical section through a precipitating cloud. Measurements show, broadly speaking, two different types of echo-pattern in displays of this type. The first is commonly associated with clouds whose tops are much colder than freezing, and consists of one or more bright horizontal bands of high echo-intensity. This pattern is associated with rain initiated by the Bergeron or ice-crystal process, and we shall defer discussion of it to a later chapter. The second type of pattern consists of a vertical column of high echo-intensity in which no horizontal bands are visible. This type of echo is typical of precipitating clouds which have not reached the freezing level, and in which precipitation is forming by a coalescence mechanism. An example of this type of pattern is shown in Plate II, from the paper by Day (1953). Very similar patterns have been observed on airborne radar sets by Styles and Campbell (1953) under appropriate cloud conditions.

We should point out, however, that columnar echoes are not confined to rainstorms produced by coalescence processes and in fact many large thunderstorms reaching well above the freezing-level show columnar echoes with no visible band structure (Day, 1953). In fact there is a general correlation of columnar echoes with convective clouds and band echoes with layer-clouds.

From the intensity of the radar echo the quantity Σnd^6 can be found and if the size-distribution of the drops is known (either absolutely or as a general function of precipitation intensity) then the precipitation intensity can be deduced. The relation between measured precipitation intensities and those determined from radar data has been investigated by Wexler (1948) and by Twomey (1953a). They conclude that because of uncertainties in the size-distribution the radar values may be in error by as much as a factor of two in either direction, assuming that instrumental errors are negligible. Whilst such an uncertainty leaves much to be desired, the echo-intensity is still a valuable guide to the intensity of the precipitation.

Drop size-distributions

The distribution of raindrop diameters at the surface of the earth has been measured by many observers since the time of Lenard (1904). Most of the observations have taken no particular note of the process by which the rain originated, and measurements on precipitation from non-freezing clouds are rare. We shall therefore comment briefly upon the general features of the distributions which have been obtained, before considering the few observations strictly relevant to our present discussion.

The size-distributions ordinarily observed have in common the feature that the number of droplets in a given size-range decreases rapidly with increasing size. A similar decrease in the number of very small droplets is also often observed, but many measuring techniques are limited to drop sizes too large for this effect to be seen. Several observers have fitted empirical curves to their data, and we shall examine some of these briefly.

Best (1950) has examined the data of a considerable number of earlier workers and shown that it can be fitted by an empirical relation of the form

$$1-F = \exp[-(x/a)^n], \qquad (7.16)$$

where F is the fraction of liquid water comprised by drops of diameter less than x, and a and n are constants, the average value of n being about 2·25. The value of a depends upon the precipitation rate I and has the value

$$a = AI^P, \qquad (7.17)$$

when x and a are measured in centimetres and I is in metres per hour. The constants A and P have mean values 0·13 and 0·232 respectively.

Expressing (7.16) in terms of droplet numbers so that $n(x)dx$ is the number of droplets with diameters in a range dx about x, we find for the average spectrum

$$n(x) \approx Bx^{-2} \exp(-Cx^2), \qquad (7.18)$$

where B and C are constants.

A different relationship was found by Marshall and Palmer (1948) for their own observations and those of Laws and Parsons (1943). Their empirical relation is

$$n(x) = n_0 \exp(-\lambda x), \qquad (7.19)$$

where x is the drop diameter in centimetres, n_0 has the value 0·08 cm^{-4}, and for rainfall of intensity I mm/hr

$$\lambda = 41 I^{-0.21} \text{ cm}^{-1}. \qquad (7.20)$$

The fit of this curve to the experimental points was not very good for drops less than a millimetre or so in diameter.

A somewhat different approach has been adopted by Levin (1954 b) who remarks that in fact all observed drop size-distributions are very close to the log-normal distribution typical of most collections of particles having a common origin (see chapter 4). He further pointed out that both the Marshall–Palmer distribution (7.19) and that used by Khrgian, Mazin and Cao (1952), namely

$$n(x) = ax^2 \exp(-bx) \qquad (7.21)$$

are close approximations to the log-normal distribution at values of x rather greater than the median value. The log-normal distribution has the further advantage of limiting the number of very small drops in the distribution, a feature found in the observations (Laws and Parsons, 1943; Marshall and Palmer, 1948) but not shown by either of the distributions (7.18) or (7.19).

Turning now explicitly to measurements on the rain falling from non-freezing clouds, we find that almost the only values available are those of Blanchard (1953) for rain from orographic clouds over the island of Hawaii. His results are shown in the curves of fig. 7.10. The results are seen to be in at least qualitative agreement with those found by other workers for rain of undetermined origin. The distributions are peaked at diameters rather less than 1 mm and show a relative scarcity of very large and very small drops, as would be expected for a log-normal distribution.

The content of large drops was also found to increase with increasing rain intensity, as found by other workers, while the content of small droplets at the same time decreased.

Blanchard also determined curves for rain of non-orographic origin, probably originating by an ice-crystal process. The curves in this case had very high concentrations of small droplets as well as a more typical distribution of larger droplets, and followed a law somewhat like (7.18), due to Best, rather than a log-normal curve. No detailed explanation has yet been given of this distinction.

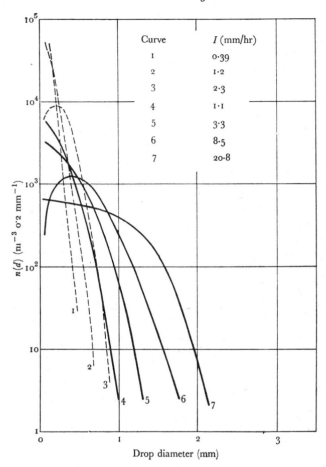

Fig. 7.10. Average raindrop size-distributions for Hawaiian orographic rain. Curves 1–3 are for measurements made at or near the dissipating edge of the cloud, while curves 4–7 represent samples taken at cloud base well in from the cloud edge (Blanchard, 1953).

Preferred drop sizes

Several of the earlier investigators to measure drop size-distribution in rain (Defant, 1905; Köhler, 1925 a, b; Niederdorfer, 1932; Landsberg and Neuberger, 1938) found what appeared to be specially preferred drop sizes whose volumes bore the relationship 1:2:4:8:16. Whilst many more recent measurements have shown no such periodicities, this may be because the drop size intervals considered in the analysis were too broad to show the effect if it did exist.

The effect has been explained by most of these workers (e.g. Defant, 1905; see also Horton, 1948) on the assumption that the rain consists of droplets, initially of closely the same size, which coalesce with each other by some sort of transverse interaction when falling at their common terminal velocity. This explains the observed peaks, and the mode of interaction shows why drops of mass 3, 5, etc., do not occur.

On the basis of the more detailed studies of coalescence which we discussed in chapter 6 we can describe this interaction more precisely. The calculations of collision efficiency by Pearcey and Hill (1957) are thought to be fairly accurate for large drops (though as we have seen there is some uncertainty when drop radii are less than 30 μ). These calculations show that for large drops of nearly the same size the collision efficiency is very large, while it decreases sharply as the drop sizes diverge. This, then, provides a detailed mechanism favouring collisions between droplets of equal size.

Whilst a reasonable mechanism thus exists favouring droplets with masses 2, 4, 8, 16, times the unit drop mass, we must still inquire to what extent this effect is of importance in actual rainfalls. This depends greatly upon the homogeneity of the original raindrop size-distribution, and is likely to vary considerably with details of the situation. Niederdorfer's peaks represent as much as 50 % modulation of the average smooth distribution curve, whilst other curves show negligible evidence of peaks.

Whilst it has been suggested (Takahasi, 1934) that the origin of the peaks may be psychological and simply related to the measuring process, belief in their reality is still fairly widespread. It remains to be seen whether they are a significant property of drop size-distributions in general.

Modification of the distribution below cloud base

In view of the many uncertainties in our present understanding of the exact mechanism by which precipitation elements develop in non-freezing clouds, it is not possible to give

The Characteristics of Rain

any detailed discussion of the means by which the size-distribution of raindrops at cloud-base level is developed. Our discussion earlier in this chapter has shown how the size of a typical raindrop depends upon various cloud parameters, and we must expect a considerable spread about this mean size. It is physically reasonable that the distribution should be lognormal, but as yet no more detailed description can be given.

Since typically the base of a cloud may be some thousands of feet above ground level, it becomes important to consider any modifications which may occur in the drop size-distribution as the drops fall through this region of cloud-free air. Fortunately this problem can be treated without too much difficulty, so that if the drop spectrum at cloud-base level is assumed, its form near the ground can be deduced to a reasonable approximation.

Three main factors influence the raindrops in their fall to the ground. In the first place, since the air is unsaturated, there is a tendency for the drops to evaporate. We have already discussed the equations governing this evaporation in chapter 6 where we considered the growth of droplets in a supersaturated environment. The argument in the present case is exactly the same, except that the sign of the supersaturation is changed.

The second factor which may influence the size-distribution is the coalescence of the drops with each other due to their differing terminal velocities. The treatment is very similar to that for the coalescence of cloud droplets except for the numerical magnitudes involved.

If we are considering a shower rather than steady rain, a third factor may enter. Since the larger drops fall with the greatest speed they tend to reach the ground before smaller drops, so that there is a drop size history during the shower as well as a size-distribution at any time. This effect may be further complicated by the presence of wind which gives a spatial distribution to the drops such that smaller drops are carried further downwind than large drops. All these effects must be considered when interpreting the results of experiments.

The variation of drop spectrum with height has been measured by Adderley (1953) using the balloon-borne microphone counter of Cooper (1951), whilst several workers (for example, Okita, 1958b) have used measuring devices located at various heights on a mountain side. Adderley's values actually apply to raindrops still within the cloud, but, by examining these together with the curves of Okita, one concludes that the drop size-distribution tends to broaden as the drops fall. Okita's results further show that below the cloud base the smallest drops disappear rather quickly, so that the median drop size shifts to larger values. As well as this, of course, there are numerous radar observations of echo-strength as a function of height, but these are difficult to interpret unambiguously.

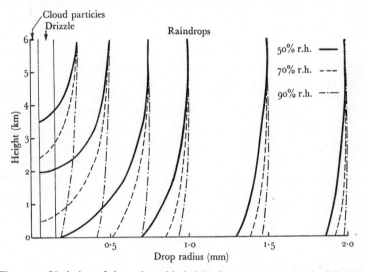

Fig. 7.11. Variation of drop size with height due to evaporation in I.C.A.N. standard atmosphere of the relative humidity shown (Best, 1952b).

Measurements of drop size history during a shower are more easy to make, since only one observing post is required. Typical examples are given by Bowen and Davidson (1951) and by Atlas and Plank (1953).

The evaporation of water-drops during their fall through unsaturated air has been made the subject of a detailed experimental and theoretical study by Kinzer and Gunn (1951), and Best (1952b) has carried out similar calculations for drops falling through typical atmospheres at various relative humidities. The results of his calculations are shown in fig. 7.11 for drops falling through the I.C.A.N. standard atmosphere, which is typical of temperate regions, and in fig. 7.12 for a summer tropical atmosphere.† It can be seen from these figures that the sizes of small drops can be reduced greatly by falls of a few thousand feet, even through an atmosphere of high relative humidity, whilst in hot arid climates with high cloud bases all but the very largest drops may evaporate before reaching the ground.

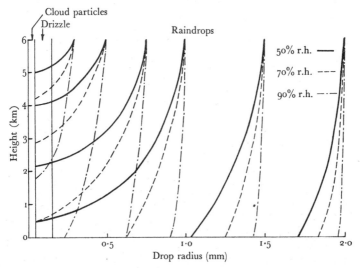

Fig. 7.12. Variation of drop size with height due to evaporation in summer tropical atmosphere of the relative humidity shown (Best, 1952b).

Rigby, Marshall and Hitschfeld (1954), Mason and Ramanadham (1954) and Okita (1958b) have combined the effects of evaporation and coalescence to study how an assumed size-distribution will vary with distance below cloud base. Starting with an exponential distribution of the Marshall-Palmer type (see 7.19), they found that the number of small droplets was progressively depleted, whilst the proportion of large drops increased with distance fallen. The distribution thus became flatter (that is, broader), whilst the median drop size shifted to

† The I.C.A.N. standard atmosphere is defined by a pressure of 1013·2 mb and a temperature of 15° C at the surface and a constant lapse-rate of temperature up to 11 km of 6·5° C per km. The summer tropical atmosphere has the same surface pressure and temperature lapse-rate as the I.C.A.N. standard atmosphere, but has a surface temperature of 41° C.

larger values as found in experiment. The changes produced, however, were not very great, and it seems likely that much of the variability in drop spectra arises from the generation processes within the cloud, rather than from any modification of the precipitation between the cloud base and the ground.

The behaviour of water-drops at terminal velocity

To conclude this chapter we shall make brief mention of some of the miscellaneous properties of raindrops falling freely through still air. The first quantity of interest is, of course, the terminal velocity as a function of drop size. For drops less than about 20 μ radius it is a valid approximation to use Stokes' law for the viscous drag and to assume the drop to remain spherical. This leads to

$$u = 1 \cdot 19 \times 10^6 \, r^2 \text{ cm/sec} \qquad (7.22)$$

for drops falling through air near sea level at 20° C, where the radius r is measured in centimetres. Above this size calculation becomes more difficult, and when deformation of the drop becomes appreciable an experimental determination of the terminal velocity becomes most satisfactory. Such measurements have been made by Lenard (1904) and more recently by Laws (1941), by Spilhaus (1948), and by Gunn and Kinzer (1949). The results of these later measurements are shown in graphical form in fig. 7.13.

Other phenomena of interest occur in the mechanical behaviour of rather large drops falling at terminal velocity. Lenard (1904), who suspended large drops in a vertical wind-tunnel, observed that they became flattened (rather than elongated as is often supposed) and that for very large drops the deformation became so extreme that rupture occurred. More detailed observations of the same type have since been carried out by Blanchard (1950). A theoretical discussion of the factors responsible for the drop shape has been given by McDonald (1954), who came to the conclusion that the only important factors were surface-tension, hydrostatic pressure gradients within the drop and external aerodynamic pressures. He found that the internal circulating currents postulated by Lenard are too small in velocity to have appreciable effect on the drop shape. Electric charges normally present on raindrops similarly seem too small to be important in determining drop shape, though exceptions may exist in thunderstorm rain.

Blanchard (1950), who used a stroboscopic photographic technique to examine the oscillations of large drops supported in a vertical wind-tunnel, found that drops as large as 9 mm in

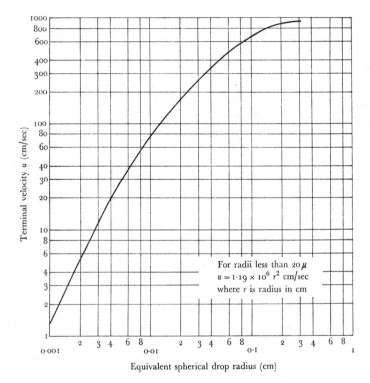

Fig. 7.13. The terminal velocity of water-drops in air at 760 mm pressure and temperature of 20° C (from data of Gunn and Kinzer, 1949).

diameter could be supported without disintegration, but that these drops were extremely sensitive to turbulence, a sudden increase in air-speed of only 6% being usually sufficient to cause disintegration. Lenard, using less elaborate apparatus, concluded that drops exceeding about 5·5 mm in diameter are unstable, and this may be a more typical figure for the conditions holding in the atmosphere.

The temperature of rain

The final property of rain which we shall consider is the temperature of typical raindrops as they reach the ground. Measurements of this quantity have been made by Byers, Moses and Harney (1949) and by Maulaud (1950), whilst Johnson (1950) and Kinzer and Gunn (1951) have conducted laboratory experiments on suspended and on freely falling water-drops.

When the falling drops have reached a steady temperature, they act as ideal ventilated wet bulbs, and their temperature appears to agree with that of the ventilated wet bulb to within $\pm 0.3°$ C in laboratory experiments (Kinzer and Gunn, 1951). The same behaviour is often found in natural rain (Maulaud, 1950) where the drop temperature is found to follow fairly closely the local ventilated wet-bulb temperature.

On other occasions, however, all observers have found marked departures from this relation, particularly at the beginning of a storm, when the rain temperature may be either higher or lower than that of the wet bulb. This effect seems to spring primarily from the fact that there is an appreciable time-lag in the temperature behaviour of a drop, so that its temperature at a given moment depends upon its past history. This effect has been examined theoretically and experimentally by Kinzer and Gunn (1951). Typical relaxation times are of the order of 10 sec, but of course vary very widely with drop size, temperature and humidity. A particularly important case occurs when the raindrops originate as ice-particles; the latent heat makes the time constant so long that the drops may be many degrees below their equilibrium temperature when they reach the ground.

CHAPTER 8

NUCLEATION OF THE ICE-PHASE

INTRODUCTION

In this chapter we shall consider some details of the phase-change from water or water-vapour to ice. Unfortunately, the theory of this transition is rather less fully developed than for condensation, and much further work is needed. The experimental evidence too, is very confused, though some sort of coherent picture is now beginning to emerge which we shall discuss in terms of the theory as it has so far been developed.

After an initial discussion of the structure of water and ice we shall first examine the homogeneous nucleation of ice by freezing and sublimation, not so much because of any practical importance of these processes, but rather to gain some insight into the problems arising in discussions of these phase transitions. We shall then go on to consider the role of nuclei in aiding the nucleation of ice-crystals and discuss certain experimental results which bear on the problem.

THE STRUCTURE OF WATER AND ICE

Water has, in the preceding chapters, been simply considered to be a liquid having a moderate vapour-pressure at ordinary temperatures, and we have had no need to inquire further into its structure and properties. Now, however, that we come to consider the ice-phase, and particularly the phase-change from water to ice, it becomes necessary to examine the structure of water in its different phases from a molecular point of view.

When we begin such an investigation we find that water is far from being a simple liquid in its molecular properties, and that the solid forms of ice present many interesting peculiarities. Most of these features can be traced back to the water molecule itself, which is far from the simple spherically symmetric structure considered in the elementary theory of liquids and solids. The water molecule can be regarded roughly speaking as a regular tetrahedron with an oxygen atom at its centre. Two of the vertices of the tetrahedron are occupied by positive hydrogen ions, while the remaining two are regions of high negative charge. The molecule as a whole is, of course, electrically neutral, but its charge distribution is very asymmetric and it possesses a large dipole moment. When water molecules become bonded together in a liquid or solid structure it is these positive hydrogen ions and negative vertices which act as the bonding centres. Since these vertices are four in number, the preferred co-ordination number for a water molecule is 4 and this is the value exhibited in ice. In water, as we shall see later, the picture is rather more complicated. We shall therefore begin this brief discussion with a description of what is known of the structure of ice, and then lead on to a consideration of water.

The structure of ice

From the point of view of the physicist ice is by no means a simple solid. At normal pressures three different forms of ice exist

(1) Normal, hexagonal ice having a crystal structure similar to that of tridymite (a form of SiO_2).

(2) A cubic ice with a structure like that of cristobalite (another form of SiO_2), which can be obtained over a range of temperatures near $-100°$ C.

(3) An amorphous ice stable below about $-140°$ C.

In addition Bridgman (1912, 1935, 1937) has found six further forms of ice which exist only under very high pressures. Whilst little is known of the molecular structure of these high-pressure modifications, the three forms stable at ordinary pressures have recently been investigated in considerable detail, and reviews of our present knowledge have been given by Lonsdale (1958) and by Blackman and Lisgarten (1958).

In cloud physics we are only interested in pressures of the order of normal atmospheric pressure and in temperatures above $-100°$ C. For our present purposes, therefore, ice may be regarded as a substance of unique crystal structure, the only form of interest being the hexagonal, tridymite-like ice-I. It is this structure which we shall imply when we speak simply of 'ice'.

The Structure of Water and Ice

The crystal structure of ice has been investigated by the methods of X-ray diffraction, and more recently by electron and neutron diffraction and there is now fairly general agreement as to this structure. The positions of the oxygen atoms, which are the principal scatterers of X-rays, were located by Bragg (1922), and Barnes (1929) determined the space group to be D_{6h}^4 (or $P6_3/mmc$). This is a hexagonal lattice as shown in fig. 8.1 (where black spheres represent oxygen atoms), and each

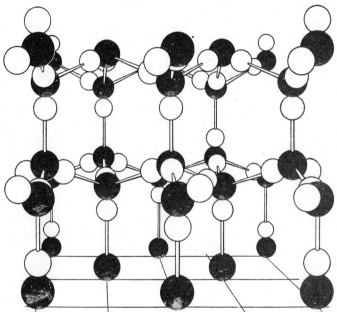

Fig. 8.1. An expanded model of the structure of ordinary ice. Black spheres represent oxygen atoms and white spheres hydrogens, of which there are two attached to each oxygen. The rods represent hydrogen bonds (Buswell and Rodebush, 1956).

oxygen atom is surrounded tetrahedrally by four other oxygen atoms. The crystal may be regarded as consisting of crinkled sheets of atoms lying in (0001) planes, these sheets being joined together by relatively fewer bonds along the direction of the c-axis.

Because protons (H^+) do not scatter X-rays to any extent, it is very difficult to determine the positions of the hydrogen ions by X-ray measurements. Barnes (1929) proposed that there was a hydrogen ion on the middle of each bond between an oxygen and its four neighbours, thus preserving the space-group symmetry and giving an essentially ionic structure. Bernal and Fowler (1933), on the other hand, proposed an essentially molecular structure in which each oxygen had two hydrogens close to it, the hydrogens thus being asymmetrically situated upon the bonds. Pauling (1935) further suggested that apart from the requirement of two hydrogen atoms close to each oxygen atom and one hydrogen atom on each bond the configuration was entirely random, giving a statistical structure having a large number of possible configurations, and hence a finite entropy at absolute zero, in agreement with experiment. It is this structure which is shown in fig. 8.1, the white spheres representing one particular random arrangement of the hydrogen ions. The statistical distribution of the hydrogens has since been verified by electron and neutron diffraction (Blackman and Lisgarten, 1958; Owston, 1958) and by infra-red spectroscopy (Ockman and Sutherland, 1958).

Table 8.1. *Lattice parameters and density of ice at atmospheric pressure* (Lonsdale, 1958)

Temperature (°C)	Lattice a (Å)	c (Å)	Density (g/ml.)
0	4·523	7·367	0·916
−30	4·518	7·360	0·919
−60	4·512	7·352	0·923
−90	4·505	7·344	0·927

Most of the unusual properties of ice can be explained in terms of this crystal structure—its low density, high dielectric constant and high proton mobility, for example. We shall have no cause to consider most of these further here,† but should point out the open structure of the molecular arrangement compared with a close-packed structure, the four-fold coordination of the molecules and the hydrogen bonding which constitutes the intermolecular force. Table 8.1 (from Lonsdale, 1958) gives data on the density and lattice parameters of ice at atmospheric pressure.

† For excellent summaries of our present knowledge of the structure and properties of water and ice the reader is referred to the reports of two recent symposia on 'The Physics of Water and Ice' to be found in *Advances in Physics*, **7** (1958), 171–297, and in *Proc. Roy. Soc.* A, **247** (1958), 421–538, respectively.

The structure of water

Water cannot properly be said to possess a structure; rather it has an instantaneous structure and an average structure. Most measurements that can be performed give only information about the average structure, and, whilst this is physically important, many phenomena can only be discussed in terms of the particular variations in structure which take place from one instant to another.

Since no means have yet been devised of obtaining information about the instantaneous structure of the liquid, the approach has usually been to propose various models for this structure, to compare their overall properties with the experimentally determined average structure, and to see how well they can explain the other observed properties of the liquid. As can be imagined there is little agreement upon details of the model.

One of the most satisfactory general structural models is that proposed by Bernal and Fowler (1933), who considered water to consist of fluctuating transient groupings of molecules into three different types of configurations: (i) a tridymite ice-like structure present only to a very small extent and only at temperatures below $4°$ C; (ii) a quartz-like structure predominating at ordinary temperatures; (iii) a close packed ideal-liquid structure predominating at temperatures near the critical point. These are not to be regarded as different forms of water, but only as fluctuating microstructures describing the immediate environment of a proportion of the molecules at a given instant. This model gives good agreement with the radial distribution function found by X-ray scattering and ascribes the decrease in density of water below $4°$ C to the increasing number of molecules in ice-tridymite-like environments. This in itself is in agreement with Frenkel's theory of heterophase fluctuations which we discussed in chapter 3. Frenkel (1946) has also discussed from a general theoretical viewpoint the existence and properties of 'cybotactic groups' (as these transient quasi-crystalline microstructures are sometimes called). The existence of increasing numbers of water molecules in ice-like structures in supercooled water has been confirmed by the X-ray diffraction measurements of Dorsch and Boyd (1951).

That the subject of liquid structure is, however, as yet only in its infancy is emphasized in a recent paper by Bernal (1959).

The surface structure of water also merits detailed attention, though we have as yet little knowledge of this field. Molecules at the water surface are in an asymmetric environment and because of their polar nature it is not surprising that they have a preferred orientation relative to the plane of the surface. The orientation appears to be that with the hydrogen ions directed into the body of the liquid and thus gives rise to a dipole layer with negative side directed outwards.

Such a preferred orientation at the surface reduces the entropy of surface molecules (Good, 1957) and the orientation may extend a depth of several molecular layers into the liquid (Henniker, 1949). Some of the effects of the existence of these surface-layers have been considered by Weyl (1951), and though many of his conclusions would not be accepted by the majority of workers, there is little doubt that surface effects are of importance in much of the behaviour of small droplets.

HOMOGENEOUS NUCLEATION OF FREEZING

The amount by which a small water droplet can be supercooled depends upon its content of foreign nucleating particles. When all these particles are removed it is to be expected that freezing should occur by homogeneous nucleation, and this process should set a limit to the amount of supercooling possible. Various experiments, which we shall discuss in detail later, suggest that the supercooling-limit temperature for small pure water droplets lies in the range -33 to $-41°$ C depending upon the droplet size, and it is, therefore, natural to suppose that in this range freezing is due to homogeneous nucleation. The validity of this assumption will now be examined.

The theory of the homogeneous nucleation of ice-crystals within a volume of supercooled water is formally almost identical with the theory of nucleation of water droplets from a supersaturated vapour, which we considered in chapter 3. Within the supercooled liquid there is a population of water molecules which have come together to form embryonic groups of ice-structure, the number of groups of a given size depending exponentially upon the free energy required for their formation.

Homogeneous Nucleation of Freezing

Because ice is a crystalline structure it is no longer generally sufficient to consider these embryos to be spherical in shape, and a more general polyhedral form must be considered. The exposed crystal faces of the embryo will be those of lowest free energy, and the equilibrium habit can be determined by minimizing the total free energy of an embryo of given size with respect to its habit. In practice we usually have no definite information about the free energies of different crystal faces and must content ourselves with the use of an average free energy σ_{SL} per unit area.

If we assume a particular crystal habit for the embryo then it is convenient to consider an inscribed sphere of radius r so that the volume of the embryo is $\frac{4}{3}\pi r^3 \alpha$ and its surface-area is $4\pi r^2 \beta$, where α and β are both greater than unity, but approach unity as the polyhedron tends to spherical shape.

The free energy of formation of an embryo inside which a sphere of radius r can just be contained is

$$\Delta G = -\tfrac{4}{3}\pi r^3 \alpha n_S kT \ln(p_L/p_S) + 4\pi r^2 \beta \sigma_{SL}, \qquad (8.1)$$

where n_S is the number of molecules per unit volume of solid, σ_{SL} is the free energy per unit area of ice-water interface, and p_S, p_L are the saturation vapour-pressures over ice and water respectively at the temperature considered. Except for the factors α and β, this equation is the same as (3.11) referring to condensation.

The size of a critical embryo is found as before by setting

$$\frac{\partial \Delta G}{\partial r} = 0 \qquad (8.2)$$

and the resulting critical free energy is

$$\Delta G^* = \frac{16\pi \sigma_{SL}^3 \xi}{3[n_S kT \ln(p_L/p_S)]^2}, \qquad (8.3)$$

which is identical with the expression for a spherical embryo (3.16) except for the factor $\xi = \beta^3/\alpha^2$, which is slightly greater than unity but approaches unity as the shape of the embryo tends towards that of a sphere.

An alternative expression for (8.3) may be obtained by use of the Clausius-Clapeyron equation (Glasstone, 1947, p. 227) to express $\ln (p_L/p_S)$ in terms of supercooling ΔT. The result is

$$\Delta G^* = \frac{16\pi \sigma_{SL}^3 \xi}{3(\Delta S_V \Delta T)^2}, \qquad (8.4)$$

where ΔS_V is the entropy of fusion of ice per unit volume, averaged over the range ΔT, and has the approximate value $(1.13 - 0.004 \Delta T) \times 10^7$ erg $°C^{-1}$ cm^{-3}. This form of ΔG^* is particularly convenient for calculation.

Discussion of the rate of nucleation now follows exactly as in chapter 3. The number of embryos of critical size per unit volume of liquid is approximately

$$n(r^*) \approx n_L \exp(-\Delta G^*/kT) \qquad (8.5)$$

and the nucleation rate is the rate at which these embryos gain a further molecule to reach a size from which they can grow freely. This rate cannot be worked out simply by kinetic theory, as we did for a supersaturated vapour, because the motion of the liquid molecules is hindered by binding to their neighbours and an activation energy associated with the diffusion process is involved. Turnbull and Fisher (1949) have evaluated this rate constant to be approximately

$$B = \frac{kT}{h} \exp(-\Delta g/kT), \qquad (8.6)$$

where h is Planck's constant and Δg is the activation energy for self-diffusion in the liquid (or more properly across the liquid-solid boundary). Thus the complete expression for the nucleation rate is

$$J \approx \frac{n_L kT}{h} \exp(-\Delta g/kT) \exp(-\Delta G^*/kT). \qquad (8.7)$$

The factor $n_L kT/h$ is of order 10^{35} cm^{-3} sec^{-1}, whilst, in a limited temperature range below 0° C, $\Delta g \approx 4 \times 10^{-13}$ erg/molecule (McDonald, 1953a). The coefficient of $\exp(-\Delta G^*/kT)$ is thus of order 10^{30} cm^{-3} sec^{-1} and the nucleation rate may be expected to have a very sharp threshold, as was the case with homogeneous condensation.

Calculation of the supercooling ΔT for which J becomes appreciable presents several difficulties, though it is simple in principle. The chief obstacle is the evaluation of the free energy σ_{SL} of an ice-water interface. Attempts have been made by Krastanow (1941), Mason (1952a), McDonald (1953a) and

Ouchi (1954), but none of these approaches is entirely satisfactory. Calculated values diverge so widely that one can only conclude that σ_{SL} probably lies between 10 and 25 erg/cm^2. This is rather larger than Frenkel's (1939) general estimate of the magnitude of such interfacial free energies, but as we shall see the range of calculated values is reasonably consistent with the results of experiment. Since an uncertainty of 10% in the value of σ implies an uncertainty of about 15% in ΔT, however, calculations are on little more than an order of magnitude basis.

The second difficulty is to decide upon the habit of the ice embryo. There are clearly two general possibilities—either the embryo surface will be molecularly rough, in which case the embryo should be approximately spherical, or else smooth close-packed planes will develop giving a polyhedral form. In the second case it is necessary to decide upon the shape and upon the parameters α and β of (8.1) before a calculation can be made. Mason (1952a) and McDonald (1953a) considered the embryo to be a hexagonal prism, but did not give an explicit treatment of the habit parameter ξ. Roulleau (1958), too, has given a different treatment of the same habit with results again differing from our argument above. Within limits, however, uncertainty in this factor is not important because of the greater uncertainty in the value of σ.

Finally, there are somewhat less important uncertainties in the appropriate values for Δg and ΔS_V at temperatures well below 0° C. These points have been discussed in some detail by McDonald (1953a), and his values, extrapolated from measurements at higher temperatures, are probably sufficiently accurate for this sort of calculation.

With all these uncertainties it can only be said that the calculation indicates a nucleation threshold for freezing of small droplets somewhere between -20 and -50 °C. Even this broad estimate, however, is of value, as we shall see below.

Comparison with experiment

As with all experiments on homogeneous nucleation, the difficulty in the present case is to eliminate the disturbing effects of foreign surfaces or suspended particles. As far as cloud droplets are concerned this is most easily achieved by forming them in clean air and freezing them while they are still in suspension. Such experiments have been performed by Cwilong (1947), Schaefer (1948a) and Fournier d'Albe (1949), who all found critical temperatures in the vicinity of $-40°$ C, at which large numbers of ice-crystals were formed. Cwilong and Fournier d'Albe used an expansion chamber to produce a fog of small clean droplets at a particular temperature, and found a critical freezing temperature at $-41°$ C. Schaefer dropped small pellets of mercury (melting-point $-38·9°$ C) through clean moist air producing a fog of droplets in their path. If the mercury was frozen the droplets were observed to grow as ice-crystals, locating the critical temperature at $-38·9°$ C. Weickmann (1947), on the other hand, reported the presence of water droplets as well as ice-crystals in clouds at a temperature as low as $-50°$ C.

For work on larger drops of water some means must be found to support them, and it is important that the support should not influence their freezing behaviour. Dorsch and Hacker (1950) and Hosler (1954) supported drops on a metal plate, whilst Jacobi (1955) used a film of collodion. Meyer and Pfaff (1935) and Wylie (1953) used glass tubes containing rather large amounts of water, while Bigg (1953a) avoided the use of solid surfaces altogether by suspending the water-drops at the interface of a pair of immiscible liquids of different densities. These last methods have the additional advantage that they protect the liquid from contamination by airborne particles.

The results of these and other experiments have been collected by Langham and Mason (1958), and observed freezing-temperature as a function of drop-size is shown in fig. 8.2. Since even the best water samples are likely to contain some impurities, it is the lowest results which are of significance for the study of homogeneous nucleation, and it appears that measurements such as those of Bigg were performed on water samples containing appreciable numbers of foreign nuclei.

It now becomes useful to work back from the experimental data with the aid of (8.7) to find the interfacial free energy σ_{SL} on the assumption that $J \times$ (volume) ≈ 1 sec.$^{-1}$. Once again, however, the calculated value of σ_{SL} depends rather seriously upon the values assumed for Δg, ΔS_V and ξ, and different calculations on essentially equivalent experimental data have yielded values of σ_{SL} at $-40°$ C ranging from 15·5 erg/cm² (Jacobi, 1955) to almost 20 erg/cm² (Roulleau, 1958). In addition Jacobi found a considerable temperature variation of σ_{SL} between -33 and $-40°$ C, whilst Langham and Mason (1958) obtained good agreement with experiment on the assumption that σ_{SL} had a constant value of 17·2 erg/cm² over this range.

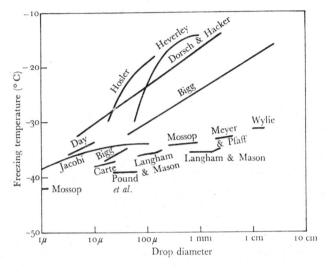

Fig. 8.2. Critical freezing-temperature as a function of size for purified water-drops, as found by various workers.

Whilst, then, theoretical work on homogeneous nucleation requires some refinement before it is much more than semi-quantitative, the theoretical predictions are sufficiently close to the results of experiment that we can with reasonable confidence attribute the freezing of pure water-drops in the range -33 to $-41°$ C (according to size) to a process of homogeneous nucleation. We can further assume, for the purposes of calculations to follow, that the interfacial free energy between water and ice is about 17 erg/cm² near $-40°$ C, with its value perhaps increasing to 20 erg/cm² or more at temperatures near $0°$ C. Such a variation with temperature seems reasonable on the basis of the variation of the structure of water over this temperature range.

HOMOGENEOUS NUCLEATION OF ICE BY SUBLIMATION

Just as water droplets can be nucleated homogeneously from supersaturated vapour, so we might expect that under suitable conditions and at low temperatures it might be possible to nucleate ice-crystals directly from the vapour. Theory shows this expectation to be justified and there is certain experimental evidence in its favour as we shall see presently. Because the nucleation of droplets and of crystals represent competing processes, interest centres upon the conditions under which one or other of these is dominant.

The formal theory of the homogeneous nucleation of ice-crystals from supercooled, supersaturated vapour has essentially been covered by our discussions of condensation and freezing. The ice-embryos may be considered to have a polyhedral habit as for freezing (though the crystal shape is not necessarily the same and will vary, in general, with temperature). This results in a factor ξ, analogous to that in (8.3), being introduced into the free energy ΔG^* of a critical embryo, which now takes the form

$$\Delta G^* = \frac{16\pi\sigma_{SV}^3 \xi}{3[n_S kT \ln(p/p_S)]^2}, \qquad (8.8)$$

where σ_{SV} is the free energy per unit area of an ice-vapour surface, p is the vapour-pressure of water in the environment and p_S the saturated vapour-pressure over a flat ice-surface. This equation is, except for the factor ξ, formally identical with (3.16) which applies to homogeneous nucleation of water-drops from saturated vapour. Similarly, the kinetic coefficient is analogous to the condensation case so that as in (3.19) the nucleation rate is

$$J = [p/(2\pi mkT)^{\frac{1}{2}}] 4\pi r^{*2} \beta n(1) \exp(-\Delta G^*/kT), \qquad (8.9)$$

where $4\pi r^{*2}\beta$ is the surface-area of the critical embryo and $n(1)$ is the number of vapour molecules per unit volume. A Becker–Döring correction can be applied as for (3.19).

We must now examine the implications of (8.8) and (8.9) and particularly of their similarity to the corresponding equations for condensation, (3.16) and (3.19).

The surface free energy of water is a well-known quantity at ordinary temperatures and can be extrapolated with reasonable confidence to temperatures below 0° C, the numerical value being roughly 80 erg/cm². The surface free energy of ice has not been measured directly, but various theoretical estimates (Mason, 1952a; McDonald, 1953a; de Reuck, 1957) place its value in the range 100–120 erg/cm². The values of r^* for liquid and solid embryos are therefore approximately equal, and the shape factor β in (8.9) is not much greater than unity. The whole pre-exponential factors in (3.19) and (8.9) are thus equal within about a factor of two and for our purposes may be considered to be equal.

Because of the rough similarity in the surface free energies of ice and water the processes of homogeneous condensation and of homogeneous sublimation thus proceed under similar conditions, and when the temperature is less than 0° C the question of interest is which process occurs more readily. The question can be examined in the following way.

If we write equation (8.9) in the form

$$J_S = A_S \exp(-\Delta G_S^*/kT), \qquad (8.10)$$

where the subscript S indicates nucleation of a solid crystal, then the corresponding equation (3.19) for condensation can be written in the same form with subscripts L, and we have already remarked that

$$A_S \approx A_L \qquad (8.11)$$

to within an order of magnitude. Comparing the rates of formation of crystals and droplets then, we see that crystals predominate if

$$\Delta G_S^* < \Delta G_L^*. \qquad (8.12)$$

Substituting from (8.8) and (3.16) this condition can be put in the form

$$\frac{\ln(p/p_S)}{\ln(p/p_L)} > \xi^{\frac{1}{2}} \frac{n_L}{n_S} \left(\frac{\sigma_{SV}}{\sigma_{LV}}\right)^{\frac{3}{2}}. \qquad (8.13)$$

The right-hand side of (8.13) has a constant value greater than unity; the left-hand side is greater than unity at temperatures below 0° C, but its value depends upon the water vapour-pressure p. For the results to be physically significant we require that nucleation take place at an appreciable rate, so that we must assign some finite value to J_S in (8.10), and this fixes p as a function of T. Equations (8.10) and (8.13) can now be solved to find the temperature and saturation ratio for which direct sublimation becomes important.

Krastanow (1940) investigated this question on the assumption that σ_{SV} and σ_{LV} were equal, and that the ice-crystals had cubic shape ($\xi = 1\cdot 91$); he found that direct sublimation would occur for temperatures below about $-65°$ C, the required saturation ratio being, with respect to water, about 14. The initial assumptions are not, however, fully justified, since ice-crystals are not cubic and it would seem to be a better approximation to take $\sigma_{SV} \approx 1\cdot 1\sigma_{LV}$ at low temperatures. The value of ξ cannot be determined without more detailed knowledge of the surface free energy of ice for different crystal-planes. If this free energy is the same for hexagonal and prism faces then $\xi = 1\cdot 65$, but values ranging from near unity (for nearly spherical shapes) to well over 2 (for crystals tending towards disk or needle shapes) cannot be ruled out *a priori*. These values suggest a sublimation threshold below $-100°$ C, but the uncertainties are so great that little reliance can be placed upon these estimates.

When we seek this threshold experimentally a difficulty immediately arises, since, as we have seen, water-drops formed at temperatures below $-40°$ C freeze spontaneously. Sander and Damköhler (1943), in a low-temperature expansion experiment, found a discontinuity at $-62°$ C in their plot of $\ln(p/p_S)$ against $1/T$, observing further that below this temperature the presence of ions had no effect on the measurements, and the precipitated cloud glittered. The saturation ratio required was about 15 with respect to ice or 8 with respect to water. Very similar results were obtained by Pound et al. (1951), who located a discontinuity at $-63°$ C. It seems possible that this discontinuity is the sublimation threshold.

HETEROGENEOUS NUCLEATION OF ICE-CRYSTALS

The formation of ice from a supersaturated or supercooled environment may be aided by the presence of foreign surfaces or of suspended particles, as was the case with condensation. Thus water in contact with foreign solid materials or small suspended particles will usually freeze at a temperature considerably warmer than $-40°$ C, and sublimation can occur at supersaturations much less than that required for homogeneous sublimation. In this section we shall investigate the properties

Heterogeneous Nucleation of Ice-crystals

required of efficient ice-forming nuclei and study their mode of action.

Let us start from the simplest case and suppose that ice is being nucleated from the vapour on to a flat, perfect crystal face. We shall also suppose that ice-embryos forming upon this surface have the form of spherical caps as shown in fig. 8.3. The form of such an embryo can be specified in terms of r, its spherical radius, and $\mathfrak{m}$, the cosine of the angle of contact. Though $\mathfrak{m}$ has

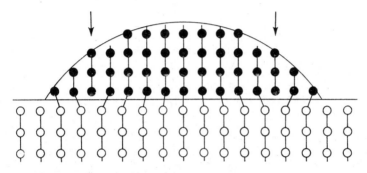

Fig. 8.3. An ice-embryo growing upon a crystalline substrate with a misfit of 10 %. The interface is dislocated, dislocations being indicated by arrows. In the case of coherent epitaxy the ice-embryo would be deformed elastically to eliminate the dislocations.

little macroscopic significance for solid systems it can be specified uniquely in terms of surface free energies as

$$\mathfrak{m} \equiv \frac{\sigma_{\mathrm{CV}} - \sigma_{\mathrm{CS}}}{\sigma_{\mathrm{SV}}}, \qquad (8.14)$$

where subscripts V, S and C refer to vapour, solid and catalysing substrate respectively.

It can be seen that there are two distinct ways in which the embryo can grow upon the substrate. Either the ice will retain its normal lattice dimensions right to the interface, in which case there will be dislocations in the sheets of atoms, or else the ice lattice will deform elastically to join coherently to the lattice of the substrate. Clearly there will always be a dislocated boundary when the substrate differs greatly from ice in its symmetry or lattice parameter, whilst coherent growth will be favoured upon substrates which are very similar to ice in crystal structure. We may in general expect a combination of these two possibilities, the interface possessing some dislocations and the ice-embryo being strained to some extent. This situation has been considered in detail by Frank and van der Merwe (1949; also van der Merwe, 1949) and its implications for the nucleation process worked out by Turnbull and Vonnegut (1952).

Dislocations at the interface will have the general effect of raising σ_{CS} and hence of lowering the value of $\mathfrak{m}$, i.e. increasing the contact angle. Elastic strain within the embryo, on the other hand, raises the bulk free energy of the ice-molecules. Both these effects are in a direction to lower the nucleating efficiency of the substrate as the number of dislocations and the elastic strain increases. An ideal nucleating agent should, therefore, have a surface-structure identical with that of ice from the point of view of crystal symmetry and lattice dimensions, and any departure from this ideal will lower the nucleation efficiency.

This is not, however, the sole requirement, since the interfacial free energy σ_{CS}, and hence the value of $\mathfrak{m}$, depends upon the way in which water molecules are bound on to the substrate surface. Because of the polar nature of the water molecule this binding is largely electrostatic, so that ice nucleation is likely to be preferred on surfaces with relatively intense ionic fields. Fletcher (1959d) has suggested that entropy effects require equal numbers of exposed positive and negative ions at surfaces which are good nucleating agents, and we shall mention this requirement again later. These requirements suggest that ionic crystals may in general be better nucleating agents than less polar compounds.

To be more quantitative, let us suppose that the misfit between the ice and substrate lattices is given by

$$\delta = \left| \frac{a - a_0}{a_0} \right|, \tag{8.15}$$

where a is a lattice vector of the substrate surface and a_0 the corresponding vector in the ice-lattice. Further, let ϵ be the average elastic strain produced within the ice-embryo. Then, following Turnbull and Vonnegut, the concentration of dislocations in the interface, and hence $\mathfrak{m}$, depends linearly upon

($\delta-\epsilon$), and instead of a free energy difference per unit volume of embryo of $-n_S kT \ln(p/p_S)$ we have the value

$$\Delta G_V = -n_S kT \ln(p/p_S) + C\epsilon^2, \quad (8.16)$$

where C is a constant depending upon the elastic properties of ice. If nucleation from the vapour is considered then p is the ambient vapour-pressure of water-vapour, whilst if freezing is involved p is equal to p_L, the saturation vapour-pressure over water at this temperature. Turnbull and Vonnegut estimate a value of 1.7×10^{11} dyn/cm² for C in the case of ice nucleating upon a basal plane at a temperature only slightly below 0° C. For substances like β-silver iodide, which have lattice constants differing from those of ice by only about 1%, it is probably correct to assume coherent nucleation, so that $\epsilon = \delta$, whilst for substances very different in structure from ice it is reasonable to assume $\epsilon = 0$.

The formal theory of nucleation rates can now be carried through just as for the case of condensation, and if we make the approximation of neglecting crystal anisotropy in both the ice and the nucleating substance this theory can be applied to the case of spherical nucleating particles. For a particle of radius r causing elastic strain ϵ within the ice because of misfit δ, and having contact angle $\cos^{-1} \mathfrak{m}$, the critical free energy is

$$\Delta G^* = \frac{16\pi \sigma_{SV}^3 f(\mathfrak{m}, x)}{3[-n_S kT \ln(p/p_S) + C\epsilon^2]^2}, \quad (8.17)$$

where
$$x = r/r^* \quad (8.18)$$

and $f(\mathfrak{m}, x)$ is given by (3.34).

When nucleation from a supersaturated vapour is under consideration, the nucleation rate per particle is given by an equation analogous to (3.36), namely,

$$J \approx [p/(2\pi mkT)^{\frac{1}{2}}] 4\pi r^2 r^{*2} n'(1) \exp(-\Delta G^*/kT), \quad (8.19)$$

where $n'(1)$ is the density of single molecules adsorbed on the nucleating surface.

On the other hand, if we consider nucleation from a supercooled liquid by freezing, then we must replace the subscript V by subscript L in (8.14) and (8.17) and instead of an ambient vapour-pressure p in (8.17) write the saturation vapour-pressure p_L over water. Then by an extension of (8.7) the nucleation rate for freezing, per particle, is approximately

$$J \approx \frac{kT}{h} n'_c \, 4\pi r^2 \exp\left(-\Delta g/kT\right) \exp\left(-\Delta G^*/kT\right), \quad (8.20)$$

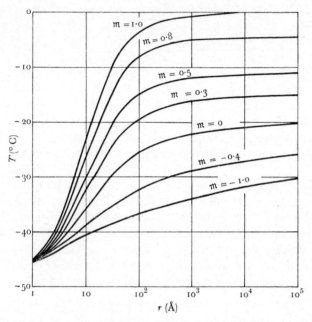

Fig. 8.4. Temperature T at which a spherical particle of radius r and surface parameter $\mathfrak{m}$ will nucleate an ice-crystal from water in one second by freezing. The effect of elastic strain is not included. On the basis of the values used in this calculation homogeneous freezing of micron-size droplets would occur about $-30°$ C (Fletcher, 1958a).

where n'_c is the number of liquid molecules in contact with unit area of catalyst surface. In the case of water $n'_c \approx 3 \times 10^{-8} n_L$ so that the pre-exponential factor is of order $10^{28} r^2$ sec^{-1}.

In discussing calculations based upon (8.19) and (8.20) it is convenient to separate out the effects of elastic strain and treat them separately. Since this strain merely contributes an additive term to the bulk free energy, as shown by (8.16), its inclusion simply adds a constant to $\ln(p/p_S)$ in the case of sublimation, or, by comparison with (8.4), to the supercooling ΔT for freezing.

Heterogeneous Nucleation of Ice-crystals

Calculations based on (8.19) and (8.20) for the effects of particle-size and contact angle (that is $\mathfrak{m}$-value) upon nucleation efficiency have been made by Fletcher (1958a) for the important cases of freezing, and of sublimation from an environment at water saturation at a temperature below 0° C. These curves assume surface free energy values $\sigma_{SV} = 100$ erg/cm^2 and $\sigma_{SL} = 20$ erg/cm^2 respectively, which are the approximate values suggested by our present knowledge. The curves are shown in figs. 8.4 and 8.5.

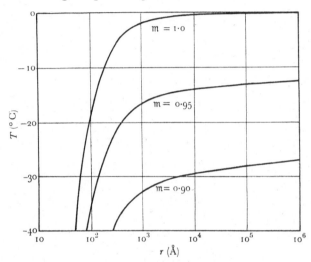

Fig. 8.5. Temperature T at which a spherical particle of radius r and surface parameter $\mathfrak{m}$ will nucleate an ice-crystal in 1 sec by sublimation from an environment at water saturation. The effect of elastic strain is not included (Fletcher, 1958a).

From these curves it can be seen that in the case of sublimation, size is relatively unimportant for particles more than about 0·1 μ radius, while below this size the nucleation efficiency falls sharply. In the case of freezing the size-effect becomes comparably important only below a radius of about 0·03 μ. In both cases the contact parameter $\mathfrak{m}$ has a very large effect on nucleation efficiency.

The effects of elastic strain can be taken into account to a good approximation by writing

$$\Delta T = \Delta T_1 + \Delta T_2, \qquad (8.21)$$

where ΔT_1 is the supercooling required by size and contact angle as shown in figs. 8.4 or 8.5 and ΔT_2 represents the contribution of elastic strain. ΔT_2 as calculated by Turnbull and Vonnegut (1952) is shown in fig. 8.6. In most cases of small misfit it is probably correct to assume $\epsilon = \delta$, though there is some evidence (Newkirk and Turnbull, 1955) that either ϵ may be considerably smaller than δ or that the constant C used in computing fig. 8.6 is too large. This figure should therefore only be regarded as semi-quantitative.

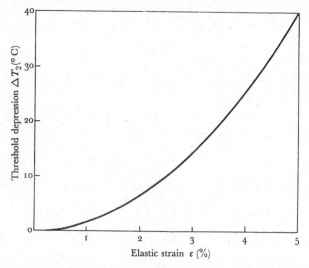

Fig. 8.6. Additional depression of nucleation threshold due to the effects of elastic strain within the ice-embryo. For coherent nucleation ϵ is equal to the misfit δ between the two lattices (Turnbull and Vonnegut, 1952).

One assumption which has been made in the discussion above should perhaps be made explicit. We have assumed the nucleating surface to have a fixed configuration and not to be affected by the nucleation process. This explicitly leaves out of consideration the action of appreciably soluble substances as freezing-nuclei, and indeed no treatment of their action has yet been attempted.

We must also recognize that the treatment which has been given represents only the behaviour of a very simplified model. In practice many small particles may have crystalline shapes differing greatly from simple spheres, and similarly it may not be a very good approximation to assume an ice-embryo in the form of a spherical cap. However, the difficulties to be faced in setting up a more general theory are considerable, and this more elementary treatment seems to be adequate for most purposes. Turnbull (1950) and Fletcher (1960) have examined some of the consequences of particular deviations from simple nucleus geometry, whilst Fletcher (1959d) has discussed some aspects of nucleation theory from a molecular point of view.

One further approach which should receive mention is the study of the adsorption of water-vapour on the surface of the nucleating particle. This is clearly closely related to the nucleation of ice-crystals upon the surface and a study of adsorption isotherms should give valuable information about surface energies. Unfortunately, the few results which have been published are in direct contradiction. Thus, while Birstein (1955) found extremely large adsorption of water-vapour on to silver iodide and lead iodide, Karasz, Champion and Halsey (1956) found the amount of adsorption to be undetectably small and Sano and Fukuta (1956a) found adsorption with a time constant of hundreds of hours. Papée (1959), using an indirect microcalorimetric method, obtained an equilibrium adsorption of about one monolayer of water on to lead iodide. When these disagreements are resolved the method may be able to give valuable information about the state of the nucleating surface.

Sublimation and freezing nuclei

When we were examining the homogeneous nucleation of ice-crystals from the vapour we found that it was necessary not only to determine under what conditions sublimation could be initiated, but also to decide whether sublimation actually took place at a greater rate than did condensation. We must make a decision of exactly the same type in the case of heterogeneous nucleation.

Many experiments have been conducted with the aim of finding out whether certain particles act as sublimation or as freezing-nuclei, but most of these have been inconclusive because of lack of understanding of the distinctions involved. In the following paragraphs we shall attempt to clarify the situation and to fit the distinction between sublimation and freezing-nuclei into the background of theory we have built up.

If we limit ourselves for the moment to consideration of completely insoluble nucleating particles in a water-vapour atmosphere, then it is clear that no condensation can occur provided the vapour-pressure remains below water saturation. Fig. 8.5 suggests that, for particles of most materials having sizes greater than $0 \cdot 1\,\mu$, sublimation will occur at vapour-pressures near water saturation provided the temperature is sufficiently low, and this is an unambiguous case of sublimation.

When the environment is supersaturated with respect to water-vapour and the temperature is below $0°$ C, both condensation and sublimation can occur, the rates of each being determined by the m-values for water and ice and the nature of the lattice misfit. In general, condensation is preferred at small supercoolings and large supersaturations, while sublimation is preferred at large supercoolings and small supersaturations. Fig. 8.7, taken from a discussion by Fletcher (1959a), illustrates this point for a hypothetical substance having contact angle about $5°$ for water and m about $0 \cdot 95$ for ice. Surfaces are drawn for both condensation and sublimation below and behind which these processes occur in a time less than 1 sec. The surface shown with broken lines divides the region of preferred sublimation from that of preferred condensation on the basis of a comparison of ΔG^* for the two processes.

It should be noted, as with homogeneous condensation, that once a thin film of water has condensed upon a nucleating particle it may freeze due to nucleation of ice by the particle surface. A liquid film 100 Å in thickness may be sufficient for this purpose, so that no macroscopic droplet is developed, and as far as can be seen there is merely the growth of an ice-crystal.

When we consider more complex nuclei the situation may be different. Consider, for example, a composite nucleus containing a soluble and an insoluble particle. Then condensation can occur on the soluble component at a vapour-pressure which may be less than saturation and which depends upon the chemical nature of the material. A macroscopic droplet cannot form below saturation, but the microscopic droplet of solution may freeze upon the insoluble particle if the temperature is sufficiently low and the ambient vapour-pressure above ice-saturation. Such a composite nucleus may thus act as a freezing-nucleus at less than ice-saturation. We shall discuss the freezing of solutions further below.

Finally, we should note that particles whose surfaces contain microscopic pits or cracks may, if their surface properties are suitable, condense water in these fissures from environments at

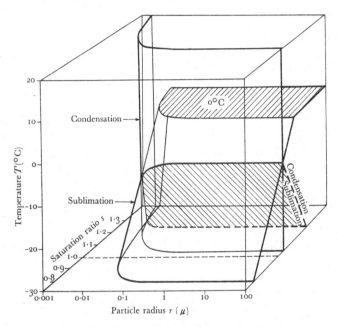

Fig. 8.7. Activity surfaces for a typical nucleating substance. Behind the condensation surface droplets nucleate at an appreciable rate, and behind the sublimation surface ice-crystals form at an appreciable rate. The remaining surface divides regions of preferred condensation and sublimation (Fletcher, 1959a).

less than water saturation. This has been discussed theoretically by Turnbull (1950), but the behaviour in condensation from the vapour is far from simple, as shown by the work of Bangham and Razouk (1937a, b).

The freezing of solutions

Since, as we have seen, the freezing of solution droplets may be of importance, let us examine what can be expected in this case.

Suppose we have an ideal solution, that is, one which obeys Raoult's law. Then, if the mole fraction of water in the solutions is **m**, the vapour-pressure of water over the solutions is

$$p' = \mathbf{m} p_\mathrm{L}, \tag{8.22}$$

where p_L is the saturation vapour-pressure of water. Since water molecules in the solution are in equilibrium with those in the vapour they must have the same chemical potential μ'_L, given by

$$\mu'_\mathrm{L} = kT \ln p' = kT \ln \mathbf{m} p_\mathrm{L}, \tag{8.23}$$

taking pure water as reference potential (see chapter 3, equations 3.5, 3.6, and 3.37).

The chemical potential of an ice-molecule at the same temperature is similarly

$$\mu_\mathrm{S} = kT \ln p_\mathrm{S}, \tag{8.24}$$

where p_S is the saturation vapour-pressure over ice. Thus the increase in free energy per unit volume when the solution freezes, assuming pure ice to be formed, is

$$\begin{aligned}\Delta G_V &= n_\mathrm{S}(\mu_\mathrm{S} - \mu'_\mathrm{L}) \\ &= -n_\mathrm{S} kT \ln(p_\mathrm{L}/p_\mathrm{S}) - n_\mathrm{S} kT \ln \mathbf{m},\end{aligned} \tag{8.25}$$

which, by use of the Clausius–Clapeyron equation as before, can be written if ΔT is small

$$\Delta G_V = -\Delta S_V \Delta T - n_\mathrm{S} kT \ln \mathbf{m}, \tag{8.26}$$

where ΔS_V is the entropy of freezing per unit volume.

If equilibrium freezing is considered then $\Delta G_V = 0$, and the depression of the freezing-point is

$$\Delta T_3 = \frac{n_\mathrm{S} kT}{\Delta S_V} \ln \mathbf{m}. \tag{8.27}$$

If the solution is dilute then $\ln \mathbf{m}$ is approximately equal to $-\mathbf{m}'$, where $\mathbf{m}' = 1 - \mathbf{m}$ is the mole fraction of solute.

If nucleation is considered then the effect of the extra term in (8.26) is to reduce the effective supercooling by an amount ΔT_3. Nucleation phenomena will therefore require an additional supercooling of this amount before they can proceed, and this ΔT_3 can be added to the other terms in (8.21). This discussion is, of course, much simplified and applies only to the case of

Heterogeneous Nucleation of Ice-crystals

ideal solutions. The same final conclusion is, however, reached in the more general case.

This conclusion has been verified experimentally in a general way by Bigg (1953b), but in addition he found that some substances, such as HI, NH_4I, Li_2SO_4 and several others, actually appeared to raise the nucleation temperature when their concentration was about 1 %, though in larger concentrations they depressed it in the usual manner. No good explanation has yet been put forward for this phenomenon.

EXPERIMENTS ON HETEROGENEOUS NUCLEATION

The number of experiments in which heterogeneous nucleation has played a part is extremely large, but at this juncture we shall only be interested in nucleation by well-defined crystalline substances, and shall postpone discussion of natural nuclei to the next chapter.

The experiments we shall discuss fall into three well-defined groups. We can examine the nucleation behaviour of large single crystals of various materials by watching the growth of microscopic ice-crystals on their surfaces, or we can study the behaviour of large numbers of small particles, suspended either as an aerosol or within small water droplets.

Single crystal experiments

Although some work had been done on the growth of ice-crystals on a freshly cleaved mica surface as early as 1939 (Nakaya, Hanaxima and Dezuno, 1939), most researches of this type have been carried out during the last few years (Montmory, 1956; Jaffray and Montmory, 1956a, b, 1957; Montmory and Jaffray, 1957; Kleber and Weis, 1958; Bryant, Hallett and Mason, 1960). The apparatus used in these experiments was similar to that used by Shaw and Mason (1955) for the study of ice-crystal growth (see fig. 10.8 on page 272). The vapour-pressure within the enclosure is at a fixed value in equilibrium with a film of ice on its floor, and the temperature of the crystal surface under study, which rests on the end of a cooled copper rod in the centre of the enclosure, may be varied in any manner desired.

The crystal surfaces used in these studies were mica, β-silver iodide and lead iodide, which were chosen because of the close similarity between their crystal symmetry and lattice parameters and those of ice. Because flat crystal surfaces of AgI and PbI$_2$ were found difficult to prepare, these materials were in some cases deposited by vacuum evaporation on to freshly cleaved mica surfaces, where they were found to produce flat, essentially monocrystalline films. When the substrate is cooled down, large supersaturations are produced and ice-crystals nucleate and grow upon its surface. The pattern of this growth is typically as shown in Plate III.

The most striking feature of these results is that the ice-crystals grow upon the substrate with their axes parallel to corresponding directions in the substrate lattice. On the basis of our discussion this type of epitaxy is clearly favoured because it gives a minimum number of interface dislocations and hence a minimum interfacial free energy. Less frequently, epitaxy is observed with ice-lattice directions parallel to other low index directions in the substrate. This is clearly possible if misfit is low for this relative orientation, but the number of possible relative orientations is strictly limited because of the increasing number of unmatched atoms across the interface as epitaxies of lower symmetry are considered. Epitaxy of ice has been considered in some detail by Weickmann (1951) and by Montmory (1955), though they restrict their arguments to purely geometric grounds.

As yet experiments have only been carried out upon predominantly hexagonal substrates, but one would expect the same sort of behaviour to occur upon surfaces making a good match with the prism faces of ice-crystals. Indeed Fletcher (1959d) has suggested that nucleation does not in fact occur easily upon hexagonal faces but only upon small exposed steps of prism face. This is supported to some extent by the observations of Schaefer (1954b), Mason (1958), and Bryant, Hallett and Mason (1960).

To illustrate the close general resemblance between the structure of ice, silver iodide and lead iodide, we show in fig. 8.8, approximately to scale, atomic models of these three crystals. Because AgI is polar it is shown viewed from two directions. The misfit between ice and AgI amounts to only about 1·4% on either face while that for ice and PbI_2 is about 0·5% on the hexagonal face and 3% on the prism faces.

It is, of course, possible to have oriented overgrowth of ice upon substrates which differ greatly from it in crystalline form.

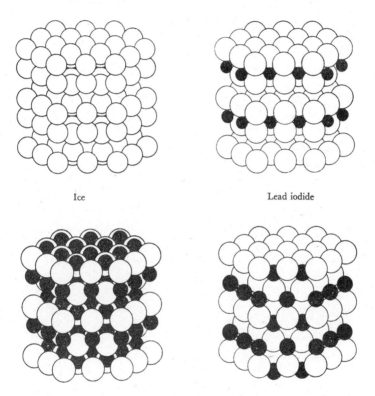

Fig. 8.8. Models of the crystal structures of ice, lead iodide and silver iodide. In the latter white spheres are iodide ions and black spheres metallic ions (Fletcher, 1959d).

Indeed we should always expect one particular relative orientation to give the lowest interfacial free energy, but this minimum will be sharpest for substrates closely resembling ice, and in other cases may be so shallow as to have negligible orientating influence.

Small particle experiments

We shall consider first experiments in which large numbers of small particles are introduced into a volume of cold, moist air and their ice-nucleating behaviour examined. We shall postpone considerations of details of the apparatus used in these experiments to the next chapter and concentrate here upon the results which have been obtained.

There are two different sets of conditions under which these experiments can be performed and they might be expected to lead to rather different results. In the first case a thermal gradient or diffusion chamber (see chapter 4) is used to produce a cold moist environment and the water-vapour pressure may be considerably above water saturation. Particles introduced into such a chamber may serve initially as either condensation or sublimation nuclei depending upon their properties and details of the supersaturation and temperature. The second type of experiment uses a chamber at an approximately uniform temperature containing a supercooled fog. This has the advantage of maintaining the environment at water saturation and simulating the conditions prevailing in supercooled clouds.

Solid insoluble particles with finite contact angles for water can only act as sublimation nuclei, unless they collide with water droplets, for example by the mechanism of Facy (1957), or are caused to act as condensation nuclei by the effects of some extraneously produced supersaturation. Hygroscopic or mixed nuclei can, of course, produce ice-crystals by condensation followed by freezing. Most experiments have taken no particular note of the conditions of the experiment and it is, therefore, not usually possible to say whether sublimation or freezing is the operative mechanism.

Particularly extensive determinations of nucleation thresholds have been made by Hosler (1951), Pruppacher and Sänger (1955*a*), Birstein and Anderson (1955), Mossop (1956*b*), Sano and Fukuta (1956*b*) and Fukuta (1958), while other workers have made more detailed measurements on limited numbers of substances.

The disagreement between the results of these measurements is so great, however, that it does not seem possible at the present time to assign reliable threshold temperatures to many materials.

PLATE III

Photomicrograph of ice-crystals growing epitaxially upon a substrate of lead iodide at − 15° C. The field shown is about 1 mm across (Fletcher, 1960).

PLATE IV

Ice-crystals growing in supercooled sugar solution. The form of the crystals depends upon the concentration of the solution and the degree of supercooling. Approximately natural size (Bigg).

There are several reasons for these disagreements. In the first place many widely differing criteria have been used to define the nucleation threshold and these alone would give differences of several degrees, assuming the same materials were used. Even this is not generally the case, and particle suspensions having widely differing size-distributions and chemical purities have been used by various workers. This has a particularly marked effect on the threshold temperature, depending as it does upon the action of the minute fraction of most efficient particles. Indeed Mason and Hallett (1956) suggest that the reported activity of many silver salts and of many iodides may be primarily due to contamination with silver iodide.

Table 8.2. *The principal ice-nucleating crystals. Basal misfit is defined as* $(a-a_0)/a_0$ *and prism misfit as* $\frac{1}{2}[(a-a_0)/a_0 + (c-c_0)/c_0]$. *Nucleation thresholds are only approximate*

	Crystal form	a in Å	c in Å	Basal misfit	Prism misfit	Nucleation threshold
Ice	Hexagonal	4·52	7·36	—	—	—
AgI	Hexagonal	4·58	7·49	1·4%	1·6%	−4° C
PbI$_2$	Hexagonal	4·54	6·86	0·5%	3·6%	−6° C
CuS	Hexagonal	3·80	16·43	2·8%†	7·1%‡	−7° C†

† Ice lattice $[\bar{2}110]$ along $[\bar{1}010]$ in CuS surface.
‡ Two ice unit cells against one of CuS.

The ordering of nucleation efficiency in terms of lattice misfit has been investigated by Montmory (1955), supplementing earlier work by Weickmann (1951) and others. He found nucleation efficiency among materials of similar crystal structure to depend upon lattice misfit against ice, in qualitatively the manner we have discussed. Pruppacher and Sänger (1955b) started from another viewpoint and ordered nucleating crystals in terms of the relative polarizability of anion and cation. They found that in general good nuclei had a large polarizability difference though there were some out of order in the list. It seems reasonable that some sort of electrostatic criterion such as this must reflect the strength of the binding forces across the ice interface. In general it seems that both these factors, and probably others, must be taken into account in assessing nucleation efficiency.

Table 8.2 gives some details of the physical constants of some of the most active simple nucleating substances. Among these silver iodide is pre-eminent, and a great deal of work has been done upon its properties. Because of this, and because of its practical importance in cloud-seeding operations we shall discuss its properties in detail in a later chapter.

It can be seen from this table that these substances all have small lattice misfits against ice (there is experimental evidence by Bryant *et al.* (1960), on the orientation of ice grown on CuS), and that their structures are relatively polar. Both AgI and CuS are almost insoluble, and PbI_2 is not very soluble (this slight solubility of PbI_2 may explain why it is not a very good freezing-nucleus, as noted by Schaefer, 1954).

Most experiments of this nature have simply determined the nucleation threshold and not sought any further information. Sano, Fujitani and Maena (1956), however, have examined the size-distributions of aerosols produced in various ways, and by varying the size-distribution have been able to find nucleation threshold as a function of particle-size for several materials. Their curves for AgI, HgI_2 and PbI_2 look very much like the theoretical size-effect curves for sublimation shown in fig. 8.5, the nucleation efficiency falling sharply for particles less than about 0·1 μ in radius. Curves for smokes of substances like NiO and MgO show an even greater size-effect, but this may be attributed to deviations from a spherical shape in these particles.

An interesting experiment of a different nature was performed by Fournier d'Albe (1949) during an investigation of the properties of cadmium iodide. He found that cadmium iodide particles were inactive as ice-forming nuclei, but, being soluble, acted as condensation nuclei, the resulting droplets freezing below $-41°$ C. When these ice-crystals were warmed and dried off, the temperature remaining below about $-9\cdot5°$ C, the cadmium iodide particles were found to act as sublimation nuclei at a little more than ice saturation. Re-examining this phenomenon Mossop (1956a) later found that iceland spar and Na-Bentonite clay behaved in a similar manner up to even warmer temperatures, and attributed the effect to the retention of small amounts of ice in cavities in the particles. The list of substances active in this way was extended by Mason and Maybank (1958) while Gourley and Crozier (1955) have reported the same phenomenon in air cleaned as far as possible of nuclei. G. A. Day (1958), however, has shown that in all cases investigated the activation was of a temporary nature, a sojourn of 10 min at 90 % ice saturation and a temperature of $-9°$ C being sufficient to destroy the enhanced activity.

Droplet-freezing experiments

The interpretation of droplet-freezing experiments is more difficult than that of cloud-chamber measurements for several reasons. In the first place most substances have appreciable solubility so that dilute suspensions may in fact disappear by solution; in the second place the distribution of suspended particles among the droplets to be measured is statistical, so that the results will depend markedly upon the size of the drops and the concentration of the suspension.

For this reason, though many experiments have been performed on the freezing of nominally pure water droplets or those containing only atmospheric nuclei, few results are available on the behaviour of drops containing suspended particles whose properties are known.

Because of its prime importance as a nucleating agent most experiments have been performed upon suspensions of silver iodide. Bayardelle (1955) and Roulleau (1957) passed suspensions of AgI through fine filters and then examined their freezing as droplets to investigate the effect of particle-size on freezing temperature. The curves obtained differ considerably from those of fig. 8.4 if the pore size of the filter is taken to define maximum particle-size. They do show, however, a decrease in freezing threshold temperature with smaller particle-size as predicted by the theory.

The effect of repeated freezing of the same drops was investigated by Itoo and Hama (1956) and by Roulleau (1957), who reached the conclusion that repeated freezing tends to deactivate the nuclei. The reason for this effect has not yet been found.

CONCLUSIONS

Though the theory of ice-crystal nucleation is far from complete, we have seen that most of the general features seem to be fairly well understood. The maximum supercooling which can be supported by pure water is about 40° for small droplets and about 35° for volumes of the order of a cubic centimetre, and it seems fairly clear that freezing begins by a process of homogeneous nucleation.

Freezing and sublimation can occur at temperatures much warmer than this in the presence of foreign nuclei, and the action of the simpler crystalline nuclei seems to be fairly well understood. Their activity depends upon their surface structure and their interaction with adsorbed water molecules. Most efficient nuclei are hexagonal in structure with polar molecular bonding and have only slight solubility in water. The activity of nucleating particles decreases sharply as their size falls below about $0 \cdot 1\ \mu$.

Many experimental measurements of nucleation thresholds for simple substances have been made but there are wide divergences between the results. Some of the general reasons for these disagreements are now known, but as yet the true thresholds for many substances are uncertain. There is, however, fairly general agreement that silver iodide is the most efficient ice-forming nucleus known, and a consideration of the criteria for nucleation efficiency makes it seem unlikely that a better material will be found. Our knowledge of the behaviour of silver iodide is quite extensive because of its practical importance, and we shall consider this in detail in a later chapter.

CHAPTER 9

ICE-FORMING NUCLEI IN THE ATMOSPHERE

INTRODUCTION

THE importance of the ice-phase in precipitation processes was not realized until it was pointed out in Bergeron's classic paper in 1935 (Bergeron, 1935), and interest in atmospheric ice-forming nuclei has only developed since that time. For this reason most of the literature on this topic is quite recent, and many basic points are still the subject of controversy.

In this chapter we shall first consider means by which the properties of ice-forming nuclei (or 'freezing-nuclei' as they are generally rather loosely termed) in the atmosphere may be measured, and then go on to discuss the results of such measurements and the general conclusions which can be drawn. Finally, we shall consider the origin of these particles with particular reference to the meteor hypothesis.

There are two basically different methods by which the properties of atmospheric freezing-nuclei may be investigated. The more direct and clearly relevant method is to isolate a parcel of air, cool it until a supercooled cloud is produced, and then to observe the nucleation of ice-crystals as the temperature is further lowered. The second method makes the initial assumption that the ice-forming nuclei are in fact freezing-nuclei and reside in the droplets of the supercooled cloud. The production of ice-crystals must therefore commence with the freezing of cloud droplets, and the method consists of observing the freezing of small water droplets formed by condensation of atmospheric air and isolated for study upon a suitable support.

The study of droplet freezing is essentially straightforward from an observational point of view, and we have already discussed some of the means which have been used to support the droplets so that the support has a minimum effect on their behaviour. The problem is in this case rather one of interpreting the results in terms relevant to cloud physics, and we shall return to discuss this later.

The application of results obtained in cloud-chamber studies to large-scale cloud physics is much more direct, and the question as to whether the nuclei act by freezing or sublimation does not need to be answered. The method, however, involves various experimental difficulties, and we shall discuss in the next section some of the means which have been used to overcome them. Since most currently used freezing-nucleus counters are cloud-chamber types we shall discuss several of these in some detail.

EXPERIMENTAL METHODS

A supercooled cloud may be formed by simply chilling an air-sample, provided it contains sufficient water-vapour. The cooling process may be by adiabatic expansion, convectional cooling from a refrigerated container, or a combination of the two. Ideally an apparatus should cool the air at a rate comparable with the cooling rate found in natural clouds, but this is difficult to achieve in the laboratory with containers of a convenient size. In fact much faster cooling rates have been used in the majority of experiments.

Having formed the supercooled cloud, any ice-crystals appearing within it must be detected and counted. In principle, the simplest method which has been used is to shine a light-beam into the cloud when, viewed from a suitable angle, crystals may be distinguished from water droplets by their brighter appearance and the scintillations accompanying their fall. Their instantaneous concentration can be estimated visually from their number or spacing.

A more sensitive detector used by Cwilong (1947) is a pool of supercooled water which freezes rapidly when an ice-crystal falls into it. This method, however, can only detect the threshold for ice-crystal production without yielding a crystal count at lower temperatures. The method was made more practical by Schaefer (1948b), who used bubble-forming agents to slow the rate of crystallization in the detecting liquid. The pool of liquid was replaced by a thin film, stretched on a ring, which could be supercooled readily. Ice-crystals falling on to the film soon

Experimental Methods

grew to visible size and were easily counted. A supercooled solution of cane sugar in water has been found by Bigg (1957 b) to be a more suitable detector in some types of cold-box.

Most combinations of these methods of cloud production and ice-crystal detection have been used in measurements, and the following more detailed discussion of a few of the experimental arrangements will illustrate their application.

Large cloud chambers

Almost the sole attempt that has been made to cool by expansion at a rate comparable to that occurring in clouds was made by Findeisen and Schulz (1944). They used a cylindrical tank with a volume of two cubic metres and made measurements at expansion rates corresponding to uplift velocities of 5 and 20 m/sec, ice-crystals being detected visually.

An even larger container was made use of by aufm Kampe and Weickmann (1951) who employed a small room cooled by refrigeration. The advantage of such a large enclosure might be expected to be a freedom from wall effects, but as we shall point out later the large untreated wall area may actually prove a prolific source of frost splinters, and lead to erroneous results.

Both these installations were far too big to be portable and were therefore restricted to sampling the air in the building in which they were constructed. We shall now turn our attention to more easily portable instruments suitable for more general use.

Controlled-cooling counters

Counters can be divided into two classes. In one type a single measurement at a fixed temperature is made on each aerosol sample and a spectrum is deduced from a series of such measurements. The other type cools an air-sample continuously and derives a complete spectrum from a single sample of aerosol. The counter we are about to describe belongs to the second class and was developed primarily for aircraft use by Smith and Heffernan (1954).

The apparatus is shown schematically in fig. 9.1. It consists of a copper tank of volume 76 litres, cooled by means of alcohol circulating through copper pipes, the alcohol in turn being cooled by dry ice in another reservoir. Since the water-content of the atmosphere at high altitudes is often too low to produce a cloud at the temperature of measurement, provision is made for humidifying the chamber by heating distilled water in a small vacuum flask.

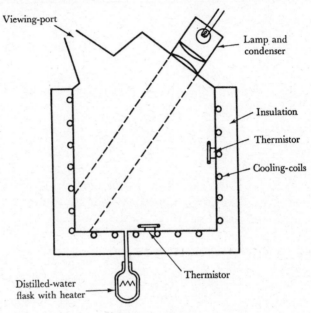

Fig. 9.1. Large visual cold-box (Smith and Heffernan, 1954).

In operation a sample of air is blown into the chamber, sufficient being used to flush it out completely. The chamber is then humidified and the temperature lowered at a rate of about 4° C/min, the appearance of ice-crystals being watched for in the beam of light traversing the cloud.

There is, as we shall discuss later, some difficulty in the interpretation of results obtained from such equipment because of the continual fall-out of crystals as they nucleate and grow. Smith and Heffernan estimated correction factors to allow for this effect under typical measuring conditions.

Constant-temperature counters

The majority of measurements of freezing-nuclei have been made with chambers operating at fixed temperatures. Such measurements have the advantage of simplicity of interpretation, the count being simply the total number of nuclei active at or above the chamber temperature, but against this a series of measurements must be made if a spectrum is to be determined.

The simplest type of counter consists simply of an open refrigerated box—Schaefer (1948b) used a commercial deep-freeze chest—maintained at the observation temperature. The air-sample is introduced along with sufficient water-vapour to form a cloud, and is retained stably by the temperature inversion at the top of the box. The ice-crystals are then counted visually in a light beam.

Such an open box is, however, subject to contamination from the immediate environment, and a more satisfactory apparatus uses a completely enclosed box, which also has the advantage of being more nearly isothermal throughout. Such an instrument has been described by Bigg (1957b) and is shown in fig. 9.2. The enclosed space is within a multi-walled cylinder cooled by dry ice in a glycol-water mixture. The enclosed volume is 1 litre in the original version of this instrument and 10 litres in a later model. Completely covering the floor of the chamber is a tray of sugar solution (1·2 kg of sugar in 1 litre of water) held at a temperature of $-12°$ C. All interior surfaces of the chamber are carefully wiped over with glycerol before each measurement to inhibit the formation of frost which will give rise to spurious counts.

When a humid air-sample is drawn into the chamber it cools rapidly to a temperature near that of the walls, a cloud forms, and any ice-crystals nucleated within it begin to grow. These ice-crystals fall into the sugar solution and grow much more rapidly, attaining a diameter of 2 mm in about 2 min. At this stage the crystals have the appearance shown in Plate IV, and are readily counted. It has been found that the sugar solution responds only to ice-crystals falling into it, and is not affected by unactivated nuclei.

The immediate advantage of this system is that it replaces the

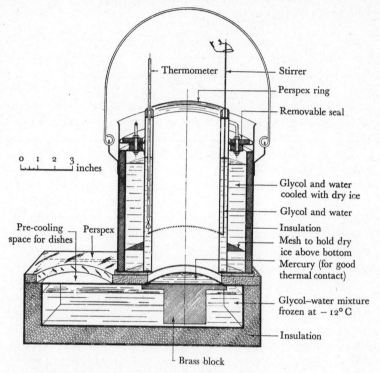

Fig. 9.2. The cold-box used by Bigg. A constant cold temperature is maintained using dry ice in a glycol-water mixture, and ice-crystals are collected in sugar-solution in a dish on the floor of the chamber (Bigg, 1957*b*).

rather subjective results obtained by the visual method by a definite objective count. The sugar solution also integrates the number of ice-crystals which fall into it from the time the cloud is injected into the chamber until the time the tray is removed. This measurement is, therefore, very easily interpreted in terms of the concentration of nuclei in the air-sample.

Expansion-chamber counters

The advantage of using expansion as a means of achieving a low temperature is the flexibility of the equipment and the ease with which the measurement temperature may be changed. Against this must be set the fact that the minimum temperature

Experimental Methods

is maintained for only a few seconds, after which the air begins to warm up by conduction from the walls. The effect of such a small measurement time has been investigated experimentally by Warner and Newnham (1958) and theoretically by Fletcher (1958b). They conclude that the count is a valid one but that the number of nuclei counted is less, by a factor between 2 and 5, than the number that would be counted with a 20-min integrating time.

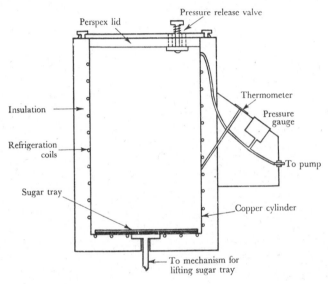

Fig. 9.3. Expansion-type ice-nucleus counter (Warner, 1957).

A small expansion chamber having a volume of only a few hundred cm^3 was used by Cwilong (1947) to study natural aerosols, but because of its small volume only the more abundant nuclei active at temperatures below about $-30°$ C could be studied. More recently, Warner (1957) has described a chamber of volume 10 litres in which the air-sample is initially cooled to $-12°$ C by conduction from the chamber walls, and is then further cooled to the measurement temperature by expansion. The supercooled sugar method developed by Bigg is used to count the ice-crystals produced. This apparatus is shown in fig. 9.3.

Continuous cold-boxes

For many studies of atmospheric nuclei it is clearly an advantage to have a continuous record of concentration. Attempts have, therefore, been made to develop automatic counters based on the measurement methods we have discussed.

Luttermoser and Brown (1959) have developed an automatic version of a visual counter in which supercooled fog flows steadily through a chamber and the scintillations produced by ice-crystals are detected photo-electrically. Electronic means are then used to sort out pulses due to fog droplets and other extraneous particles. The method is particularly designed for aircraft use.

Bigg, on the other hand, has adapted his sugar-solution method for continuous operation by replacing the tray by a flexible belt, coated with the supercooled solution and driven slowly past the bottom of the chamber. Ice-crystals fall on to the belt, grow, and are photographed as the belt emerges from beneath the enclosure. Means are also provided, of course, for continuous injection of air (Bigg and Meade, 1959).

Thermal precipitation counter

An ice-nucleus counter of an entirely different type has been described by Nathan and Hill (1955, 1957) and by Fenn and Weickmann (1959). The principle of this instrument is to precipitate all the particles in a given volume of air by means of a thermal gradient, and then to examine them for ice-nucleating ability by subsequent humidification and cooling.

The hot-plate of the thermal precipitator shown in fig. 9.4 consists of a piece of electrically conducting glass whose temperature is raised to about 100° C. The cold-plate is a thin sheet of rhodium-plated copper cemented to another sheet of electrically conducting glass, which is in turn in good thermal contact with a large block of dry ice. During collection the temperature of the cold-plate is maintained just above the ambient dew-point so that the strong temperature-gradient removes all suspended particles from the air as it passes through the 0·1 mm gap between the two plates.

After a fixed volume of air, of the order of a litre, has been drawn through the precipitator with a constant-speed pump, the current to the hot-plate is turned off and the temperature of the cold-plate gradually lowered. When the dew-point has been reached the cold-plate is exposed momentarily to the air to provide a moisture supply in the form of condensed droplets, and then the appearance of ice-crystals is watched as the temperature is further lowered. In this way a complete freezing-nucleus spectrum may be determined by measurements on a single sample.

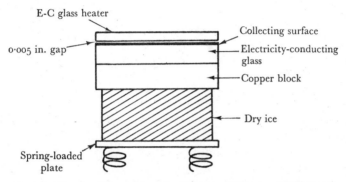

Fig. 9.4. The thermal-precipitator cell used by Nathan and Hill (1955) in their freezing-nucleus counter.

Discussion

Simple as the measuring process is in principle, many practical difficulties arise which may make the measurements unreliable. The worst practical difficulty, which has undoubtedly produced many errors in published results, is that of frost-formation on the walls of the chamber. Even extremely small frost-deposits shed ice-splinters at a very rapid rate, and it becomes imperative either to take positive steps to eliminate frost, or to verify by careful tests that no frost-formation is taking place. Frost-formation may be particularly troublesome in large chambers where adequate treatment of wall surfaces is difficult. In smaller enclosures rubbing the walls with an antifreeze solution such as glycerol in water is effective, but care must be taken that any vapours thus introduced into the system do not affect the nucleation rate of ice-crystals on atmospheric nuclei.

The problem of frost-prevention applies equally to all types of counter. We shall mention briefly certain other difficulties which depend more explicitly on the particular techniques used. The visual counting method suffers from the great disadvantage that it is not objective but depends upon the recognition of certain scintillations as characteristic of ice-crystals, and upon estimation of the concentration of these scintillating particles. In addition, as the ice-crystals grow they fall out of the cloud so that some sort of empirical integration method must be used. Furthermore, the fog itself becomes more dense with time and settles out, so that nuclei may be removed from view before they become active. These troubles are serious in any quantitative instrument, but become particularly acute in the case of counters in which a series of measurements are made as the temperature falls. Whilst measurements can certainly be made, the interpretation of these in terms of physically relevant quantities is extremely difficult.

The use of a supercooled solution as an ice-crystal detector overcomes many of these problems. It responds only to ice-crystals, and yields an integrated count which is easily interpreted. On the other hand, it requires careful use to avoid excessive supercooling (in which case stray crystals may nucleate upon the floor of the dish) or insufficient cooling before the measurement is made (in which case some ice-crystals may melt on the still warm tray). The use of sugar and glycerol solutions within small chambers may also bring their own problems through desiccation of the fog, but again with careful handling these can be overcome.

The thermal precipitation counter is so widely different from the others described that some misgivings might be entertained about the applicability of the results to processes in clouds. The nucleation process is, however, fundamentally unaltered and provided that there are no nucleation sites on the substrate and that contamination is avoided the results should be valid.

A comparative study of several of these counters was conducted recently in a co-operative programme (Meteorology Research Inc., 1957), the equipment involved being Bigg's static cold-box, Warner's expansion chamber, the thermal precipitation counter and three other constant temperature cold-boxes of various designs. Frost on the walls was a serious

Experimental Methods

problem in some cases, but when tested on the same aerosol sample average counts agreed to within an order of magnitude at a given temperature, or, alternately expressed, to within about 4° C for a given concentration. The shape of the activity spectrum of the aerosol was very much the same, independent of the equipment used.

ACTIVITY SPECTRUM OF NUCLEI IN THE ATMOSPHERE

Several investigations have been made of the activity spectrum of ice-forming nuclei in the atmosphere. The measurements were all made at approximately water saturation, in the presence of a fog of water droplets. From the point of view of results as well as of method they can conveniently be divided into two groups.

In the first group may be placed the investigations of Findeisen and Schulz (1944), of Smith and Heffernan (1954) and of Murgatroyd and Garrod (1956). These have in common the use of a relatively large chamber in which the temperature is reduced fairly slowly, counting of ice-crystals being done by visual estimation. The derived spectra have the general form shown in fig. 9.5, each curve typically rising as the temperature is lowered, reaching a plateau, and then increasing very sharply at a temperature below $-30°$ C.

Experiments in the second group are more numerous and have the common characteristic of making a single measurement at a fixed temperature, though a fast expansion might be used to reach this temperature. The count was in some cases made visually, and sometimes with the aid of a supercooled sugar solution. Among such measurements are those of Palmer (1949), Workman and Reynolds (1949), aufm Kampe and Weickmann (1951), Rau (1954), Warner (1957) and the co-operative measurement programme (Meteorology Research Inc., 1957).

With the exception of Rau, who obtained a curve which was similar to those of Findeisen and Schultz, all other measurements indicate a steady exponential increase in nucleus count with decreasing temperature, as shown in fig. 9.6. The slopes vary considerably from observer to observer, and the count at a

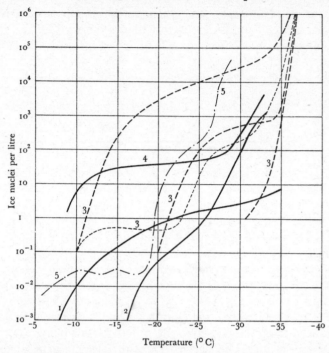

Fig. 9.5. Natural ice-nucleus spectra exhibiting complex structure: 1, Findeisen-Schulz 5 m/sec; 2, Findeisen-Schulz 20 m/sec; 3, Smith and Heffernan; 4, Rau; 5, Murgatroyd and Garrod.

given temperature differs by several orders of magnitude in different cases.

In attempting to come to some conclusion about the average nucleus spectrum it must be recognized that this may vary widely from day to day and from place to place. Further, though the spectrum determined by most of the fixed temperature counters is usually exponential, and thus a straight line on the graph, Warner (1957) has noted the occasional occurrence of curves and even humps in the spectrum. Thus whilst it seems likely that much of the characteristic shape of the curves of fig. 9.5 may be due to instrumental effects such as ice-crystal fall-out and homogeneous freezing (Bigg, 1953c), we cannot dismiss the possibility that these are real features characteristic of the particular place and time of the measurement.

Activity Spectrum of Nuclei in the Atmosphere

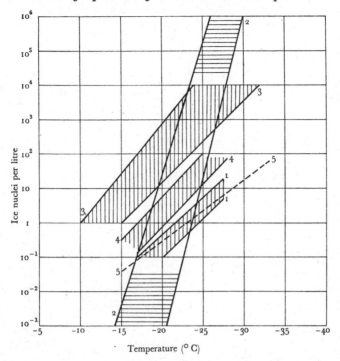

Fig. 9.6. Natural ice-nucleus spectra exhibiting simple exponential form: 1, Palmer; 2, Workman and Reynolds; 3, aufm Kampe and Weickmann; 4, thermal precipitator; 5, Warner.

The average spectrum of ice-forming nuclei in the atmosphere is thus probably best represented, in the region $10° < \Delta T < 30°$, by an exponential of the form

$$n(\Delta T) = n_0 \exp(\beta \Delta T), \qquad (9.1)$$

where ΔT is the supercooling, $n(\Delta T)$ is the number of nuclei active at supercoolings less than ΔT and n_0 and β are parameters. The usual value of β is about 0·6 with values between 0·4 and 0·8 being common. n_0 is more variable, typically being about 10^{-5} litre^{-1}, with variations of several orders of magnitude sometimes occurring. This gives typically a concentration of one active nucleus per litre at about $-20°$ C, the concentration changing by a factor of ten for about $4°$ C temperature change.

Droplet-freezing experiments

Further indirect confirmation of the form of the spectrum comes from experiments on the freezing temperatures of nominally pure water droplets, such as those of Heverly (1949), Dorsch and Hacker (1950), Bigg (1953a), and Hosler (1954) which are shown in fig. 8.2. It may be assumed that since the water droplets froze at temperatures well above those for homogeneous nucleation, they contained freezing-nuclei, the most likely source of which was atmospheric contamination.

Assuming contamination by a random selection of nuclei from an atmospheric spectrum $n(\Delta T)$, the probability that a drop of volume V contains at least one nucleus active above temperature ΔT is (Bigg, 1953a)

$$p = 1 - \exp(-V\alpha n), \qquad (9.2)$$

where $\alpha n(\Delta T)$ is the spectrum per unit volume of water, α being a constant. At the median freezing-point $\Delta T'$, p is 0·5 and (9.2) becomes

$$n(\Delta T')V = \text{const.} \qquad (9.3)$$

But the results of Bigg and of Dorsch and Hacker have the form

$$\ln V = a - b\Delta T', \qquad (9.4)$$

where b is approximately unity, so that, combining this with (9.3) we have

$$n(\Delta T) \propto \exp(b\Delta T), \qquad (9.5)$$

which is exactly the form of (9.1). The curves of Heverly and Hosler give somewhat similar results, though the distribution is now curved.

Langham and Mason (1958) measured freezing curves parallel to that of Bigg, which simply moved up and down with the degree of purity of the water sample. This is in itself somewhat surprising as it implies that all the nuclei behave in exactly the same manner towards the purification process. It suggests, however, that the nuclei in the droplets may have a spectrum very similar to the atmospheric spectrum from which they are derived.

Rau (1950, 1953) also studied the freezing of droplets on a polished metal surface and concluded that natural nuclei had preferred activation temperatures near -4, -12 and $-20°$ C, these being maxima in his plot of fraction of drops freezing at a given temperature. This result has not been confirmed by other workers.

Spatial variations

Because of the fact that different types of counter give considerably differing results when measuring the number of nuclei active at a given temperature, it is necessary to consider the counting method used when comparing nucleus counts in different places. Fortunately, several sets of measurements are available using virtually identical equipments and techniques in widely scattered stations so that intercomparison is simple.

Considering first the large visual chamber of Smith and Heffernan (1954), measurements have been made in Australia and over the South-western United States (Smith and Heffernan, 1954; Smith, Kassander and Twomey, 1956; Kassander, Sims and McDonald, 1957) and in South Africa (Mossop, Carte and Heffernan, 1956), both near sea-level and from aircraft at heights up to 7000 m. Despite relatively large day-to-day variations in the count there is a fairly close agreement between the average concentrations of active nuclei found in all places and at all altitudes. Whilst there may in fact be statistically significant differences between the counts they amount to only two or three degrees difference in the average temperature at which the observed nucleus count reached 1 per litre, namely, about $-30°$ C, corresponding to less than a factor of ten in concentration. This very low temperature was presumably due to some lack of sensitivity in this type of apparatus or observing technique.

More extensive sets of measurements using the cold-boxes developed by Bigg (1957*b*) and by Warner (1957) which are known to give results in close agreement with each other have since been reported. These measurements were made at ground level at ten widely scattered stations in Australia, the United States and Norfolk Island during the month of January in 1956 and 1957 (Bigg, 1956*a*; Bracewell, 1956; Bowen, 1958). The average temperature at which the count was 0·1 crystals per litre varied between about -15 and $-24°$ C (temperature for

1 per litre being about 4° C lower), some stations reporting consistently lower counts than others. There does not, however, appear to be any obvious common factor linking stations with low or high counts. From these more extensive measurements then, it appears that the average concentration of freezing-nuclei at the earth's surface may vary by as much as a factor of 100, from place to place, when the measurement is made at a temperature near $-20°$ C.

We have already mentioned the fact that little variation of nucleus count with height was found. Smith and Heffernan (1954) reported that nucleus concentrations tended to be somewhat higher above inversions than immediately below them, while the measurements of Kassander et al. (1957) show a slightly higher nucleus concentration near the ground than aloft, though counts at 1600 m and 5000 m were, on the average, the same. The measurements of Murgatroyd and Garrod (1956), similarly, show little systematic variation either in count or in spectrum up to 7000 m. Recently Telford (1960a) has made measurements of ice-nucleus concentration above and below the tropopause with the conclusion that though the ratio of counts above and below varies from day to day, the two concentrations are, on the average, about the same.

Temporal variations

Although the average freezing-nucleus concentration varies by a factor of 10 to 100 from place to place on the earth's surface, the fluctuations of count at one place and different times may be even more marked. Schaefer (1954a), in a series of three-hourly readings taken over a period of six years at the summit of Mount Washington, discovered an annual trend towards low counts in the summer and high counts in the winter. A similar result was found by Rau (1954) in a shorter series of measurements near Ravensburg. On a shorter time-scale Schaefer reported individual counts varying by a factor of 10^6. Some very high counts may be spurious, however, and due to ice-fragments, rather than to freezing-nuclei. Bigg's measurements (Bigg, 1956a) show maximum variations of the order of a factor of 10^3. Some of these variations persist for several days, as we shall discuss later, while Bigg and Meade (1959) have recently found very large fluctuations over times of the order of a few hours.

These fluctuations are very significant features of the occurrence of natural freezing-nuclei, and might be expected to give some clue as to their origin. We shall return to this important question in the following sections.

ORIGIN OF NATURAL ICE-FORMING NUCLEI

The most natural assumption to make about the source of atmospheric freezing-nuclei is that they originate at the surface of the earth in much the same way as do condensation nuclei, and are then distributed throughout the atmosphere by the action of winds. One important difference should, however, be noted, and that is that whilst efficient condensation nuclei are, in the main, composed of small hygroscopic particles, we should expect efficient freezing-nuclei to be solid insoluble particles.

There are three main ways in which information can be obtained about the origin of atmospheric freezing-nuclei. In the first place, acting upon the assumption that they are particles generated at the earth's surface, samples of naturally occurring dusts and smokes may be examined for nucleation ability and then their abundance and dispersion assessed. The second approach is to collect ice-particles from the atmosphere, either in the form of snowflakes or of crystals produced in a cold-chamber, and to endeavour in some way to identify the nuclei on which they have formed. The third method is to examine the spatial and temporal variations of the nucleus content of the atmosphere and correlate these with various physical and meteorological phenomena. We shall examine these methods in turn to see what results they give.

Activity of naturally occurring substances

The most obvious substances to examine for ice-nucleating activity are naturally occurring soils from the earth's surface. Various clays and soils in finely divided form were investigated by Schaefer (1949), with the conclusion that some clays have thresholds of activity as warm as $-12°$ C and reach a high activity level by $-24°$ C. Many other types of soil were found to have thresholds in the range -18 to $-24°$ C. A more detailed investigation on substances of higher purity was later carried out by Mason and Maybank (1958) who found nineteen minerals, mostly in the clay and mica groups with thresholds above $-18°$ C. Seven substances had thresholds above $-10°$ C, the most abundant being kaolinite ($-9°$ C). A large number of these materials could be pre-activated by prior freezing, but it was not determined how long this activation could persist under unsaturated conditions. Several other investigators have examined more limited classes of materials; thus Manson (1957) found vaterite ($\mu - CaCO_3$) to be active below $-15°$ C, while Isono and Komabayasi (1954) found volcanic dust from Mount Asama to have a threshold of $-14°$ C.

The sizes which these dust-particles must have to be active ice-forming nuclei can be estimated from the theory of chapter 8. If the particles behave as sublimation nuclei, then they will only have optimum activity if they are greater than about $0.2\ \mu$ in diameter, whilst if they act by freezing they must be larger than about $0.05\ \mu$. The smaller particles will, of course, still exhibit activity but their threshold activities will be lower. The observed exponential spectrum could be explained approximately if the average size of the particles is considerably less than the values quoted above (say $0.01\ \mu$). This seems, however, to be rather small, and it appears more likely that the approximately exponential distribution is due to the relatively greater abundance of materials having lower thresholds, the mean particle-size being somewhat greater than $0.1\ \mu$. Particles very much larger than this fall out rather quickly under gravity, whilst much smaller particles tend to be lost by coagulation.

Another possibility is that freezing-nuclei, like Aitken nuclei, are produced prolifically as combustion products near cities or industrial sites. Whilst it has generally been observed that the concentration of freezing-nuclei is not markedly greater in cities than in country air, Soulage (1958a) has found a copious source of such nuclei in the exhaust gases of electric steelworks, which he found to have an output per furnace of as many as 3×10^{15} nuclei per day with activity threshold $-9°$ C. Telford (1960b) has also found high concentrations of freezing-nuclei in smoke-plumes and down-wind of large industrial areas.

The way in which such particles might be distributed throughout the atmosphere has not been considered in detail. A comparable calculation has, however, been made by Junge (1957 b) for the vertical distribution of particles of sea-salt produced continuously at the surface and distributed by turbulent diffusion below an inversion at 2000 m. This calculation showed that after about 16 hr the concentration varies by only a factor of 3 throughout this region.

Woodcock (1951) and Twomey (1955) have measured the concentration of large salt-particles in the atmosphere at various altitudes and found that, whilst the concentration decreases sharply with height over the ocean, the distribution becomes much more uniform after the air-mass has spent several days over land. Byers, Sievers and Tufts (1957) made measurements of micron size chloride particles over the central United States and found no general decreases of concentration with altitude up to 3000 m.

Whilst therefore we can reach no quantitative conclusion it seems possible that after some days dust-particles originating at the earth's surface could be fairly uniformly distributed in the atmosphere up to the tropopause. Penetration of the tropopause does not ordinarily take place by diffusion, but the tropopause is itself often not well defined, and its irregular vertical motion could provide a capture mechanism for particles from lower levels of the atmosphere. In addition, of course, large thunderheads regularly penetrate the stratosphere and these must carry up air from lower levels. No quantitative estimate of these effects has yet been made.

The lack of any extreme geographical variations in nucleus concentration would seem to require that the nucleus material is a common constituent of the earth's surface, rather than being confined to small regions like active volcanoes. Many clay minerals should fulfil this condition.

Nuclei in snow crystals

The collection and examination of snow crystals has long been of interest, and several studies have been made of particles contained within them. Of course, the finding of a particular particle within a snow crystal does not necessarily mean that it was the nucleus responsible for its formation, since many particles may be collected by the crystal during its growth. However, the finding of a single solid particle at the centre of a snow crystal makes it very likely that this was the active nucleus.

Weickmann (1947) was among the first to examine the nuclei in natural ice-crystals, though he did not attempt specific identification. He collected crystals from cirrus clouds and on observation found that about one-third of the solid nuclei present were greater than a micron in diameter. Later aufm Kampe, Weickmann and Kedesdy (1952) used an electron microscope to examine similar cirrus crystal nuclei and identified some of them tentatively as kaolin, clay and quartz particles.

In this they agreed with the findings of Kumai (1951), who examined forty-three central nuclei from snow crystals, most of which were considered after electron-microscope examination to be soil particles. Kumai further found that these central nuclei ranged from 0·5 to 8 μ in greatest dimension, this distinguishing them from the condensation nuclei collected on other parts of the crystals, which had diameters of 0·05–0·15 μ. Isono (1955) later confirmed these results and made more positive identification of the nuclei as clay particles by electron diffraction.

Soulage (1955a, 1957) followed a rather different method to examine the residues from evaporated snowflakes. He dissected these under an optical microscope and in the larger nuclei found both soluble and insoluble components, most of the nuclei examined being 7–25 μ dia. He then examined the nucleation behaviour of the various fragments and found activity, usually near $-15°$ C. However, if freezing were repeated the threshold was depressed, probably due to some contamination. Soulage therefore concluded that most atmospheric ice-nuclei are mixed particles, of which the solid components are the active freezing-nuclei. He was unable to identify positively the nature or origin of the particles involved.

Correlation with meteorological variables

Because a relatively long series of measurements is required for any statistical significance, only a few attempts have been made to find correlations between freezing-nucleus concentration and other meteorological variables. These few cases are, however, worth considering.

We have already mentioned the seasonal variation of nucleus concentration found by Schaefer (1954a) and Rau (1954). In addition Schaefer found that high counts were usually associated with westerly winds and strong jet-streams, and suggested that freezing-nuclei may originate in dust-storms in the continental interior. Rau made a more detailed analysis of his results from Ravensburg and classified them in terms of the origin of the prevailing air-mass. This classification shows that, with the exception of maritime polar air, maritime air-masses tend to contain fewer nuclei than do continental air-masses.

More recently, Georgi and Metnieks (1958) have carried out a careful though brief series of measurements at Valentia on the coast of Ireland. They measured sea-salt nuclei, Aitken nuclei and freezing-nuclei, as well as wind-speed and direction, and concluded that there was a strong positive correlation between freezing-nucleus count and Aitken count. Since Aitken nuclei are primarily of continental origin, and there was also a correlation with winds from the mainland it was again concluded that freezing-nuclei are primarily of continental origin.

Bigg (1956a, 1957b) on the other hand, in measurements at Carnarvon in Western Australia, found no significant correlation between nucleus count and wind direction, and most other sets of measurements have been too brief to give any useful data on this question. The development of continuously recording counters should help greatly in this study.

Summary

Before proceeding to the next section it may be well to summarize the evidence so far discussed, and to see towards what conclusions it points.

The observational data indicates that the concentration of freezing-nuclei, averaged over a time of about a month, differs from place to place on the earth's surface by a factor of 10–100 when measured at $-20°$ C. This is comparable with the fluctuations in average Aitken count (table 4.1) and rather greater than the fluctuations in the number of large condensation nuclei in the atmosphere at different places. On the other hand, the concentration of freezing-nuclei does not vary greatly with height up to, and at least for a limited height above, the tropopause. This differs greatly from the behaviour of the Aitken count, which, as we saw in chapter 4, falls sharply with altitude. On the other hand, it is similar to the behaviour of the concentration of large salt-particles in the atmosphere, at least up to altitudes of 3000 m.

Measurements made on the ground have shown that many soil particles and industrial products provide efficient ice-forming nuclei, and studies of natural snow crystals indicate that the large nuclei upon which they appear to have formed are probably clay particles. The diameter of these particles is usually in the micron range.

It therefore seems likely that just as the sea provides giant salt-particles in substantially constant concentration over both land and sea and up to high altitudes, so the soil and to a lesser extent industrial smokes may provide solid insoluble particles which act as ice-forming nuclei.

We have yet to decide, however, whether all the ice-forming nuclei in the atmosphere come from the earth's surface, and as we shall see in the next section there is a strong possibility that an entirely different source is of importance.

THE METEOR HYPOTHESIS

We have so far discussed the behaviour and distribution of atmospheric freezing-nuclei on the basis of the hypothesis that they are dust-particles borne aloft from the earth's surface. We shall now consider an entirely different hypothesis proposed by Bowen (1953) to the effect that the most active freezing-nuclei in the atmosphere are particles of meteoritic dust captured by the earth in its orbit, and falling freely through the atmosphere. We shall indicate the lines of argument used by Bowen in developing this hypothesis.

Rainfall on the earth's surface is influenced by many factors, some of which are periodic (like the seasons) and many of which on a shorter time-scale are much more random, though there may be serial correlations due to such things as the movement of pressure systems. It is therefore to be expected that

The Meteor Hypothesis

when a sufficient amount of data from different places and different years is combined to a single curve of rainfall as a function of day of the year, a smooth curve should result.

The difficulty lies, of course, in estimating the degree of smoothness to be expected, since rainfall is very far from normally distributed. Putting aside then, for the moment, the question of the statistical significance of deviations from the mean, the daily rainfall of Sydney during January, summed over the period 1859–1952 shows distinct peaks on 13 and 22 January and on 1 February. Peaks of comparable magnitude occurring on exactly the same days were found in the daily rainfall totals of twenty-three stations in the state of New South Wales for the period 1871–1946.

Whilst these peaks are interesting, they come from a single area and could be explained on the basis of widespread cyclonic rainfall. However, when the data from fifty stations in New Zealand (1900–52), four stations in South Africa (1900–53), forty-eight stations in the U.S.A. (1869–1950), thirty-two in Japan (1876–1950), five in the Netherlands (1901–50), and the whole of the British Isles (1919–49) is analysed (Bowen, 1956a), and each group shows rainfall peaks within one or two days of those for Australia, such a local explanation becomes impossible.

Similar peaks were found near the same calendar days by Dmitriev and Chili (1955) who examined rainfall records for six stations in the U.S.S.R. over a period of 22 years. For the whole year they found thirty peaks which they considered to be statistically significant.

Combining all this data Bowen (1956b) produced a world rainfall curve incorporating data from approximately 300 stations distributed over the whole globe and each contributing records for a period of about 50 years. This curve is shown in fig. 9.7. Deviations from the mean are as large as $\pm 15\%$ and the three peaks on 12 and 22 January and 1 February are well marked.

A further analysis of this data (Bowen, 1953, 1956a) shows that in general a peak in the rainfall curve is produced not by one or two exceptionally heavy falls on that day, nor yet by an extraordinarily large number of rainy days on that date; but rather by a somewhat greater than normal number of rainy days on most of which the rainfall was consistently higher than average.

The wide geographic and temporal extent of these rainfall peaks argues against any purely local meteorological causes, and Bowen therefore concluded that the cause was extra-terrestrial, such an agent having the desired characteristics of yearly periodicity and a virtually simultaneous action over the whole surface of the earth.

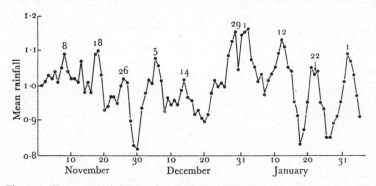

Fig. 9.7. The mean rainfall by days for November, December and January for three hundred stations distributed over the globe (Bowen, 1956*b*).

Of all extra-terrestrial mechanisms, that of injection of meteoritic particles into the atmosphere to serve as ice-forming nuclei seems most likely. Meteor streams recur annually and have durations of only a few days. If they were to provide efficient freezing-nuclei these might increase the rainfall immediately by effectively seeding clouds whose tops were not yet cold enough to contain appreciable numbers of ice-crystals. In addition, the widespread release of latent heat caused by glaciation of cloud-tops might have a more important influence on the development of cloud systems as a whole.

In order that this hypothesis be acceptable, two conditions must be fulfilled. It must be shown that meteoritic dust-particles can act as efficient ice-forming nuclei, and there must be a strong correlation between meteoritic dust concentration and rainfall peaks. Postponing consideration of this first requirement, Bowen examined the correlation between known meteor showers and rainfall peaks. It was found that this

The Meteor Hypothesis

correlation was indeed strong if a time-lag of about 30 days was allowed for penetration of the dust from the upper stratosphere to the troposphere. The relevant dates and meteor showers for the prominent peaks shown in fig. 9.7 for January are given in table 9.1. Similar correlations have been given by Bowen for the months August–December (Bowen, 1956 b, c, 1957), and by Dmitriev and Chili (1955) for the whole year.

Table 9.1. *Correlation between meteor showers and rainfall peaks (Bowen, 1956 a)*

Meteor shower	Date	Date of rainfall peak	Time difference (days)
Bielids I†	15–27 November	29 December	—
Bielids II	2 December	2 January	31
Geminids	13 December	12 January	30
Ursids	22 December	22 January	31
Quadrantids	3 January	1 February	29

† The date of appearance of this stream retrogresses about one day in every five or six years.

To strengthen the correlation, several meteor showers which have well-defined periodicities (the Perseids in September, the Giacobinids in October and the Bielids in December) have been found to be associated with rainfall peaks having the same periodicity and phase.

Additional evidence of a similar nature has been provided by study of the variations in the date on which snow first covered the ground in Tokyo (Bowen, 1956d), a phenomenon which might be expected to be related to freezing-nucleus concentration, and for which the records extend back for over 300 years. This study again exhibited a correlation in periodicity with the Bielid shower at the beginning of December.

Evidence of a somewhat different kind comes from an examination of the occurrence of ice-crystal clouds. Bowen (1953) showed that the occurrence of noctilucent clouds, which are probably clouds of ice-crystals produced by sublimation on to nucleating particles at altitudes of 80–100 km, and which can be observed during summer months in high northern latitudes, is correlated with meteor showers in June and July, there being in this case no time-lag involved. A similar result has been found for the occurrence of dust-clouds at heights of about 80 km by Bigg (1956b), the dust being detected by its optical scattering properties at sunset.

Somewhat similarly, Bigg (1957a) has found a correlation between the occurrence of cirrus clouds over Australia and meteor showers occurring 30 days previously, this time-lag being expected since the clouds are below the tropopause.

The time-delay

An important feature of the meteor hypothesis is the time-delay of 30 ± 2 days between the date of a meteor shower and the date of the corresponding rainfall peak. This has been interpreted as the time required for meteoritic particles, slowed to terminal velocity at a height of about 80 km, to fall to the level of the tropopause. The particles in a meteor shower cover a range of sizes from about 1 to 1000 μ. Smaller particles are removed from the solar system by the repulsive effects of radiation from the sun, and most larger particles are removed by gravitational effects. When these particles enter the upper atmosphere the larger particles burn up, those of intermediate size melt, and the smallest simply slow down to terminal velocity. Whipple (1950, 1951) calculated that iron particles less than about 10 μ dia. will be unmelted by air resistance; while Öpik (1951), taking into account the earth's angular velocity, fixed the melting-limit for 50% of incident particles at about 20 μ. Most particles greater than 100 μ dia. will burn up rather than merely melt.

The behaviour of these particles has been reviewed by Junge (1957c), and the result of calculations by Link show that particles of diameters 1–10 μ should take from 1000 to 30 days respectively to fall to the tropopause. The time-delay experienced by the largest unmelted particles is thus of the order required by the meteor hypothesis.

This time-delay aspect is, however, one of the most unsatisfactory features of the meteor hypothesis. If melting destroys the nucleating power of the particles then a reasonably sharp increase in freezing-nucleus count might be observed, but there should be a long tail. In the case of rainfall it has been suggested by Bowen that the arrival of the nuclei might trigger off rainfall and several days might be required to re-establish an unstable situation. This has, however, not been carefully investigated.

Freezing-nucleus peaks

Stimulated by controversy over the meteor hypothesis, widespread measurements of atmospheric freezing-nucleus concentration have been made during the last few years. Measurements have, for the most part, been confined to January, in which month Bowen's rainfall curve shows well-defined peaks which have been related to prominent meteor showers. Summaries of the results of such measurements have been given by Bowen (1956*e*, 1958). Though curves plotted for individual stations often show no marked agreement with each other, a composite curve incorporating all the available data has, for both 1956 and 1957, shown peaks in fairly good agreement with the positions of the prominent peaks in the rainfall curve.

Criticism

Many objections have been raised to the meteor hypothesis, some of them of a serious nature. We shall discuss some of these below. Most have not yet been resolved.

In the first place the statistical significance of the rainfall peaks themselves has been questioned. Swinbank (1954), Martyn (1954), Newmann (1954), and Oliver and Oliver (1955) have made criticisms of this nature and shown with the aid of sample calculations with data from other stations that large peaks can be produced from random reshuffling of the data. Whilst this criticism might be valid for limited sets of data it hardly applies to the large mass of observational data given in Bowen's later papers. Analysis of the significance of the peaks is, however, difficult because of the non-normal distribution of rainfall.

A somewhat similar question can be asked about meteor showers—is the rate of collection of small particles by the earth significantly greater during meteor showers than the rate of collection of random micrometeorites at other times? Evidence from correlations with noctilucent clouds (Bowen, 1953) and atmospheric transparency (Zacharov, 1952) suggests an affirmative answer, but these correlations have been attacked by Millman (1954) and by Whipple and Hawkins (1956). These latter authors show that the accretion rate for shower meteorites detected by radio or photographic means is less than twice the normal sporadic rate and can therefore have little effect upon terrestrial phenomena. They suggest, however, that corpuscular erosion of meteorite fragments in space may produce large numbers of shower particles which are too small to be detected by optical or radio means. Though there is no positive evidence to support this concept, it may prove to be an important process.

The chief arguments against general acceptance of the meteor hypothesis are, however, on physical grounds. The postulated delay of 30 ± 2 days between the incidence of meteor showers and the appearance of rainfall and freezing-nucleus peaks requires either some constant convection mechanism or else a high degree of size uniformity among the active particles. No physical mechanism has been suggested for either of these alternatives. We have seen that if the delay is attributed to a time of free fall then this requires a particle diameter a little less than $10\,\mu$, a size in good agreement with the nuclei found in cirrus crystals.

Estimates of the concentration of such particles in the atmosphere, deduced from studies of the accretion rate of meteoritic material have been made by several authors. Bowen (1953) gives a figure of $1\,\text{m}^{-3}$ in the stratosphere while Junge (1957c) surveys various accretion estimates to derive results in the range from 10^{-4} to $1 \cdot 5$ particles per m^3 in the troposphere. Recently, Pettersson (1958) has found much higher accretion rates and his data gives 300 particles per m^3 at the earth's surface. These values are, however, upper estimates assuming the dust to consist entirely of $10\,\mu$ spheres. If most of this material provides active freezing-nuclei, then these values are high enough conceivably to have an effect on rainfall. They are, however, hardly adequate to explain the peaks in freezing-nucleus concentration found, for example, by Bigg (1956a) on the days predicted by Bowen. Bigg's results would require meteor particles to be present at a concentration greater than 10^4 per m^3 on peak days, and some of the data of Mossop, Carte and Heffernan (1956) suggests concentrations of $10^6\,\text{m}^{-3}$. If

Pettersson's results are correct then such peak values may possibly occur in particle concentration, but the disagreement between the various estimates has not yet been resolved. The alternate possibility of considerably smaller particle sizes coupled with some other type of transfer mechanism from the high stratosphere has not been explored.

Finally, and rather crucially, it has been questioned whether in fact meteoritic particles are active as ice-forming nuclei. It is hard to obtain experimental information on this point, since large meteorites reaching the earth's surface will certainly not have surface properties exactly the same as those of small particles stopped high in the atmosphere, and the identification of other material as meteoritic is uncertain.

Schaefer (1957) has examined small magnetic particles collected from snow and rain samples and found them inactive above $-25°$ C. He was also unable to produce active nuclei by vaporizing pieces of an iron-nickel-cobalt meteorite. Mason and Maybank (1958) tested the nucleation properties of dust from a stony meteorite, finding it only slightly active, rather less so than many terrestrial mineral dusts.

Since noctilucent clouds give some evidence of water-vapour at greater than ice-saturation in the high stratosphere, it may be possible that meteoritic dust-particles become activated by formation of ice upon them during passage through this layer. Mason and Maybank (1958), and also Day (1958), found that in fact such pre-activation may occur with meteor dust, but Day showed that for all the samples tested the activation was destroyed after a few minutes in an unsaturated environment.

There is thus little direct evidence to confirm the meteor hypothesis, though no evidence has yet been brought forward which completely disproves it. The statistical arguments cannot, of course, be conclusive, and a decision on the truth of the hypothesis must await further experiments explicitly designed to test it.

CONCLUSIONS

Our survey has shown that, as might be expected, the soil contains numerous particles which can act as ice-forming nuclei in the atmosphere, of these some of the clay minerals probably being most important. Details of the distribution of these particles through the atmosphere are not known, but it seems probable that turbulent mixing could distribute them fairly uniformly up to high altitudes.

On the other hand, it has been proposed that some at least of these atmospheric freezing-nuclei enter the earth's atmosphere as meteoric dust which then falls slowly to lower altitudes. There appears to be a correlation between meteor showers and the rainfall and freezing-nucleus counts observed approximately 30 days later, but the statistical significance of the observed deviations from the mean, and the interpretation of the correlation coefficients obtained is still the subject of controversy. No positive tests of the meteor hypothesis have yet been devised, and the mechanism is not yet clearly formulated in detail so that critical experiments have not been possible.

We must, therefore, conclude that, whilst the majority of freezing-nuclei in the atmosphere probably originate at the earth's surface as wind-blown soil-particles of particular types, an appreciable contribution may be made by dust of meteoric origin.

CHAPTER 10

THE GROWTH OF ICE-CRYSTALS

INTRODUCTION

The study of crystal growth has long been popular for aesthetic reasons, and the intricately patterned hexagonal forms of snow-flakes are among the most pleasing geometrical designs found in nature. Scientific studies of the growth of ice-crystals have been undertaken within the last few decades, notably by members of the Japanese school, and in the last few years considerable experimental and theoretical work has been done. Despite this the physical mechanisms underlying the growth of intricate snow-crystal patterns are largely unknown, though, as we shall see, some general qualitative conclusions can be drawn.

From the point of view of cloud physics principal interest lies in the rate of growth of an ice-crystal within a cloud. This growth occurs both by direct vapour transfer from cloud droplets to the crystal because of the difference in vapour-pressure over ice and water at the same temperature, and by a process of accretion of cloud droplets very similar to that which we discussed in connexion with the formation of non-freezing rain. Ice-crystal habit is important in both these processes.

In this chapter then, we shall first give an account of our present knowledge of the factors determining the habit of ice-crystals growing from the vapour, and then go on to discuss growth by sublimation and by accretion.

SHAPES OF ICE-CRYSTALS

If the growth of an ice-crystal were a simple equilibrium process, then the resulting crystal habit could be determined by application of Wulff's theorem (Wulff, 1901), which states that in equilibrium the distance of any crystal face from the centre of the crystal is proportional to the free energy per unit area of that face. Whilst no exact values of such free energies are known for ice, approximate values can be found by assuming that molecules interact only with their nearest neighbours, and counting the concentration of nearest-neighbour pairs linked across any crystal plane. Because of its hexagonal structure the planes having lowest energy in ice are basal planes {0001} and prism planes {10$\bar{1}$0}. Using the crystal-structure model and lattice parameters discussed in chapter 8 a simple counting of nearest-neighbour pairs then shows that the energy of a prism face is about 6% greater than that of a basal face, per unit area. Wulff's theorem then shows that the equilibrium form should be a hexagonal prism, the ratio of whose axial length to hexagonal diameter is 0·82. This result was obtained by Krastanow (1943).

Whilst ice-crystals formed in nature in snow clouds or grown in the laboratory do in fact have primarily the shape of hexagonal prisms, the dimensional ratio may vary from less than 0·1 to more than 10. In addition many complex forms such as stars, hollow prisms, scrolls and dendritic plates commonly occur. Whilst it is hard to imagine surface-energy changes adequate to explain the variations of simple prismatic habit, it becomes quite impossible to explain the more complicated forms on the basis of any equilibrium theory. We therefore conclude that the forms of ice-crystals are not equilibrium structures, but represent kinetic effects due to thermal or diffusion gradients, surface accommodation coefficients and similar mechanisms, which have become 'frozen-in' before attaining equilibrium.

Several attempts have been made to find the dependence of ice-crystal shape upon the variables involved in crystal growth. Parameters which almost certainly affect crystal habit are the temperature and the ambient pressure of water-vapour, while the presence of fog droplets and the fall velocity of the growing crystal may also be involved. Finally, it is possible that foreign particles or vapours may exert appreciable influence on the growing ice-crystal.

Early experiments were mainly confined to examination of the forms of snow crystals falling from clouds under conditions when reasonably reliable information could be obtained about the temperature of the cloud in which growth took place. Particularly beautiful sets of photomicrographs of such snow crystals are to be found in the books of Bentley and Humphreys (1931) and Nayaka (1954). As Nakaya (1951) has pointed out, however, there is often a natural tendency to concentrate

Shapes of Ice-crystals

attention upon snowflakes of almost perfect symmetry to the neglect of the, often more numerous, less regular crystals.

Mason (1957, pp. 173–9) has reviewed these results and summarized them in the form shown in table 10.1. Because of the uncertainties attending such experiments these results are principally of use for comparison with more accurate laboratory work to verify that conditions in natural clouds are being validly approximated. This data refers only to individual crystals; we shall defer to a later stage any discussion of the aggregation of crystals to form snowflakes of large size.

Table 10.1. *Temperature ranges for the origin of snow crystals in the atmosphere according to several observers (Mason, 1957)*

Temperature range (° C)	Crystal type
-3 to -8	Needles
-8 to -25	Plates, sector stars
-10 to -20	Stellar dendrites
< -20	Prisms, single crystals, twins
< -30	Prismatic clusters

Laboratory experiments

Laboratory experiments on ice-crystal growth can be divided into two general categories; in the first type ice-crystals are grown under conditions close to water saturation in a droplet fog, while in the second type the crystals grow in a droplet-free vapour whose temperature and supersaturation can be varied independently. The droplet fog environment simulates more closely conditions in a cloud in the atmosphere, but gives more limited information about the growth process.

Typical of studies of growth from a fog are those of aufm Kampe, Weickmann and Kelly (1951) and of Mason (1953). In both these experiments ice-crystals were formed by seeding with dry ice or silver iodide a supercooled fog formed in a room-sized cold-chamber. Ice-crystals were either collected on glass plates and photographed immediately, or collected on glass plates coated with formvar solution, which on drying retained permanent replicas of the crystals. These experiments gave very similar results, indicating a habit sequence: plates–prisms–plates–prisms, as shown in table 10.2, as the temperature is lowered. The transition points found between the different habits are seen to agree quite well.

Table 10.2. *Transitions between different crystal forms*

	Plates (° C)	Prisms (° C)	Plates (° C)	Prisms (° C)
aufm Kampe et al. (1951)	0 to −4	−4 to −10	−10 to −20	−20 to −40
Mason (1953)	0 to −5	−5 to −10	−10 to −25	−25 to −40

Perhaps the best-known experiments on snow-crystal growth are those of Nakaya (1951, 1954), who suspended growing ice-crystals upon a fine filament in a cylindrical cold-chamber. In the bottom of this chamber was a beaker of water maintained at a warm temperature to produce a fog which rose slowly and was supercooled well before reaching the crystals. The ice-crystals were thus growing in an environment which was probably above water saturation but which also contained fog droplets. The habit of growth of the ice-crystals was studied as a function of ice-crystal temperature and of the temperature of the water-bath. Unfortunately, the presence of the droplets makes it very difficult to estimate the true supersaturation of the environment so that the results are rather hard to interpret (Marshall and Langleben, 1954). The crystal habit appeared to depend primarily on temperature as in table 10.2, but was also a function of supersaturation.

Shaw and Mason (1955), in experiments which we shall discuss in more detail later, examined the growth-rate of single ice-crystals growing on a cooled metal plate under conditions of controlled temperature and supersaturation. They concluded that crystal habit was primarily a function of temperature, the supersaturation having a much smaller, non-systematic effect. The transition temperatures between the various forms were substantially those found by Mason (1953) and shown in table 10.2.

Kobayashi (1957) devised a useful technique for studying the growth of ice-crystals over a wide range of temperatures and supersaturations. Using a 'diffusion' or 'thermal gradient' chamber (see chapter 4) at a low temperature he grew ice-crystals upon a fine thread or rabbit hair and studied their habit photomicrographically. By varying the position of the crystals in the chamber and changing top and bottom temperatures a wide range of growth environments was studied. Later Hallett and Mason (1958) performed a similar experiment with a more elaborate diffusion chamber incorporating movable ice-plates to aid in the study of growth at low supersaturations. Both these experiments confirmed that habit depends primarily upon temperature, the predominant forms and boundary temperatures agreeing closely with those of Mason (1953) in table 10.2. Some dependence of habit on supersaturation was found, and

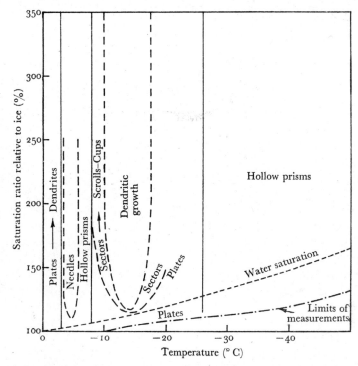

Fig. 10.1. The growth habits of ice-crystals, grown in a diffusion chamber, in relation to temperature and supersaturation (Hallett and Mason, 1958).

this agreed, at least qualitatively, with Nakaya's results. Apart from slight shifting of boundaries there is a marked tendency towards dendritic growth at high supersaturations. This is shown diagrammatically in fig. 10.1.

A conflicting view has been taken by Marshall and Langleben (1954), who reanalysed Nakaya's results and attempted to express them in terms of true supersaturation instead of the hybrid 'supersaturation' employed by Nakaya. They found that on such a diagram loci of constant excess of vapour density over that existing over a growing ice-particle delineated the various crystal forms almost as well as did Nakaya's empirical boundaries. They therefore concluded that habit may be determined primarily by the active vapour excess rather than any temperature effects. Whilst they did not carry out any actual experiments to test this hypothesis, Marshall and Langleben devised an ingenious electrical analog experiment in which the distribution of current to a crystal-shaped electrode immersed in an electrolytic tank was studied. They found that habit changes could be reasonably explained if an inhibiting potential (corresponding to a higher vapour-pressure) were associated with edges and corners. These sites would then only grow under conditions of high vapour excess, giving rise first to plates and then to dendrites.

Whilst this theory introduces many ideas which are undoubtedly of importance in habit determination, the weight of experimental evidence appears to be in favour of a strong temperature dependence of crystal habit. The case for vapour-excess dependence is further weakened by an uncertainty (acknowledged by Marshall and Langleben) in the interpretation of Nakaya's original data. The true situation is obviously, however, a combination of the two dependences, and we shall now discuss a recent attempt at a synthesis.

This attempt was made by Kobayashi (1958) who studied the growth of ice-crystals in a diffusion chamber at pressures much below atmospheric. The effect of such reduced pressure is to decrease the effective supersaturation over a growing ice-crystal by its combined effect on thermal conductivity and diffusion constant. In this way effective supersaturations as low as $0 \cdot 1 \%$ with respect to ice were attainable and growth could be studied under these near-equilibrium conditions. Kobayashi found that at these low supersaturations the habit of the ice-crystals grown was in fact close to the equilibrium form predicted by theory over a large temperature range. With increased vapour excess, $\Delta\rho$, the temperature demarcations previously discussed emerged, and for large $\Delta\rho$ the dendritic forms of Marshall and Langleben became predominant. The approximate positions of the demarcations between these regions are shown in fig. 10.2.

Shapes of Ice-crystals

Whilst these experiments have brought some order into the discussion of the habit of ice-crystals, it cannot be said that we have more than the most rudimentary understanding of the causes of these habit variations. Anything approaching a complete discussion must take into account the crystal structure of ice, the diffusion of molecules across the surface of the growing crystal and their accommodation at the points of growth, as well as the properties of the enveloping vapour diffusion field.

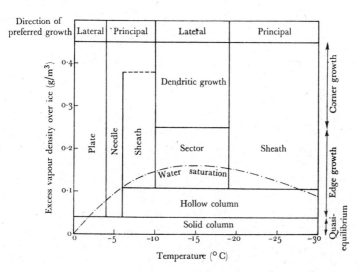

Fig. 10.2. Ice-crystal habit as a function of temperature and vapour density excess over that of ice (Kobayashi, 1958).

One other aspect of the study of ice-crystal habit should perhaps be mentioned, not so much because of any significance it might have in cloud physics, as because it often presents a complication in laboratory experiments. This is the effect of impurities upon the growth and habit of ice-crystals. Nakaya (1955) reported that aerosol particles may profoundly affect ice-crystal habit, but this has not been confirmed, and the most active contaminants have been found to be organic vapours such as isobutyl alcohol (Vonnegut, 1948). The mechanism of the effect of such 'poisons' on crystal growth has been discussed by Sears (1958), but no direct application can be made to ice until more is known of its normal growth.

GROWTH-RATE OF ICE-CRYSTALS

Ice-crystals in sub-freezing clouds are initially formed as extremely small embryos by sublimation on to active nuclei, or as somewhat larger crystals by the freezing of cloud droplets containing active freezing-nuclei. In the first case the ice-embryo may be less than a tenth of a micron in diameter when steady growth begins, whilst frozen cloud droplets are of the order of 10 μ dia. Such crystals are at this stage volumetrically unimportant, but because the cloud air is at approximately water saturation, the environment is considerably supersaturated with respect to ice, and the crystals grow rapidly. In this section we shall consider the early stages of this growth, that is, the region in which growth is from the vapour rather than by any accretion processes.

The overall theory of the growth of an ice-crystal from the vapour is very similar to that discussed in chapter 6 for the growth of a water droplet. There are, however, two complications. The first is that the ice-crystal is a polyhedron rather than a sphere so that solution of the diffusion equation is more difficult. The second is that vapour molecules cannot condense in a haphazard manner on to the crystal surface, but must be accommodated into the regular structure of the crystal.

Treating the more straightforward part first let us postpone consideration of molecular effects at the crystal surface and discuss the effect of crystal habit on growth-rate. This follows on closely parallel lines to our discussion of droplet growth in chapter 6. Since the concentration n of vapour molecules near the growing crystal obeys Laplace's equation

$$\nabla^2 n = 0, \qquad (10.1)$$

there is a close analogy between the vapour field and the electric potential near a charged conductor. Jeffreys (1918) followed this analogy and showed that the rate of growth of a crystal whose electrostatic capacity is C is

$$\frac{dM}{dt} = 4\pi CDm(n_0 - n_c), \qquad (10.2)$$

where D is the diffusion constant of the vapour, m the mass of a vapour molecule, n_0 the concentration of vapour molecules at a large distance and n_c the concentration at the surface of the growing crystal. This is a direct generalization of (6.4) in which C replaces the capacity r of a sphere of radius r.

The development now follows exactly as in chapter 6. The latent heat released by condensing molecules raises the temperature of the growing crystal and affects n_c, the extent of this temperature rise depending upon the thermal conductivity κ of the air. If the effects of surface free energy and of ventilation are neglected an equation closely analogous to (6.14) is obtained:

$$\frac{dM}{dt} = 4\pi C G's, \qquad (10.3)$$

where s is the supersaturation relative to a plane ice surface and

$$G' = \frac{D\rho_v}{\rho_s}\left[1 + \frac{DL^2\rho_v M_0}{RT^2\kappa}\right]^{-1}. \qquad (10.4)$$

Here ρ_v is the density and M_0 the molecular weight of the vapour and ρ_s the density of ice.

With the aid of (10.3) and (10.4) the rate of growth of a crystal of given shape can be found, provided the electrostatic capacity C of the crystal form is known. Three simple cases serve as approximations for the majority of ice-crystal forms:

(i) Sphere of radius r
$$C = r. \qquad (10.5)$$

(ii) Circular disk of radius r
$$C = 2r/\pi. \qquad (10.6)$$

(iii) Prolate spheroid, major and minor semi-axes a and b respectively, $e = (1 - b^2/a^2)^{\frac{1}{2}}$

$$C = 2ae/\ln\{(1+e)/(1-e)\}. \qquad (10.7)$$

The fact that these are not very good approximations to real ice-crystal shapes is of no great consequence since, as we shall see, the process of crystal growth is much more complex than simple equations like (10.3) would suggest.

This approach to ice-crystal growth was first followed by Houghton (1950), though he only obtained an approximate solution to (10.2) instead of completing the analysis leading to (10.3) and (10.4). Mason (1953) later carried through the discussion in the way we have outlined above and arrived at equations equivalent to (10.3) and (10.4).

Mason calculated the expression $G's$ as a function of temperature for the case of a crystal growing in an environment at water saturation, and this curve is shown in fig. 10.3. From this it can be seen that the growth-rate is a maximum at a temperature near $-14°$ C and decreases rather sharply at warmer or colder temperatures. This variation is largely a consequence of

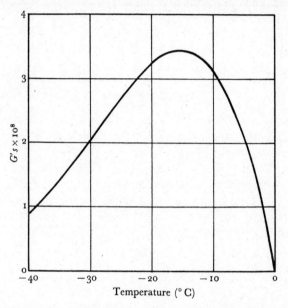

Fig. 10.3. The function $G's$ plotted as a function of temperature for an ice-crystal growing in an environment at water saturation (Mason, 1953).

the variation of the absolute vapour-density difference between water and ice as a function of temperature as is clear from (10.2). The other terms entering the final equation, however, do have a considerable effect.

Molecular basis of crystal growth

The growth of a crystal differs from the growth of a droplet in that the molecules of a crystal are arranged in a regular array and any added molecules must also lie in this pattern. The mechanism of growth generally involves adsorption of a vapour molecule on to the crystal surface, followed by migration of this adsorbed molecule over the surface until it reaches a position where it can be fitted into the crystal structure with a net decrease in the free energy of the system.

Fig. 10.4 shows pictorially a section of the surface of a growing crystal at a moderate temperature. The sites at which adsorbed molecules can be most favourably incorporated into the crystal structure are the kinks in the growth step shown.

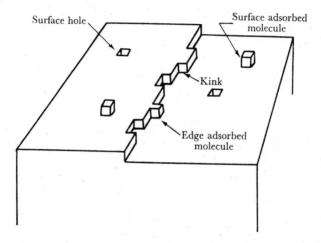

Fig. 10.4. A model of a step on a crystal surface at a moderate temperature (after Burton and Cabrera, 1949).

Molecules will not tend to cluster together into isolated islands on the crystal surface because the free energy associated with the edges of such islands forms a large free energy barrier when the island is small. Easy growth therefore requires the existence of at least one step upon the growing crystal face.

If the internal structure of a crystal is undistorted, growth upon surface steps will rapidly lead to the development of a crystal bounded by perfect, unstepped faces of low index. Any further growth requires the development of new crystal layers, and the only process by which these can form is by the clustering together of adsorbed molecules to form islands upon the perfect faces. This is the two-dimensional analogue of homogeneous nucleation, and the energy barrier to be surmounted is quite large. Under typical conditions the supersaturation required for an appreciable nucleation rate is of the order of 50 %. Since crystals are observed to grow at much smaller supersaturations than this, there must clearly be some other mechanism acting to facilitate the addition of new crystal layers to a growing crystal.

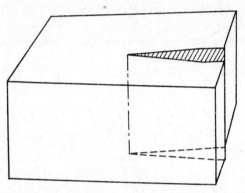

Fig. 10.5. Model of a screw dislocation. The surface step, shown shaded, is an integral number of atomic layers high.

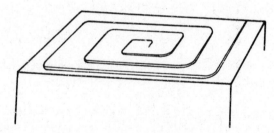

Fig. 10.6. Growth terraces on the face of a crystal from which emerges a single screw dislocation.

Such a mechanism was proposed by Frank (1949), who showed that if the crystal lattice was distorted, as shown in fig. 10.5, by the inclusion of a screw dislocation, then a step would always be present upon the face of emergence of the dislocation. The behaviour of such a dislocated crystal was studied in detail by Burton, Cabrera and Frank (1951) who showed that as molecules were added to the step it would twist into a polygonal spiral as shown in fig. 10.6. Growth patterns of this

Growth-rate of Ice-crystals

kind have been revealed by delicate microscopic techniques in a variety of crystals (Verma, 1953), verifying the essential correctness of the theory.

The growth-rate of such a dislocated crystal surface requires careful study, and this has been done by Burton et al. (1951). They find that for small supersaturations the distance between the turns of the growth spiral is approximately inversely proportional to the supersaturation s, so that the number of steps per unit area of surface is proportional to s. While each of these steps is sufficiently far from its neighbours to be uninfluenced

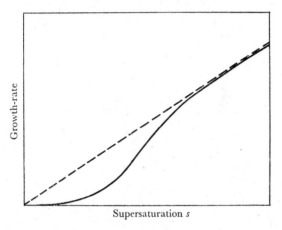

Fig. 10.7. Dependence of growth-rate of a crystal face containing screw dislocations upon supersaturation. Initially, variation is quadratic, but becomes linear at large supersaturations.

by them its rate of advance is proportional to s, so that the growth-rate of the crystal face is proportional to s^2. When the supersaturation is high the steps are close together and compete for the adsorbed vapour molecules, giving a rate of growth simply proportional to s. The overall behaviour is thus as shown in fig. 10.7.

In general, of course, since screw dislocations occur in quite high concentrations in all but the most perfect crystals, we must expect that rather than a single dislocation there will be a whole group of dislocations emerging on most crystal faces. Except for inactive close pairs of opposite sign, these dislocations all give rise to growth spirals which interact with one another. In the

s^2 region the presence of extra growth terraces increases the growth-rate, but it cannot exceed the linear rate because of interactions between adjacent steps.

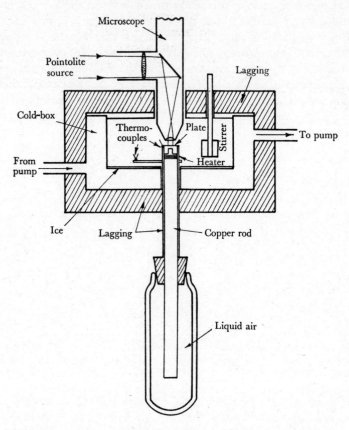

Fig. 10.8. Apparatus for study of ice-crystal growth under conditions of controlled temperature and supersaturation (Shaw and Mason, 1955).

Experimental results

The most careful experiments which have yet been made to examine the growth-rate of ice-crystals are those of Shaw and Mason (1955). Using the apparatus shown in fig. 10.8 they supported a small ice-crystal on the top of a cooled copper rod in an enclosure maintained at ice saturation at a slightly higher temperature. The supersaturation and temperature could thus be easily varied and the growth followed through a microscope.

They found that for a crystal of diameter $2a$ and axial height $2c$ the growth-rates in the two directions were given by

$$d(2a)^2/dt = U_a; \quad d(2c)^2/dt = U_c, \qquad (10.8)$$

where U_a and U_c were constants. This is in agreement with what might be expected from the diffusion theory which we outlined above, though the theory makes no predictions about the way in which the added mass is distributed over the crystal. Equations (10.8) show that the habit of the crystal tends to a limiting form given by

$$\Gamma = |c/a|_{\text{lim}} = (U_c/U_a)^{\frac{1}{2}}. \qquad (10.9)$$

Shaw and Mason found, as we have discussed before, that Γ is a definite function of temperature but seems to have only a small, non-systematic dependence upon supersaturation.

Examining growth-rate as a function of supersaturation s, Shaw and Mason found that both U_a and U_c varied as s^2 both for growth and evaporation in the range $-0.3 < s < 0.3$. If the crystal is relatively large, this behaviour is in agreement with the calculations of Burton, Cabrera and Frank for small growth-rates. The values of U_a and U_c were found to vary from crystal to crystal, and with time on a single crystal face. This is reasonable since growth-rate is a function of dislocation density in this region. It should also be remembered that minute traces of impurities may have a large effect upon the growth-rate if they become adsorbed on the active growth steps.

This variation of growth-rate with s^2, rather than with s as suggested by (10.3), has not yet been considered in practical applications of the theory to processes of ice-crystal growth in clouds, and indeed it must first be verified that ice-crystals growing in free air show the same behaviour as those supported on a metal plate as in the experimental arrangement.

GROWTH OF ICE-CRYSTALS IN CLOUDS

When ice-crystals grow in clouds the situation is more complicated than in the simple growth from the vapour which we have been considering. In particular the interaction between ice-crystals and cloud droplets must be taken into account, and also the interaction between individual ice-crystals to form aggregates or snowflakes. Finally, we must discuss the further interaction between these aggregates and the cloud droplets to form rimed structures or graupel. We shall consider these in turn.

Interaction with cloud droplets

Kumai (1951), in a microscopic investigation of snow crystals, found traces of condensation nuclei distributed widely over the surface of the ice-crystal replicas and suggested that the accretion of very small droplets might be an important process in the growth even of very small ice-crystals in clouds. This was contested by aufm Kampe *et al.* (1952) and later discussed in more detail by Kuroiwa (1955). Kuroiwa considered a small water droplet moving uniformly relative to a larger ice-crystal at a temperature below 0° C. Because of the diffusion gradient surrounding the ice-crystal the small droplet tends to evaporate, and Kuroiwa showed that only droplets larger than about a micron in diameter could reach a typical small ice-crystal without evaporating. His analysis, however, neglected any aerodynamic effects, and these, as we shall discuss presently, preclude collection of droplets smaller than several microns in diameter.

The effect of the evaporation of such small droplets and of the partial evaporation of large droplets near a growing ice-crystal is, however, important. Instead of the diffusion field of watervapour around the crystal extending to infinity it contains a distribution of vapour-sources which limit its range. This in turn increases the gradient of vapour-pressure near the growing crystal and so increases its growth-rate. This effect has been considered by Marshall and Langleben (1954) and shown to depend very markedly upon the size of the growing ice-crystal and upon the droplet spectrum of the cloud. Large crystals and small droplets favour an increase in the growth-rate over that given by (10.3). For crystals less than 1 mm in diameter and typical cloud droplet spectra, however, the increase is only about 10 %. There is again a statistical aspect to this effect, and crystals which experience near-collisions with water droplets may have their mass increased very substantially by

Growth of Ice-crystals in Clouds

vapour transfer. This is not of any macroscopic significance, however, since the absolute number of crystals so favoured is not appreciable.

Mass and terminal velocity of ice-crystals and aggregates

Before any discussion of the accretion of water droplets by ice-crystals or of the aggregation of ice-crystals can be entered upon it is necessary to have some quantitative idea of the terminal velocity of various crystalline structures.

Almost the only measurements which have been reported on the falling velocity of single ice-crystals are those of Nakaya and Terada (1934). Working on a mountain side where snow was generally observed to fall as single crystals, they allowed individual crystals to fall through a vertical cylinder 2 m in height, closed at each end to eliminate draughts. The time of fall was observed visually in the case of small snow crystals, but for the faster graupel pellets a photographic exposure was made and the length of the resulting streak measured. After falling through this cylinder the crystal was caught on a glass plate coated with paraffin, photographed, melted, and the diameter of the resulting hemispherical drop measured to give its mass.

Table 10.3. *Relation between mass m (mg) and maximum dimension d (mm) of various natural ice-crystals (Nakaya and Terada, 1934)*

$m = 0.065\ d^3$ for graupels
$m = 0.027\ d^2$ for crystals with water-drops
$m = 0.010\ d^2$ for 'powder' snow and spatial dendrites
$m = 0.0038\ d^2$ for plane dendritic crystals
$m = 0.0029\ d$ for needles

Two useful sets of data were derived from this investigation. The first is a set of empirical relationships between the mass m and maximum dimension d of various crystal types. These relationships are shown in table 10.3 for five general types of crystal structures. The coefficients represent average values. Graupel particles, representing an extreme form of crystal rimed with frozen water-drops, have an approximately constant density of 0.125 g/cm^3 indicating a very open structure. Crystals other than needles have $m \propto d^2$, indicating a constant thickness in one dimension, which for plane dendritic crystals was found to be approximately 0·01 mm. Needles have approximately constant cross-section and grow only in one dimension.

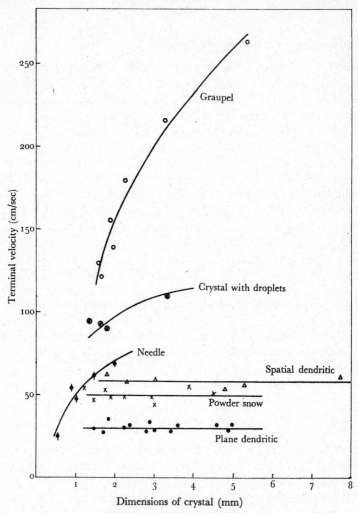

Fig. 10.9. Measured terminal velocities of different types of ice-crystals (Nakaya and Terada, 1934).

The relation between size and terminal velocity of different crystal types found by Nakaya and Terada is shown in fig. 10.9. The striking thing about these results is that while the terminal velocity of needles and rimed crystals increases with size, the fall speed of plate-like crystals is constant and independent of size. The actual terminal velocity is in all cases much less than that of a water-drop of the same mass because of the open structure of the crystal forms.

These rather striking results were later examined theoretically by Magono (1954), after first collecting additional data on crystal density and falling attitude. He found (Magono, 1953) that a crystal of plate-type falls keeping its plane horizontal. Since the thickness of the plate is constant, its mass is proportional to its area, as also is the aerodynamic drag. The falling velocity is therefore independent of crystal size, and calculation gives a value of 35 cm/sec in good agreement with experiment. Similar approximate treatments were given by Magono for the other crystal types studied by Nakaya and Terada and adequate general agreement with experiment was obtained.

Most other measurements of the terminal velocity of ice-crystals have concentrated upon aggregates of small crystals forming snowflakes, though Schaefer (1954c) made a few measurements on individual crystals which were roughly in agreement with those of Nakaya and Terada. Magono (1951) made measurements on simple crystals and also on snowflakes of diameter up to about 2 cm, timing their fall over a distance of 4·2 m. He found terminal velocities in the range 1–2 m/sec, with a considerable scatter but an approximate variation as the square root of the snowflake dimension, this being determined by photographing the falling flake.

Langleben (1954) carried out a more detailed investigation using a motion-picture camera to determine fall-speed and a dye-impregnated filter-paper (Marshall, Langille and Palmer, 1947) to find the diameter d of the water-drops obtained by melting the snowflake. He found the terminal velocity u to be given by a relation
$$u = kd^n, \qquad (10.10)$$
where the parameters k and n varied from one type of snowflake to another, depending upon the shape of the constituent crystals and the amount of riming or melting. When u is measured in cm/sec and d in cm the value of n giving best fit was usually about 0·3, while k was 160 for dendrites and 234 for

combinations of columns and plates, values of k increasing rapidly with riming or melting.

Litvinov (1956) later carried out a similar series of measurements in Russia, using a 12-m falling-cylinder and stop-watch timing. His results were again approximately described by (10.10), but the size dependence was much less marked than found by Langleben, the average value of n being 0·16. The value of k was also lower, best fit being given by values 87 and 115 in the two cases considered. This deviation from Langleben's results is almost certainly due to slightly different conditions of formation of the snow crystal aggregates.

Magono (1953) made very similar measurements on the terminal velocity of snowflakes, but instead of using an empirical formula of the form (10.10), developed an approximate theoretical expression. He considered the aerodynamic drag on a falling flake to consist of two parts: one represents drag due to air flowing around the flake and is proportional to r^2, where r is the radius of the flake considered as approximately spherical, the second part is proportional to r^3 and is the drag contributed by air flowing through the open structure of the flake. Both terms are proportional to the square of the velocity. This assumption gives an expression for the terminal velocity with the two drag coefficients as parameters. These were fitted by comparison with experiment to give

$$u = 132 \sqrt{\frac{r}{0\cdot40 + 0\cdot63r}} \qquad (10.11)$$

for non-rimed flakes, and

$$u = 194 \sqrt{\frac{r}{0\cdot45 + 0\cdot60r}} \qquad (10.12)$$

for rimed flakes, r being in cm and u in cm/sec. These expressions give a variation of u as $r^{\frac{1}{2}}$ for small flakes, as found earlier by Magono (1951), and a size independent velocity for large flakes, again in agreement with experiment.

To summarize this section we may say to a sufficient degree of approximation that the terminal velocity of dry unrimed single ice-crystals has a constant value of about 40 cm/sec, while the terminal velocity of dry unrimed snowflakes ranges from about 60 cm/sec for flakes a few millimetres in diameter to about 150 cm/sec for flakes as large as 4 cm diameter. The corresponding velocities for rimed or wet crystals or flakes are higher by about a factor 1·5 in typical cases, but can exceed this if riming or wetting is severe.

Growth of snowflakes

Whilst the initial growth of ice-crystals in a cloud is by sublimation from the vapour, this process becomes increasingly slow as growth proceeds, and large structures can only be built by some sort of collision process. We shall first consider the aggregation of individual ice-crystals to form snowflakes, the other possible collision process being the accretion of cloud droplets to form rimed structures and graupel.

Study of the aggregation of crystals poses problems of two types: the aerodynamic problems associated with the collision, and the physical problems associated with the adhesion of crystals to one another. The aerodynamic problems have scarcely even been considered, let alone solved, so that only semi-quantitative estimates based upon collision cross-sections for spheres can be made. Some empirical knowledge has been gained of adhesion processes, but physical understanding is still lacking.

When two ice-crystals are brought into contact, their adhesion will depend in detail upon the structure of the crystal surfaces. Weyl (1951) has discussed the surface-structure of ice and water to some extent, but no explicit theories have been formulated. It has been suggested that the surface-layer of ice has a pseudo-liquid structure, and certainly there must be numbers of molecules engaged in surface-diffusion, but again these suggestions remain vague. It seems clear at any rate that the surface will become more inert as the temperature is lowered and that any interfacial rearrangement required for adhesion will be most easily accomplished at temperatures only slightly below the melting point. This effect has been investigated by Goldshlak (1957), who examined the aggregation of small ice-crystals in a cloud chamber. He found that cohesion occurred at temperatures above $-25°$ C in an atmosphere at ice saturation, whilst in a supersaturated environment cohesion occurred down to rather lower temperatures.

Magono (1953), examining a group of snowfalls, found a

relation between air-temperature at the observation point and the observed average size of snowflakes as shown in fig. 10.10. A further study of the sizes and forms of crystals making up the individual flakes allowed deductions to be made about the history of growth and it was concluded that aggregation into snowflakes occurred only at temperatures above $-10°$ C. This result does not conflict with that of Goldshlak because of the different aerodynamic conditions involved in the collisions.

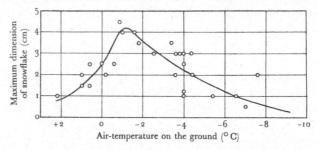

Fig. 10.10. Relation between the size of snowflakes and the air-temperature at which the snowflakes were observed (Magono, 1953).

Calculation of the growth-rate of snowflakes by accretion of ice-crystals is relatively simple if a constant collection efficiency E is assumed (Magono, 1953). The simplicity stems from the fact that to a good approximation the terminal velocities of both ice-crystals and snowflakes may be regarded as independent of size. Let us suppose the concentration of ice-crystals is n per unit volume, and that each has a fall-speed u_2 and an effective volume V_2 when incorporated into a snowflake. Let the flake be spherical and have volume V_1 and terminal velocity u_1. Then its cross-sectional area is $\alpha V_1^{2/3}$ where $\alpha^3 = 9/16\pi$ and

$$dV_1/dt = nV_2 E\alpha V_1^{2/3}(u_1-u_2), \qquad (10.13)$$

which can be immediately integrated to give

$$V_1 = [V_0^{1/3} + nV_2 E\alpha(u_1-u_2)t]^3 \qquad (10.14)$$

or

$$V_1 = [V_0^{1/3} + nV_2 E\alpha h(u_1-u_2)/u_1]^3, \qquad (10.15)$$

where V_0 is the initial volume of the flake and h is the distance through which it falls.

In a comparison with experiment Magono assumed $E = 1$ and obtained reasonable agreement, considering the uncertainty in some of the other quantities involved. Equations (10.14) and (10.15) thus seem to give a valid approximation to the growth behaviour of snowflakes.

Because the terminal velocity of snowflakes is independent of their size there should be little collision between flakes. We

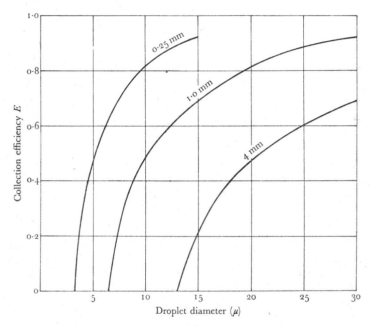

Fig. 10.11. Collection efficiency E of an ice-crystal (of diameter shown as parameter) for cloud droplets of various sizes, calculated from the theory of Ranz and Wong (1952) assuming the crystal to be a circular disk. Terminal velocity was assumed to be 50 cm/sec in each case.

might expect the same to be true of individual ice-crystals and some collision mechanism must be sought to provide the double crystals which serve as embryos for snowflakes. Two possible mechanisms come to mind. In the first place, since the ice-crystals involved are perhaps a millimetre in diameter, the 'wake' effect which Pearcey and Hill (1957) found for large cloud droplets may operate to give a large collision probability (see chapter 6). On the other hand, some of the ice-crystals may collect cloud droplets by collision, and these may increase the terminal velocity relative to unrimed crystals by a sufficient amount to cause collisions. No detailed examination of these processes has as yet been made.

Accretion of water droplets

The collection of cloud droplets by collision with ice-crystals is an aerodynamic problem of some complexity and has as yet received no detailed consideration. If the analysis of Ranz and Wong (1952) is used, as discussed in chapter 4, and the ice-crystal treated as a disk, then a collection efficiency for water droplets can be derived. Such curves are shown in fig. 10.11. There is, however, no general agreement as to the correctness of these values, and Browne and Wexler (1953) derive collision efficiencies of 0·1 and 0·8 for the same situation.

The measurements of Nakaya and Terada (1934), who examined the small frozen droplets attached to ice-crystals, show that the collection efficiency is certainly finite for droplets from 15 to 45 μ dia. when the collecting crystal is a plate about 2 mm in diameter. This is consistent with the curves shown in fig. 10.11 but by no means establishes their correctness.

The growth of graupel pellets by accretion of a large number of frozen droplets on an ice-crystal or snowflake is even more complicated because of the complex conical structure often formed. A discussion of some of the properties of graupel pellets and of their growth has been given by Magono (1953).

CHAPTER 11

RAIN FROM SUB-FREEZING CLOUDS

INTRODUCTION

In chapter 2 we discussed in perspective the Wegener–Bergeron suggestion (Wegener, 1911; Bergeron, 1935) that most if not all raindrops reaching the ground had their origin as ice-crystals nucleated in supercooled regions high in the cloud-tops. In the present chapter we shall consider this mechanism in greater detail, examine the precipitation which can be produced in this way and relate it to the growth of precipitation by coalescence between water droplets. The extreme view is no longer held that all rain originates by the Bergeron process, but it is almost certainly the major mechanism for precipitation release in temperate climates, particularly in continental regions where coalescence is inhibited by the small droplet size typical of clouds in these areas.

Four aspects of the precipitation process require study: the supply and activity of ice-forming nuclei, the growth of ice-crystals by sublimation, collision and accretion processes, and phenomena connected with melting. We have already discussed all but the last of these in some detail, and it now remains to consider their interaction in typical cloud environments. It is scarcely necessary to say that no reliable complete calculations have been made, but the work which has been done allows a semi-quantitative general picture to be built up.

ICE-CRYSTALS IN CLOUDS

From the discussion in chapter 9 it is clear that the atmosphere normally contains a considerable number of active ice-forming nuclei, so that as a convective cloud rises above the freezing-level its summit will eventually contain numbers of ice-crystals. The concentration of these ice-crystals may typically be about 0·1/litre when the cloud-top temperature has fallen to $-20°$ C, fluctuations of an order of magnitude in this concentration occurring with time and place.

It is immaterial whether these nuclei act by sublimation or by freezing; in either case the resulting ice-crystals begin to grow rapidly by sublimation, taking on the habit characteristic of their growth environment.

Whilst the initial concentration of ice-particles present in the cloud is determined by the concentration of active ice-forming

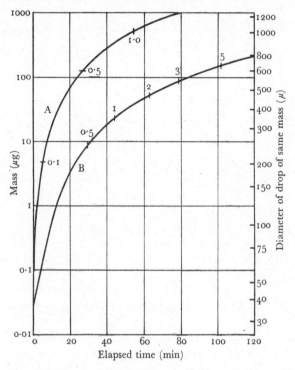

Fig. 11.1. Growth of ice-crystals by sublimation in a cloud at water saturation. A: plane dendritic crystal at $-15°$ C; B: hexagonal plate at $-5°$ C. Figures entered at intervals along curves give fall distances in kilometres (after Houghton, 1950).

nuclei in the atmosphere, there is some evidence that the population of ice-crystals may multiply through a fracture process. Thus large dendritic crystals are mechanically weak and portion of a branch may become detached and continue growth as a separate crystal. The importance of this multiplication process in typical cloud situations is not known.

Ice-crystals in Clouds

The growth process was discussed in chapter 10 and is described to a first approximation by (10.3), when molecular effects are neglected. Fig. 11.1, calculated by Houghton (1950), shows this growth on a convenient time-scale. The growth-rate is a maximum near $-15°$ C so that in any real cloud the sublimational growth-rate must be less than this because of the fall of the growing crystal through the cloud. A typical growth curve probably lies between the curves shown in the figure, tending toward the $-15°$ C curve in the early stages of growth and towards the $-5°$ C curve later. This allows for the crystal either falling to warmer temperatures or being carried by the updraught to cooler temperatures, the nucleation event having occurred in the range -15 to $-20°$ C.

These curves show that growing by sublimation alone an ice-crystal may reach a mass of a few tens of microgrammes, or assuming it to be a dendritic plate a diameter of about 2 mm, in a time of the order of 30 min. Such a crystal, from fig. 10.9, has a terminal velocity of about 30 cm/sec so that it is unable to fall against a vigorous updraught. In less active clouds such a crystal could reach cloud base and perhaps fall to the ground.

In more vigorously convective clouds such single crystals cannot fall against the updraught, and further growth by sublimation has little effect because of the virtual size independence of terminal velocity. In such situations there must be some form of aggregation before precipitation is produced.

Snowflakes

The growth of snowflakes by collision between ice-crystals is described by (10.14). In an extreme case where all the liquid water in a cloud has been converted to ice by a sublimation process, the average ice-content may amount to 1 g/m³. Assuming the mean density of a snowflake to be about 0·1 g/cm³ this gives $nV_2 = 10^{-5}$. Assuming further than the collection efficiency E is unity, (10.14) gives

$$V \approx [V_0^{\frac{1}{3}} + 6 \times 10^{-4} t]^3, \qquad (11.1)$$

where V_0 is the initial volume of the flake in cm³ and V the volume after a time t sec. A value of 100 cm/sec has been assumed for the difference in terminal velocities $u_1 - u_2$. A flake having an initial diameter of about a millimetre will, under these conditions, grow to a diameter of 1 cm in about 20 min.

The time required for the formation of a dry, unrimed snow-flake of moderate size is thus reasonable on the basis of the theory. Such growth, however, appears to require a relatively large water content in the form of ice-crystals if it is to proceed at a reasonable rate. Snowflakes formed in this way have terminal velocities of about 150 cm/sec and can fall as precipitation through moderate updraughts. The whole process of ice-crystal and snowflake growth occupies about 40 min of time, so that with an updraught of about 100 cm/sec a cloud depth of about 1500 m above the nucleation level should enable precipitation to develop.

Rimed structures

In a supercooled cloud containing water droplets as well as ice-crystals we must consider the growth of precipitation elements in the form of rimed structures—small water droplets frozen together into a porous mass of low density. Such accretion probably begins with the collection by an ice-crystal of one or more cloud droplets. Fig. 10.11 suggests that for an ice-crystal approaching 1 mm in diameter the collection efficiency is reasonably good for droplets more than about 10 μ dia. Referring back to fig. 5.3, which gave typical droplet spectra for various convective clouds, we see that in almost all cases the median droplet diameter is 10 μ or more, so that we may expect appreciable accretion.

Suppose we have an ice-crystal in the form of a flat plate 1 mm dia. falling at a terminal velocity, u, of 50 cm/sec through a cloud containing water droplets. We may reasonably take the liquid water content w to be about 0·5 g/m^3 and suppose the bulk of this liquid to be contained in droplets larger than 10 μ in diameter. From fig. 10.11 the ice-crystal will then collect this liquid water with an efficiency E of about 0·6. The rate of increase of mass of the ice-crystal due to accretion is then readily found to be

$$dM/dt = \pi R^2 E w u \approx 1\cdot 2 \times 10^{-7} \text{ g/sec}. \qquad (11.2)$$

Ice-crystals in Clouds

Taking the average density of graupel to be 0·125 g/cm³ as discussed in chapter 10, the initial rate of increase in crystal volume is of the order of 10^{-6} cm³/sec. This rate will, of course, increase as the particle grows in size, but it serves to provide a rough estimate of accretion rate. This suggests that the falling crystal could develop into a graupel pellet about 1 mm dia. in a time of the order of 10 min.

Any comparison between the growth of graupel particles in different cloud types depends in detail upon the behaviour of the collection efficiency E with droplet size, and reliable information is not yet available. Our preliminary results suggest, however, that within the usual range of droplet spectra, graupel formation will be appreciable provided that a fairly large fraction of cloud water remains in a supercooled liquid state.

Hail formation

Hail represents an extreme case of droplet accretion, but in addition there is a qualitative difference between a hailstone and a graupel or sleet pellet. Graupel is formed by a relatively slow accretion process in which each collected cloud droplet freezes almost immediately upon impact, and the resulting structure is porous and of low density. On the other hand, when the accretion rate is very large the latent heat released upon freezing of part of the accreted liquid is sufficient to prevent freezing of the remainder. The pellet thus formed is covered with a liquid skin and freezes relatively slowly to a dense, more or less transparent mass which we identify as a hailstone.

The conditions under which hail may be expected to form have been investigated in some detail by Ludlam (1950). Using the notation of chapter 6, his argument can be expressed as follows. The rate at which a spherical particle of radius r and temperature t loses heat while falling through air at a lower temperature t_0 is

$$H_1 = 4\pi r \kappa (t - t_0), \qquad (11.3)$$

where κ is the thermal conductivity of air. A ventilation factor should really be included in this equation and those that follow, but we shall omit this for simplicity. As well as losing heat by conduction, the particle loses heat through the evaporation of some of its material, the latent heat for this process being L_V. If $\rho(t)$ is the vapour density in equilibrium with the surface of the particle and ρ_0 that of the environment and D the diffusion coefficient for water vapour, then the rate of heat loss through this effect is

$$H_2 = 4\pi r D L_V [\rho(t) - \rho_0]. \qquad (11.4)$$

At the same time as the particle is losing heat by these mechanisms it is accreting water at temperature t_0 with an efficiency E close to unity. If we suppose that all the accreted water is frozen with latent heat L_F and raised to the equilibrium temperature t of the ice-particle then the rate of addition of heat is

$$H_3 = \pi r^2 E w u (L_F - t + t_0), \qquad (11.5)$$

where w is the liquid water content of the cloud, u is the terminal velocity of the particle and we have taken the specific heat to be approximately unity.

In equilibrium we must now have

$$H_1 + H_2 = H_3 \qquad (11.6)$$

and this equation determines the equilibrium temperature t of the growing particle. If the particle is to be completely frozen then t must be less than 0° C. The condition for the formation of a particle with a liquid skin is thus that t as determined by (11.6) be greater than 0° C. Fig. 11.2 shows Ludlam's calculations for a cloud with base at 900 mb and base temperature 10° C. For any given temperature and ice-particle radius the curves show the maximum water content w which can exist before a liquid surface-film develops on the falling particle. The broken line shows the adiabatic value of the water content in the cloud. Unmixed patches of cloud may attain this water content, but the average value will be much lower. The position of this curve is, of course, strongly dependent upon the assumed temperature at cloud base.

These curves show that if the liquid water content of the cloud is high, ice-particles may accumulate liquid water at temperatures below $-20°$ C if their size is large enough. At warmer temperatures hailstones may form on particles as small as a millimetre in radius. The existence of alternating layers of clear and opaque ice in hailstones reflects stages of growth where accretion takes place as liquid or as solid material and Ludlam points out that a particle may typically acquire three layers over an opaque core during its single rise, growth and fall through a dense cloud.

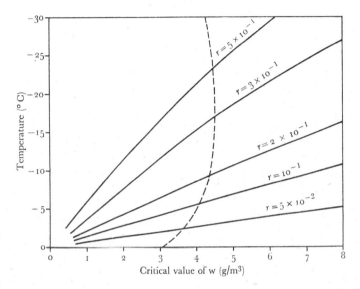

Fig. 11.2. Critical liquid water content w for coagulation elements in a cloud with base at 900 mb and base temperature 10° C. If for an ice-particle of radius r cm. at a given temperature the liquid water content exceeds that given by the curve, then liquid water is accumulated. The broken line gives the adiabatic value of the liquid water content in this cloud (Ludlam, 1950).

The overall growth process

From the discussion above it is clear that the evolution of an ice-crystal from its nucleation to the time it leaves the cloud base is a complex process. Initial growth is by sublimation, but at a diameter of the order of one millimetre accretion processes may become important. Whether accretion is of other crystals or of droplets depends in detail upon the structure of the cloud. The subject has been discussed to some extent by Ludlam (1952).

In chapter 7 and in the present chapter we have considered coalescence rain and rain originating as ice-crystals as completely separate processes. In a cloud whose top reaches above the freezing-level, however, both processes will be in competition and rain may be produced in either way. If the cloud structure is maritime and the water content high then coalescence is favoured and precipitation may commence before the cloud-top has risen sufficiently high to contain appreciable numbers of ice-crystals. On the other hand, if the microstructure is continental the cloud droplets will be predominantly below the size limit for collisions and few coalescences can occur. If the cloud is sufficiently vigorous, then its top may rise sufficiently high to contain active ice-nuclei in concentration $0 \cdot 1 - 1$/litre and these may lead to precipitation as discussed in the present chapter. Quantitative aspects of many of the processes involved are still incompletely understood, and clouds differ so much among themselves that no valid comparison of the rates of development of precipitation by these two mechanisms is yet possible. Radar techniques, however, have made possible studies of the general pattern of precipitation growth in sub-freezing clouds, and we shall now examine the information which can be obtained in this way.

RADAR STUDIES

As was pointed out in chapter 7, radar techniques provide a particularly valuable tool for the qualitative examination of precipitation elements in clouds. The extraction of reliable quantitative results from radar data, however, is a matter of extreme difficulty. The root of the difficulty lies in the fact that the intensity of the radar echo from a cloud of particles of diameter d and concentration n varies as nd^6, and this quantity is not simply related to any of the variables important in cloud physics.

When echoes from clouds whose tops are above freezing-level are being studied it must be remembered that the scattering particles may be either ice-particles or water droplets, and the scattering properties of these two sorts of particles are different. Ryde (1947) has examined theoretically the echoes produced by spherical particles of diameter d, dielectric constant ϵ and concentration n in the field of a radar beam of wavelength λ. His results have been re-expressed by Mason (1957, p. 311) to show that the echo-strength is proportional to

$$\frac{\pi^4}{4\lambda^4}\left(\frac{\epsilon-1}{\epsilon+2}\right)^2 \Sigma n d^6 \qquad (11.7)$$

provided $d \ll \lambda$. At a wavelength of a few centimetres, the dielectric constant of water is about 80, while that of ice is only about 3·3. Thus $(\epsilon-1)^2/(\epsilon+2)^2$ is about 0·93 for water-drops and 0·18 for ice-particles, representing a difference of 7 db in echo-intensity.

When radar echoes of precipitation within a cloud extending well above the freezing-level are examined on a range-height display, a pattern such as that shown in Plate V is often observed. The most noticeable characteristic is a horizontal bright band of high echo-intensity lying just below the freezing-level. Below this band normal precipitation echoes are observed, and there are fainter echoes from higher parts of the cloud.

The general explanation of this pattern is obvious. Ice-crystals growing in the upper parts of the cloud produce only weak echoes, though their concentration is considerable because of their low terminal velocities. On passing through the 0° C isotherm melting begins and the crystals become coated with a film of water. This greatly increases their radar reflectivity; indeed Labrum (1952) has shown that when 10% of a typical ice-crystal has melted its back-scattered intensity is equal to that of a water-drop of the same mass. With further melting the crystal collapses to a spherical droplet and its scattering efficiency decreases. At the same time the terminal velocity increases and the concentration of particles is reduced, giving a further decrease in echo-intensity to that of normal precipitation. The bright band is thus associated with the presence of partly melted ice-crystals within the cloud, and its presence is clear evidence for the formation of rain by the Bergeron process.

On some occasions bright bands lying higher than the melting-band have been observed (Bowen, 1951; Day, 1953). These bands usually originate at a height corresponding to a temperature of about $-16°$ C and then fall at 1–2 m/sec towards the melting-band, which is brightened on their arrival. Whilst these bands are certainly due to falling precipitation elements, the exact mechanism responsible for their formation is not clear.

The echoes produced by thunderstorms are often columnar in form, extending well above the freezing-level. The columnar echo may be accompanied by a melting-band, but this is not always the case (Byers and Braham, 1949; Day, 1953). An echo of this type appears to be associated with the strong updraughts present in the cloud. At high accretion rates, as we have seen, particles with wet surfaces are formed and these give intense echoes. The updraught is sufficiently strong to carry such particles high into the cloud and an intense echo-pattern often showing columnar cells results.

Radar observations are thus able to give general support to the mechanisms of precipitation growth which we have discussed. They give a broad picture of precipitation patterns within a cloud which is a valuable adjunct to study of precipitation development on a microphysical scale.

CHAPTER 12

ARTIFICIAL MODIFICATION OF CLOUDS

THE STABILITY OF CLOUDS

A CLOUD of water droplets of small size is, as we have seen, a system having a very large degree of stability. Any attempt to control the precipitation of clouds must, therefore, be concerned in an intimate way with this stability, with its origins and with the means whereby it can be upset. We have already covered most of these topics in the preceding chapters, but it may be as well to recapitulate briefly before we embark on a discussion of practical means of influencing cloud stability.

As well as differing in macroscopic quantities like size and updraught, clouds consisting entirely of water droplets differ from one another in two microscopic ways—in liquid water content and in droplet size spectrum. The liquid water content does not vary as much from cloud to cloud as would be expected from an adiabatic model, but seems to be limited to average values rather less than 1 g/m³ by a mixing process whose vigour is related to the existing liquid water content. Thus, while tenuous clouds of low water content certainly exist, those clouds most likely to produce precipitation usually have average water contents of 0·5–1·0 g/m³.

In droplet spectra, however, clouds differ very widely from one another, maritime clouds having small numbers of rather large droplets and continental clouds having large numbers of small droplets. Since stability in these clouds can only be upset by collision and coalescence processes between the droplets, and since collision efficiency is a very strong function of the radii of the droplets involved, details of the size spectrum are of paramount importance in determining the intrinsic stability of the cloud microstructure. If the cloud becomes unstable and precipitation develops, then it does so through the action of the small proportion of much larger than average droplets present in the spectrum, and it is the interaction of these droplets with average cloud droplets which must be considered.

Turning to clouds which penetrate above the freezing-level, the presence of ice-crystals must also be considered. The liquid water contents and droplet spectra met with are similar to those of warm clouds, with the exception that vigorous thunderclouds may have central liquid water contents approaching the adiabatic value. Such clouds, however, are so far from a state of stability that we may conveniently omit them from our present discussion.

The microstability of sub-freezing clouds is closely related to the concentration of ice-crystals in their upper parts, though the production of precipitation is also influenced by cloud geometry, liquid water content and droplet spectrum.

If then, one is to attempt to control the stability of clouds by modification of their microstructure, two important accessible parameters emerge—the droplet size spectrum and the concentration of ice-crystals present. The importance of these two quantities lies in the fact that it is only the one droplet in 10^6 which is significantly larger than its fellows or the one ice-crystal per litre of cloud which is important. The part of the microstructure important to stability is thus of a sufficiently small concentration that modifications can be contemplated without the expenditure of a prohibitive amount of energy. In the following sections we shall consider the possibility of such modifications and their effects on stability in greater detail.

WARM CLOUDS

In attempting to modify the microstructure of warm clouds, the only quantity with which we can be concerned is the size spectrum of the cloud droplets, and the only ways in which it seems possible to modify the spectrum are to add nuclei, to add droplets, or to increase the collision efficiency. We shall examine all of these possibilities.

We have seen that maritime clouds containing perhaps 50 droplets/cm^3 are unstable once they grow to depths of more than about 1000 m, so that little cause would be served by trying to enhance their instability. On the other hand, continental clouds and particularly drought-time clouds consisting of large numbers of small droplets may reach considerable depths without precipitating, and indeed the time required for natural

Warm Clouds

precipitation to develop may be so long that the cloud dissipates without having produced rain. It is upon such clouds that attention must be focused.

Salt seeding

Since it is clearly impractical to attempt to deplete the air of its nucleus content in order to decrease the droplet concentration, any modification must take the form of an addition to the nucleus population, and the only addition of value would be one which increases the number of specially large droplets. To accomplish this, rather large hygroscopic particles must be introduced into the air below cloud base in numbers sufficient to provide about one large drop per litre. Table 7.1 on page 174 suggests that nuclei producing droplets larger than about 40 μ radius might be adequate to produce precipitation from a typical moderately continental cloud, and the obvious choice for the seeding material is sodium chloride.

Here, however, one recalls our previous remarks that large sea-salt particles are usually found even in continental air-masses, and the concentration of particles greater than 10^{-10} g is typically about $0 \cdot 1 - 1$/litre, the higher values being found in the more extremely continental air-masses and in those strictly maritime in origin (Squires and Twomey, 1958). These particles are rather smaller than those in which we are interested, but the fact remains that considerable numbers of giant salt-nuclei are normally present in continental air-masses so that the proposed modification is a quantitative rather than a qualitative one, and is therefore more difficult to evaluate. It seems possible that a sufficient concentration of large salt-particles released below the base of a cloud of moderately continental character might stimulate precipitation if the vertical development of the cloud is several thousand metres and its lifetime an hour or more. Clouds of more extreme continental character are less susceptible to this treatment because of the large vertical development and long time required and because of the fact that such clouds generally possess a higher initial content of large salt-particles.

Several experiments have been performed to test this seeding technique. Davies (1954), in East Africa, sent up balloons carrying charges of finely ground salt which was dispersed explosively on reaching cloud level. A large proportion of the treated clouds were observed to rain after about 20–30 min. Augustin (1954), in Madagascar, dispersed salt-particles from an aircraft and claimed substantial rainfall to result, though much of it not until the next day. Fournier d'Albe and co-workers (1955) used blowers to disperse NaCl particles 10^{-9} to 10^{-8} g in mass from the ground in the Punjab and observed increased rainfall in the region downwind from the generators.

It is with the evaluation of the results of such trials, however, that we meet one of the crucial difficulties in the field of weather modification. The phenomena associated with clouds and weather vary in such a wide and, in the absence of complete information, capricious manner that elaborate statistical techniques of experimental design are required before reliable conclusions can be drawn. Most of the experiments which we shall describe in this chapter had no such careful design and relied mainly upon visual estimates of the results obtained when a particular technique was applied to a few chosen clouds. Whilst such methods often serve to show that some effect was achieved, the ultimate worth of a cloud modification project is measured by the increase in precipitation produced, and as we shall see later much careful planning is necessary if this quantity is to be determined.

Water-drop seeding

A more direct method of modifying the droplet spectrum, suggested by Ludlam (1951 b) and by Bowen (1952 a), is by the actual introduction of large droplets from some sort of water spray. By using suitably equipped aircraft it is possible to introduce the droplets at any level in the cloud, but as we shall see in a moment, the most suitable position is just above the cloud base.

Consider a droplet introduced at some arbitrary level into a convective cloud. If it is to give rise to a raindrop it must fulfill two related requirements—it must be large enough initially to grow to raindrop size within the life of the cloud, say 30–90 min, and it must be large enough to fall against the updraught before it reaches the cloud-top. For continental

PLATE V

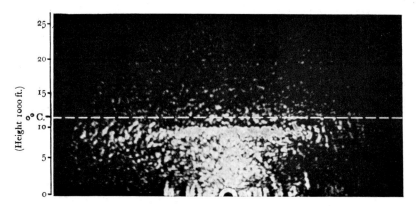

A melting-band echo as observed with a 10 cm radar system (Day, 1953).

PLATE VI

(a) A well developed cumulus cloud consisting entirely of water droplets. Note the sharp outline of the upper parts of the cloud.

(b) The same cloud about 30 min after seeding with silver iodide smoke. The upper part of the cloud is now glaciated, with ice-crystals blowing downwind in a typical anvil. Heavy rain can be seen falling from the cloud base. (C.S.I.R.O.)

clouds the first requirement suggests from table 7.1 that the droplet must exceed 30 μ in radius. The second requirement depends upon the level at which the droplets are introduced into the cloud, the cloud depth and its updraught. A 30 μ drop introduced into the base of a continental cloud with updraught of 1 m/sec will rise about 2400 m, so that only clouds thicker than this could be effectively seeded by drops of this size. If, on the other hand, seeding from the summit is contemplated, the drops must exceed 150 μ radius if they are to fall against the updraught.

In the cloud we have chosen, table 7.1 shows that the drop will have grown to 0·55 mm radius when it reaches cloud base. This corresponds to an increase of a factor 6×10^3 in mass for droplets injected at cloud base, but only a factor 50 for droplets released near the summit. It is, therefore, much more economical to release droplets near cloud base than near cloud-top level, and for this reason cloud-base injection was recommended by Ludlam and Bowen.

The multiplication factor given above is too small for practical utility but represents, of course, only a first approximation. The factor increases rapidly with updraught velocity and, for a speed of 2 m/sec, is about 10^5. There may be a further large increase if fragmentation processes become important, though this is only likely to be the case in clouds with very strong updraughts and great vertical development.

Bowen (1952a) carried out experiments to test the effectiveness of this seeding method. An aircraft fitted with a 66-gallon water tank and equipped with two horizontal spray bars each 2 m long flew through selected clouds about 300 m above cloud base. The spray gave droplets estimated to have a median radius of about 25 μ. In performing the experiments situations were chosen in which several similar non-precipitating clouds were visible, and the behaviour of the seeded cloud was compared with that of the others. In almost all cases the seeded cloud behaved in a manner different from the others and precipitation often developed in it. The initial trials of the method, whilst not yielding statistically significant results, thus supported in a general way the predictions of the theory.

More recently Braham, Battan and Byers (1957; also Braham and Battan, 1958) have described experiments in which water was released at the rate of 450 gallons per mile into the bases of cumulus clouds. In this experiment two similar clouds both showing no radar echo were chosen and one of the pair was seeded on a random basis. Varying results were obtained, but in many cases the treated cloud rained while the other remained stable. Statistical examination of the results showed that the likelihood of the effect being due to chance was less than 2 %, giving further confirmation that this seeding method produces positive results.

Electrical effects

Another method of modification of the cloud droplet spectrum which has attracted some interest involves electrical effects. In our discussion of collision efficiencies in chapter 6 we found that small droplets do not make collisions but move around each other with very small separations between their surfaces. Under these circumstances strong electric fields or charges on the droplets may sufficiently disturb the surfaces or the motion to cause collision and coalescence. Thus the presence of such fields or of opposite electric charges on colliding droplets may greatly enhance the collision rate within the cloud, building larger droplets and perhaps causing instability to develop.

Vonnegut and Moore (1958) have made some progress in this direction by investigating the way in which space charge can be introduced into the lower atmosphere. They supported a fine wire 0·025 cm dia. and about 7 km long on masts about 10 m above ground-level, and by applying a potential of about 30 kV produced a corona along its length. The ions produced rapidly became attached to Aitken nuclei in the airstream and their mobility was reduced to such an extent that they were carried downwind in detectable quantities to distances as great as 8 km.

In this experiment D.C. potential was used so that the ions were all of one sign, and their presence was detected by measuring anomalies in the normal atmospheric potential gradient. If these ions were carried into clouds they would tend to charge drops similarly with consequent inhibition of collisions. Vonnegut has proposed a different role for the ions in developing cumuli, but this work is as yet controversial.

Telford (unpublished) has carried this technique a stage

further by trailing a long wire anchored to a streamlined bomb below an aircraft flying through cloud. The wire is fed with high-voltage alternating current so that bursts of positive and negative space charge separated by about a metre in space normal to the wire are produced. These ions should become attached to cloud droplets and, as a result of the opposite charges between neighbouring regions, coalescence should be accelerated.

Though these techniques may prove useful in stimulating precipitation from warm clouds, no positive results have yet been reported.

SUB-FREEZING CLOUDS

When attention is turned to sub-freezing clouds a new possibility immediately becomes apparent, associated with the Bergeron mechanism of rain production initiated by ice-crystals. Since a sub-freezing cloud becomes unstable when ice-crystals are present in concentrations of about one per litre, it is simply necessary to devise a means of producing them at this concentration in clouds not normally cold enough to contain natural ice-crystals, in order to markedly effect stability.

Two methods have been evolved to achieve this effect. The first, discovered by Schaefer (1946), consists of dropping into a supercooled cloud pellets of solid CO_2 (dry ice, $-78°$ C) which chill the droplets formed by condensation to well below the temperature for homogeneous freezing and so produce a cloud of tiny ice-crystals in the wake of the pellets. Schaefer found that in fact pellets of any material cooled to a temperature below $-40°$ C are active in this way, but dry ice is most convenient from a practical point of view.

The second method, discussed as early as 1938 by Findeisen (1938), involves the addition to the air-mass in which clouds are forming of quantities of ice-forming nuclei which will be active at temperatures markedly warmer than those characteristic of natural nuclei. This suggestion was not taken up practically until Vonnegut (1947) discovered the great efficiency of silver iodide as an ice-crystal nucleus, but since then it has become the major technique used in cloud modification studies.

Dry-ice seeding

After Schaefer's discovery of the ice-producing effects of dry ice, Langmuir (1947) calculated that more than 10^{17} ice-crystals could be produced by a one-gram pellet falling through supercooled cloud, and almost immediately field trials showed that glaciation of clouds could be produced in this way. Some of the clouds were converted entirely to ice in these experiments in which broken pellets of dry ice were dropped into them from aircraft, and in some cases snow was observed to fall from the cloud base.

In the same year Kraus and Squires (1947) in Australia seeded cumulus clouds with dry ice and produced rain which actually reached the ground. After this the number of experiments becomes too numerous to consider in detail, a typical series being those described by Smith (1949). Useful conclusions can be drawn from the results as summarized by Bowen (1952 b). Considering the desired result to be the production of precipitation, suitable clouds are found to have summit temperatures lying between about $-6°$ C and $-15°$ C. If the cloud is warmer than this the growth-rate of ice-crystals is too slow to effectively initiate precipitation, whilst if the cloud-top is colder than $-15°$ C it will probably contain already enough natural ice-crystals to cause instability. In addition the total depth of cloud must be sufficiently great that the snowflakes or water-drops emerging from the cloud base are big enough to fall to the ground before evaporating. The experiments suggest that this will happen if the cloud depth exceeds half the terrain clearance of its base. The minimum time for the production of precipitation was found to be about 10 min, the time increasing with cloud depth. This value appears reasonable in the light of our discussion of the Bergeron process in chapter 11.

Many factors may intervene to stop the development of precipitation in seeded clouds. If the cloud is of small vertical development it may be converted entirely to ice-crystals before any of these are sufficiently large to act as precipitation elements. It may then dissipate as these crystals fall slowly into the dry air below cloud base. In towering clouds, on the other hand, wind shear may often blow the cloud-tops away from their bases so that ice-crystals fall into dry clear air instead of growing in their descent through the cloud.

An additional point emerging from the experiments is that the effective number of ice-crystals produced by dry-ice seeding must be many orders of magnitude less than the value of 10^{17} per g calculated by Langmuir. Had this value been approached then complete and stable glaciation of most seeded clouds would have resulted, since seeding rates of several pounds per mile were typical, the amount of dry ice used in seeding a single cloud amounting to perhaps 100 lb.

Laboratory experiments by aufm Kampe and Weickmann (Weickmann, 1957a) suggest as a more realistic value 3×10^{10} ice-crystals per gram of dry ice. This implies that any really large-scale operation of dry-ice seeding must of necessity involve major problems in the dispersion of dry-ice through clouds in the seeded area. As we shall see this difficulty has never really become important since dry-ice seeding has given way to vastly more efficient techniques making use of silver iodide smoke.

Silver iodide seeding

The technique of seeding clouds with silver iodide particles was discovered, as was the dry-ice method, at the General Electric Research Laboratory in Schenectady during the course of research on Project Cirrus. Vonnegut (1947) found that when silver iodide was formed into a smoke by intense heating the particles produced had a nucleation threshold near $-4°$ C, while at $-15°$ C the yield of active nuclei was about 10^{15} per gram of silver iodide burnt.

Vonnegut developed several convenient methods of producing the smoke, some of which we shall discuss in the next chapter, and found similar high efficiencies in all cases. Here, then, was a nucleating agent sufficiently active that very small amounts could modify the ice-nucleus content of the atmosphere by several orders of magnitude. Field trials immediately following this discovery showed some signs of success, though the results achieved were not as spectacular as had been obtained using dry ice. In more recent experiments, however, the technique has been amply demonstrated to be an effective one for cloud modification.

Silver iodide acts in a cloud in essentially the same manner in which the natural freezing-nuclei in the atmosphere act. It is, therefore, only necessary to distribute silver iodide smoke through the air-mass in which clouds are forming in order for it to influence their development and stability. This is a distinct contrast to the case of dry-ice seeding in which the upper parts of well-developed clouds must be seeded in a precise manner. Various means of distributing the smoke from burners on the ground or on aircraft have been developed as we shall discuss later. In cases where aircraft burners are used the seeded clouds are much more readily determined and their development studied. Warner and Twomey (1956) report a time-lag of about 20 min to be typical between the introduction of smoke at cloud base and the appearance of precipitation. This is of the order of magnitude expected from our discussion of chapter 11 and is about twice the time required for the development of rain after dry-ice seeding of cloud-tops, as might be expected. As was the case with dry-ice seeding, cloud summit temperatures below about $-6°$ C and cloud depths in excess of about 1300 m are required if there is to be a good probability of rain being produced. Plate VI shows the results of seeding a well-developed supercooled cumulus cloud with silver iodide smoke dispensed from an aircraft. The second photograph, taken after a lapse of 30 min, shows that the cloud has become extensively glaciated in its upper parts, and heavy rain can be seen falling from its base.

It is, however, very difficult to obtain good quantitative information on the overall effect of cloud seeding with silver iodide by studies of individual clouds. Braham (1960), on the basis of radar studies of a large number of clouds in Arizona, has discussed this point further. We have just seen that if a suitable cloud is seeded at its base with silver iodide, it takes about 20 min for precipitation to develop. Braham found that, at any rate for the region studied, only about 30 % of clouds judged suitable for seeding would persist for the required additional 20 min, and of those which did persist almost half rained by natural processes in any case. The statistical effect to be produced by seeding is thus small, and under most conditions the number of experiments required to give a statistically significant quantitative result is prohibitively large. For this reason, as well as because it is ultimately the rainfall reaching the ground which is important, most quantitative cloud-seeding experiments are based on large-scale programmes of the type we shall discuss in chapter 14.

Very little attention has been given to quantitative aspects of seeding conditions in clouds, as is understandable since most of our knowledge of the subject has as yet scarcely passed the qualitative stage. Since, however, it is evident that a great excess of active ice-forming nuclei may convert the whole of the upper part of the cloud into which they are introduced into ice-particles, thus inhibiting the precipitation mechanism, it is of importance to have some estimate of conditions under which this may occur.

A discussion of this topic has been given by Twomey (1958). Assuming an exponential spectrum of atmospheric freezing-nuclei and an exponential AgI spectrum saturating at $-20°$ C (see chapter 13), he calculated the rate of nucleation and growth of ice-crystals in several model clouds. A cloud may be considered to be adequately seeded, from the Bergeron point of view, when it contains about one ice-crystal per litre, and the limit of over-seeding occurs at any given level when the cloud has been converted completely to ice-crystals.

Assuming an aircraft to release AgI nuclei at a rate of 10^{14} per sec, active at $-20°$ C, into the base of the cloud, this being the output of a typical burner, Twomey calculated the amount of ice present at various levels in model clouds representing large, medium and small cumuli. The results of this calculation are shown in fig. 12.1. The full lines show the rate of glaciation due to natural nuclei and the broken lines that due to the seeding. For a large and vigorous cloud complete glaciation only occurs at about the $-35°$ C level, some 7 km above the freezing-level, and is little affected by the seeding, though of course the ice content at lower levels is increased, in this case by about an order of magnitude. In a small, weakly convective cloud, on the other hand, seeding at this rate produces complete glaciation at about $-10°$ C when the supercooled part of the cloud is only about 2 km thick; if only natural nuclei were present this would not have occurred until the $-25°$ C level. This difference is due partly to the smaller area of the base of the small cloud and to its smaller updraught giving a greater growth time for the ice-crystals.

Twomey's study shows that the minimum seeding-requirement of one nucleus per litre is always easily met by this type of seeding-operation. In medium or large clouds overseeding resulting in complete cloud-top glaciation is unlikely to be produced, but this may be possible in small clouds having areas less than 5 km² and updraughts of about 20 cm/sec.

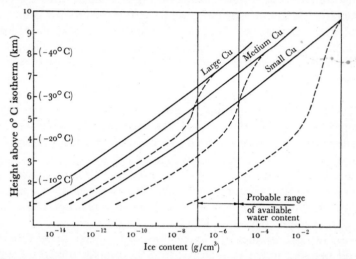

Fig. 12.1. Computed variation of ice-content with height for natural freezing-nuclei alone (solid curves) and with silver iodide nuclei added (broken curves). Natural nucleus spectrum provides 1 nucleus/litre active at $-20°$ C; silver iodide generator provides 10^{14} nuclei/sec active at $-20°$ C (Twomey, 1958).

In large-scale seeding operations it is not usual to seed individual clouds but rather to release smoke from aircraft or ground burners distributed along the upwind border of the target area. In this case the smoke is greatly diluted before it reaches the target clouds and overseeding is less likely to occur. The optimum seeding-rates in this case, however, depend greatly upon details of the experimental design and we shall postpone their discussion to chapter 14.

CLOUD DISSIPATION

If a supercooled cloud is seeded with dry ice or silver iodide, then the results may be viewed from two angles. The ice-crystals produced grow at the expense of the cloud droplets and fall with increasing rapidity, emerging from the cloud base as snowflakes or raindrops. This is the view we have so far taken, which accords importance to the precipitation elements. If instead we consider primarily the cloud droplets, then the effect of seeding is to deplete their number and, as the precipitation falls, to decrease the liquid water content of the cloud.

This latter aspect of the results of seeding is also of importance since the dissipation of clouds in the vicinity of airfields may aid materially in traffic flow. Conditions where such dissipation is necessary are mainly those in which continuous cloud-deck extends over a large area, so that seeding of such layer clouds must be considered rather than the cumulus clouds typical of rainmaking experiments.

Some seeding of clouds of this type was carried out in the early stages of Project Cirrus, and more recently Weickmann (1957a) has described experiments explicitly aimed at cloud dissipation. If about twenty ice-crystals per litre are generated in a typical cloud then each will grow to a mass of about 0·035 mg before the liquid water disappears. Such crystals have terminal velocities near 50 cm/sec and would thus fall about 1500 m in half an hour. A seeding-rate producing about this concentration of ice-crystals might therefore be expected to dissipate a supercooled cloud-deck some 1000 m in depth in reasonable time. A much greater seeding-rate would yield crystals too small to fall out in reasonable time and lead to a rather stable ice-cloud; a seeding-rate much lower would produce insufficient quantities of ice-crystals to greatly decrease the liquid water present in the cloud.

Either dry ice or silver iodide may be used as the seeding-agent, but since cloud temperatures are often warmer than $-10°$ C dry ice tends to be preferred because of its greater efficiency at these temperatures, and the ease with which comparatively large quantities may be dispersed. Since the cloud areas involved are usually not large, few problems arise in handling the quantities of material required.

Weickmann and his collaborators dispersed dry ice at the rate of about 10 lb per mile of aircraft track into supercooled layer clouds 300–1000 m in depth. This seeding produced a clear path about 2 miles wide after a time of about 30 min, longer times being required for thicker clouds.

Whilst no widespread use has yet been made of this technique, it shows promise of being sufficiently effective and economical to warrant its use in situations where supercooled stratus is a major visibility problem. The restriction to supercooled clouds is, however, often a limiting factor in the usefulness of the method. For warm clouds entirely different approaches must be used.

One such novel technique has been examined by von Straten and others (1958) at the U.S. Naval Research Laboratory. Arguing that the development of a cloud could be considerably influenced by its heat balance and in particular by its absorption of solar radiation, they sought to modify this balance by introducing large numbers of small black particles into the cloud. For this purpose they used carbon black having a free surface of the order of 10^4 m^2 per lb and seeded clouds with quantities amounting to a few pounds per cloud.

The preliminary results were said to indicate that such seeding of cumulus clouds caused them to dissipate in times of about 10 min, while seeding clear air in unstable situations resulted in the formation of clouds. As yet no careful experiments have been performed so that it is not possible to assess the effectiveness of the method nor the results which might occur in different situations. It remains an interesting new approach worthy of further study.

HAIL AND LIGHTNING SUPPRESSION

As well as modifications of cloud structure simply designed to produce precipitation, attempts have also been made to seed clouds in an effort to suppress the formation of hail and to reduce the incidence of lightning strikes.

Hail develops, as we have seen, when the rate at which an ice-particle grows by collision when falling through a supercooled cloud becomes so great that some of the water is accreted in liquid form. Hail is therefore associated with supercooled clouds of very high liquid water content, which will usually be compact clouds of great vertical development. Whilst it has been shown that the natural freezing-nucleus concentration of air in which hailstorms are forming is not significantly lower than normal (Soulage, 1958b), it has been suggested that the introduction of large numbers of artificial nuclei into incipient hailclouds might cause sufficient glaciation at warm enough temperatures to reduce the rate of liquid water accretion on precipitation elements.

To this end programmes have been designed in Europe and America to study the effects of silver iodide seeding on hail suppression. Since hail is even more variable than rainfall, and a negative result is much more difficult to detect than a positive one in small-sample experiments, any assessment of the effectiveness of the method must await the design and performance of large-scale, long-term statistical experiments. A survey of experiments which have been conducted up to the present has been given by Frank (1958).

Any attempts to reduce the incidence of lightning strikes should best be based on a detailed knowledge of the mechanism of electrification of clouds, and unfortunately such knowledge is not at present available. Partly because silver iodide seeding is the most readily available means of modifying cloud structure, and based upon the thought that glaciation may prevent the great vertical development characteristic of thunderclouds, several programmes of cloud seeding and lightning stroke evaluation have been initiated in the U.S. Of these perhaps the best known is Project Skyfire (Barrows et al. 1958).

Although the results published for Project Skyfire contained no conclusions on the effectiveness of silver iodide seeding in reducing lightning strikes, other evidence now suggests that such seeding may in fact greatly increase such strikes. This conclusion appears reasonable in the light of increasing acceptance of the view that the major part of cloud electrification is caused by interaction between ice crystals and supercooled water droplets. The seeding increases the number of ice-crystals in the cloud and hence the rate of charge separation, leading to a result directly contrary to the aims of the programme. Such conclusions are, however, as yet only tentative and much work remains to be done.

DISCUSSION

Our understanding of the structure of clouds and of the factors influencing their colloidal stability leads us to believe that it may be possible to modify the microstructure in such a way as to significantly alter the development of the cloud. Experiments have shown that such modifications are in fact feasible in practice, and may be such as to lead to precipitation.

Of the techniques which have been used, those applying to warm clouds are the least developed and it is not yet clear whether in fact they are sufficiently efficient to justify any large-scale experiments. Indeed, relying as they must upon coalescence for drop growth it seems that there may be only a small range of cloud microstructure for which any useful result may be achieved.

The seeding of sub-freezing clouds, on the other hand, has from the beginning been a field where spectacular results and over-optimistic calculations have abounded. Though many of the claims have been found to be extravagant, the simplicity and economy of silver iodide as a cloud-modifying agent remains most attractive.

The ultimate success of any cloud-seeding programme is, however, measured by the increased rainfall produced and this is a quantity not easily determined. Experiments in which even quite large numbers of individual clouds are seeded give very little information because of the enormous variety of cloud-development patterns, and the results cannot usually be expressed in terms of rainfall figures. It is necessary, therefore, to perform seeding programmes extensive both in time and area and so designed that statistically significant conclusions can be drawn, if the real effect of cloud seeding is to be found. After a brief discussion in the next chapter of the production of silver iodide smokes and of some of their characteristics, we shall therefore make an examination of such large-scale programmes and see what conclusions can be drawn from the results which have been obtained so far.

CHAPTER 13

PRODUCTION AND PROPERTIES OF SILVER IODIDE NUCLEI

INTRODUCTION

THE ice-nucleating properties of silver iodide were discovered by Vonnegut (1947), and since its initial use it has remained the pre-eminent nucleating agent in this field. Other substances such as lead iodide (Fukuta, 1958) and cuprous oxide (Pruppacher and Sänger, 1955a) have occasionally been proposed as more efficient nucleating agents, but their claims have not been generally substantiated.

In this chapter, therefore, we shall turn our attention to a fairly detailed consideration of the production of silver iodide smokes, their physical properties and nucleation behaviour. Whilst the interpretation of the experimental results in this field is in some cases still controversial, it is possible to build up a consistent theoretical description which serves to unify the discussion even though some of its details may require modification in the light of future developments. In particular, further work is required before we know whether silver iodide smoke nucleates supercooled clouds by direct sublimation or by collision with droplets and subsequent freezing.

Data on silver iodide

Silver iodide is generally considered to exist in three solid forms (Sidgwick, 1950). Their properties are briefly as follows:

α-AgI is the stable form between 146° C and the melting-point, 555° C. It is dark brown in colour and is cubic in structure, the iodine atoms occupying a body-centred cubic lattice with separations of 2·18 Å. The silver atoms are not uniquely positioned but are distributed statistically over the thirty largest spaces in the iodine lattice. On cooling below 146° C the β-modification is produced spontaneously.

β-AgI is the stable form between 137° C and 146° C, though it commonly exists in a metastable state below this range. It is greenish-yellow in colour and hexagonal in form. The atoms form an almost exact wurtzite structure at low temperatures ($-180°$ C), but at ordinary temperatures there is probably a random distribution of silver atoms in four positions which surround tetrahedrally the ideal wurtzite positions (Helmholtz, 1953).

γ-AgI, which has a cubic zincblende structure, is generally thought to be the stable form below $137°$ C. However, it does not appear to have been prepared in bulk, and evidence for its existence depends upon detection of its presence along with the hexagonal form by X-ray diffraction methods. The only difference between β- and γ-forms is that the β-form has a layer structure ABABAB..., while the γ-form is stacked in order ABCABC.... From the X-ray viewpoint at least then, a small concentration of γ in β is indistinguishable from a density of stacking faults. We shall not concern ourselves further with this question except to point out the geometrical identity between {0001} faces in the β-phase and {111} faces in the γ-phase. The lattice constants of the three forms are shown in table 13.1.

Table 13.1. *Crytallographic data for AgI*

	Temperature range (° C)	a (Å)	c (Å)
α-AgI	146–555	5·034	—
β-AgI	137–146	4·58	7·49
γ-AgI	< 137	6·47	—
Ice	—	4·52	7·37

Only a few other points need be noted about silver iodide, and these refer to the β-form or to the $\beta\gamma$ mixture found at ordinary temperatures. Silver iodide is relatively dense (5·67 g/cm^3) and is ordinarily prepared as a fine powder, though crystals several millimetres across can be grown by special methods (Helmholtz, 1953; Hoshino, 1957). It is almost insoluble (3×10^{-7} g in 100 ml.) in cold water and only about ten times more soluble in hot water. It is, however, readily soluble in liquid NH_3 and in aqueous solutions of KCN, $Na_2S_2O_3$ or other iodides such as NaI, which yield soluble complexes, or in acetone solutions of these iodides.

Introduction

Because there is some tendency for AgI to decompose, yielding elemental iodine, it has a very corrosive action on metal fittings. This decomposition is accelerated by heat and light and its effects must be taken into account in the design of equipment handling AgI solutions.

PRODUCTION OF AgI DISPERSIONS

In cloud-seeding operations employing silver iodide it is desired to produce as many active nuclei as possible from the material consumed, and this implies that a very fine state of subdivision and dispersion must be achieved. As we shall see later, there are limits to the degree of subdivision desirable, but in practice this is a secondary consideration. In this section, then, we shall consider some of the means which have been developed for the production of a very fine dispersion or smoke of silver iodide particles.

Laboratory methods

For the preparation of small quantities of silver iodide dispersions where efficiency is not a major consideration the simplest procedure is to heat a quantity of the pure salt in a stream of air. At temperatures above about 550° C the silver iodide melts and decomposes, and the vapour when quenched in the air-current yields large numbers of particles less than a micron in diameter. Whilst these particles are nominally pure silver iodide, the fact that iodine is more volatile than silver leads to variations in stoichiometry as the evaporation proceeds. Because of the amount of heat required and relatively small output rate this method has not been adapted for large-scale use.

Another method explored by Vonnegut (Schaefer, 1953) was to strike an arc between silver electrodes in an atmosphere containing iodine vapour. Again, whilst this method gave nominally pure particles of silver iodide, it was unsuitable for large-scale use because of electrical power requirements.

Ammonia solution generator

Among methods used for large-scale dispersion of silver iodide, the only one which has an output of nominally pure silver iodide particles is the ammonia solution generator developed by Tominaga and Kinumaki (1954). Silver iodide is extraordinarily soluble in liquid ammonia, 100 g of NH_3 at 0° C being capable of dissolving as much as 530 g of AgI (the solubility has fallen to about half this value at 25° C). Because of the very low boiling point of liquid ammonia ($-33°$ C at normal pressure) a feasible dispersion method is to simply spray an AgI-NH_3 solution into air at ordinary temperatures. Whilst this appears to give a good dispersion, much finer particles can be produced by spraying the solution into an electric furnace held at about 600° C and igniting the vapour. The bulk of the silver iodide is dispersed as particles a few hundred Ångstroms in diameter (though the average particle-size on a number basis is less than 100 Å).

Solid fuel burners

As a development of the simple laboratory method of vaporizing silver iodide, the idea was conceived of incorporating silver iodide in some material which would itself act as a fuel to provide the necessary heat of vaporization. Among solid materials, charcoal was found to be one of the most suitable and has been used quite widely in the field (Vonnegut, 1950, 1951).

Silver iodide is dissolved in acetone in association with about a quarter part by weight of sodium iodide and the resulting solution diluted to about 2% AgI by weight. Charcoal pellets are soaked in this solution and when dry contain about 1% AgI by weight. This impregnated charcoal is then burnt in a stream of air and gives a copious supply of nuclei. The particles are, however, no longer nominally pure silver iodide, but contain a large admixture of sodium iodide. We shall postpone discussion of the effect of this impurity to a later section.

Another arrangement sometimes used consists of a thick, impregnated cord, which, not being sufficiently combustible to burn by itself, is slowly fed into a hydrogen or propane flame. The chief advantage of this method over that of burning charcoal is that the output of the generator remains relatively constant over quite long periods of time if an automatic feed is used for the impregnated cord.

Solution burners

All these methods, however, have been largely superseded by a type of generator burning directly an acetone solution of silver iodide, which was once again suggested by the Project Cirrus group (Vonnegut, 1950, 1951). In the original model a solution of 10% AgI in a NaI-acetone base was fed into an ordinary spray gun, hydrogen being used instead of compressed air. The spray was ignited and burnt strongly, producing a fine smoke of silver iodide, again contaminated with sodium iodide. A variant of this system, designed to produce nuclei at a high rate, injected the acetone solution along with compressed air into the flame of a large kerosene burner.

Many burners based on this principle have since been developed, one of the most efficient being that made for Project Skyfire by Fuquay and Wells (1957). As shown in fig. 13.1 it consists of a flame-chamber about 12 cm in height provided with ventilating holes and an annular exit for the smoke. Silver iodide in NaI-acetone solution is drawn from a reservoir through a modified hypodermic needle and nebulized by a jet of propane gas which carries it into the flame-chamber which operates at about 1100° C. The plate at the top of the chamber serves to contain the flame and the smoke is quickly quenched as it leaves the annular exit. The burner is designed for fixed operation on the ground.

A different approach was followed by Warren and Nesbitt (1955), who set out to design a burner for use on aircraft. They argued that since the acetone solution is inflammable no other fuel should be necessary, and designed a compact burner on this principle. The construction is obvious from fig. 13.2. The burner is mounted beneath the aircraft wing and a suitable air-flow enters the combustion chamber which is about 8 cm in diameter. The acetone solution is sprayed through a swirl-type atomizer into the combustion chamber, where it is ignited initially by a spark-plug. A thermocouple serves to give a remote indication that the burner is functioning.

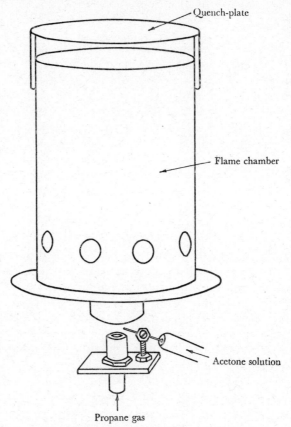

Fig. 13.1. The Project Skyfire silver iodide smoke generator (after Fuquay and Wells, 1957).

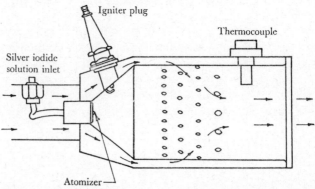

Fig. 13.2. The airborne acetone burner of Warren and Nesbitt (1955).

Chemical composition of smokes

Since the nucleation properties of a silver iodide particle may depend critically upon any impurities present, it is clearly of great importance to know as precisely as possible the chemical composition and physical form of the particles with which we are dealing. Surprisingly little work has been published on this important topic.

Manson (1955) has examined the smoke formed by evaporation of pure silver iodide at temperatures from 650 to 1000° C. He found the particles to be pure AgI in a mixture of the hexagonal β- and cubic γ-forms. In smoke produced at 650° C the β-AgI made up 73 % of the sample, while for evaporation at 800–1000° C the sample consisted of 95 % β-AgI. Lisgarten (Mason and Hallett, 1956), using electron diffraction, found in addition diffuse rings indicating the presence of some sort of amorphous structure in the particles.

Analysis of samples from charcoal or solution burners, on the other hand, may be expected to yield much more complex results. Naitô and Sugawara (1954) collected smoke from a generator burning AgI-KI-acetone solution in gasoline and subjected the samples to chemical analysis. They found undecomposed AgI amounting to between 40 % and 90 % of the amount expected together with metallic silver, free iodine, KI and carbon. The proportions varied considerably from one sample to another.

Lisgarten (Mason and Hallett, 1956) analysed smoke produced by vaporization of AgI-KI-acetone and AgI-NaI-acetone solutions, by electron diffraction. In the smoke from the AgI-KI solution he found no trace of lines associated with silver iodide and concluded that some form of mixed crystal had formed. The AgI-NaI smoke showed little or no sign of the pattern expected from hexagonal AgI, but gave lines which could be interpreted either as NaI or cubic AgI.

Unfortunately, no further work appears to have been published on the properties of these mixed smokes and the details of their action as ice-nuclei are still matters for speculation.

Size-distribution of smokes

Determination of the size-distribution of particles in AgI smokes is usually made by means of an electron microscope, the smoke particles being caught by some form of thermal or electrostatic precipitator. Several studies of this type have been made and, as might be expected, the distribution of particle sizes is found to depend considerably upon the method of preparation.

As with most dispersion processes, the size-distribution is, to a first approximation, log-normal (see fig. 4.1). The peak of the number distribution usually occurs at diameters of a few hundred Ångstroms and tends towards smaller values for higher generation temperatures. The peak in the volume-distribution curve occurs, of course, for rather larger diameters, so that most of the silver iodide is contained in particles 500–1000 Å in diameter.

Investigations of this sort have been made by Smith and Heffernan (1954), Birstein (1955), Sano and Fukuta (1956a) and Sano *et al.* (1956). Electron micrographs of the silver iodide particles often show them to have an amorphous, roughly spherical appearance. This may be due to the fact that they are probably formed first as molten droplets which later solidify, or may be evidence of a high degree of aggregation, particularly in the larger particles.

NUCLEATION BEHAVIOUR OF AgI SMOKES

Whilst many measurements of nucleation threshold for silver iodide smokes have been made, agreeing in the main to a threshold near $-4°$ C, very few complete activity curves have been determined. Those which are available apply, in the main, to the outputs of cloud-seeding generators burning either impregnated fuel, or silver iodide solution. The most convenient method of assessing burner output is to determine the number of nuclei active at any temperature produced by the burner on consumption of 1 g of silver iodide. When this quantity is known for all temperatures in the range 0 to $-30°$ C, then for practical purposes the output is specified. This leaves out of account of course the time factor which is important in practice since it determines the seeding rate. This is, however, a separate consideration which can always be met if necessary by parallel operation of several burners.

Adopting then the specification of smoke activity outlined above we turn to the results of measurements. Ignoring for the moment the broken curves, figs. 13.3 and 13.4 show the activity

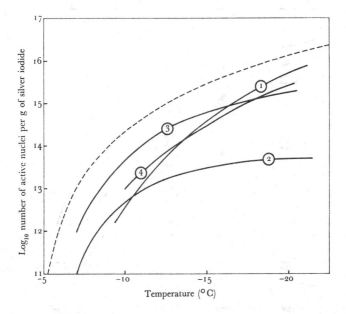

Fig. 13.3. The activity spectra of the smokes produced by various burners: 1, hydrogen burner (Vonnegut, 1949b); 2, kerosene burner; and 3, hydrogen burner (Smith and Heffernan, 1954); 4, charcoal burner (Soulage, 1955). The broken line indicates the maximum output possible at any given temperature if the smoke particles are assumed to act as sublimation nuclei (Fletcher, 1959c).

curves for several burner types in common use. All these curves are similar in that they show an initially more or less exponential rise in activity as the temperature falls from $-5°$ C towards $-10°$ C. At lower temperatures the activity rises more slowly and increases relatively little below $-15°$ C. The few measurements which have been made on pure AgI smokes suggest a very similar behaviour.

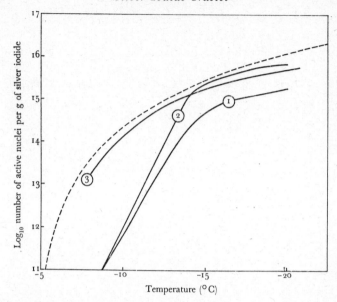

Fig. 13.4. The activity spectra of smokes produced by various burners: 1, MRI propane flame; 2, MRI oxy-propane flame; 3, 'Skyfire' solution burner (Fuquay and Wells, 1957). The broken line indicates the maximum output possible at any given temperature if the smoke particles are assumed to act as sublimation nuclei (Fletcher, 1959c).

THEORETICAL DISCUSSION

In this section we shall examine the nucleation behaviour of silver iodide smokes on the basis of the theory developed in earlier chapters. As we pointed out earlier, much of this theory is relatively new and not yet well established, and modifications may be necessary as our knowledge expands.

Pure silver iodide

Pure silver iodide is a rather hydrophobic substance, as can immediately be seen from the behaviour of millimetre-sized crystals placed in a shallow dish of water. These can be picked up by the meniscus, and float, suspended from the surface. This leads us to expect that silver iodide should be a rather poor condensation nucleus, and Fletcher (1959a) found a supersaturation of about 2·5 % to be necessary for condensation to occur on 1 μ particles. A direct measurement of contact angle

Theoretical Discussion

by Head and Sutherland (Fletcher, 1959a) yielded a value of $10° \pm 2°$ which is in agreement with the experimental condensation threshold. Fig. 13.5, drawn from the theory of chapter 3 for small spherical particles, shows the expected behaviour of AgI particles as condensation nuclei, based on these findings.

Fig. 13.5 shows that if pure silver iodide particles are introduced into the atmosphere no condensation can occur on them

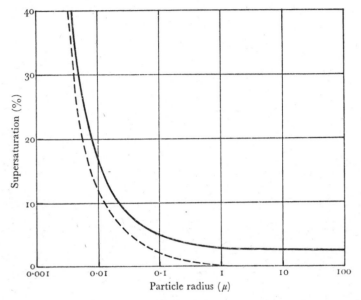

Fig. 13.5. The supersaturation at which a silver iodide particle of given size will nucleate a water droplet by condensation in 1 sec. The broken curve shows the behaviour of an ideal wettable particle (contact angle zero) (Fletcher, 1959a).

until a supersaturation of at least 3 %, and considerably more for particles smaller than $0.1\,\mu$, is achieved. These supersaturations should be compared with the values of a few tenths of a percent typical of conditions inside growing clouds.

Thus, except for the possibility of a collection effect, in which aerosol particles are captured by growing droplets, as discussed by Facy (1957), the particles will remain unwet, and any ice-formation must proceed by way of a sublimation process. The activity curve of AgI smoke should thus be one of a set of sublimation curves such as are shown in fig. 8.5. Since the values of m and ε for ice on silver iodide are not known the curve cannot be determined from first principles. However, if the nucleation threshold of −4° C at water saturation is taken into account the appropriate curve may be fixed upon as shown in fig. 13.6.

This treatment, suggested by Fletcher (1959a), is in agreement with the experimental results of Sano, Fujitani and Maena (1956) who found a curve very similar to that of fig. 13.6. Mason

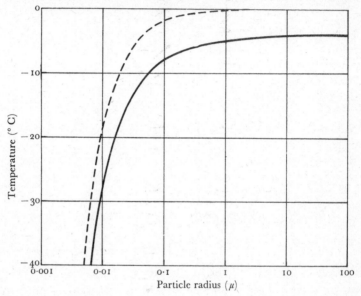

Fig. 13.6. The temperature at which a silver iodide particle of given size will nucleate an ice-crystal by sublimation from an environment at water saturation in 1 sec. The broken line shows the behaviour of an ideal sublimation nucleus (Fletcher, 1959a).

(1959), on the other hand, has suggested that the nucleation threshold for sublimation is at −12° C. Whilst a result of this kind could be explained by use of a crystal surface less active than that of a smoke particle because of its method of preparation, the question must be left open until further experiments have been performed.

The action of silver iodide as a freezing-nucleus is less easy to discuss and the situation is complicated by the small but finite solubility of the particles. The curves of fig. 8.4 applied to silver iodide suggest that the size-effect should not be important for particles larger than about 100 Å. Bayardelle (1955) and Roulleau (1957) have examined the freezing-points of droplets containing suspended particles of silver iodide which have been passed through fine filters. They find a size-effect rather intermediate between that expected for freezing and sublimation, but it is difficult to interpret the results because of the statistical nature of droplet experiments.

Whilst it is possible that an appreciable number of ice-crystals in a cloud may be nucleated by the Facy collection mechanism followed by freezing, no adequate treatment has yet been given of this process. Recent careful experiments by Edwards and Evans (1960) have, however, supported this view. They prepared closely controlled silver iodide aerosols with maximum particle radius 100 Å, and found that these particles were active in producing ice-crystals provided the experimental situation was such that the particles were quickly included in water droplets. The activity was consistent with the freezing-curves of fig. 8.4, and the particles appeared almost inactive when in a situation in which only sublimation could take place.

Whilst these results are not in disagreement with the hypothesis that larger particles act by sublimation, they pose the question why the activity of smokes is not several orders of magnitude higher than observed due to activity of these small particles as freezing-nuclei. It may be that only a minute fraction of these tiny particles are active due to the necessity for a particular type of surface configuration, but this is still in the realm of speculation.

Mixed smokes

Discussion of the behaviour of mixed smokes is especially difficult because so little information is available about their constitution. Whilst smokes based on ammonium iodide as a solution agent may consist predominantly of pure silver iodide, more common types using sodium or potassium iodides inevitably contain much of this material mixed with silver iodide in the smoke. Whether a mixed crystal or simply a mixture of crystals is formed, the resulting particle is likely to be hygroscopic and to form eventually a solution droplet containing a suspension of AgI particles, when it is exposed to high humidity.

If this is the process involved, then its analysis in terms of size-effects will prove very complicated.

Fletcher (1959c) has adopted an alternative approach. He makes the working assumption that the smoke from solution burners behaves in all respects essentially like pure silver iodide smoke. This implies that the 25 % or so of NaI or KI is separated or else enclosed in AgI so as to be rendered inert. The nucleation behaviour of the smoke particles under conditions of water saturation can then be expressed as in fig. 13.6, the particles acting by sublimation.

From this curve it is now a simple matter to determine the maximum number of nuclei active at a given temperature which can be produced from one gram of AgI, this maximum being produced by a dispersion into uniform particles having a nucleation threshold at the temperature considered. When such a number is found for every temperature the curve shown broken in figs. 13.3 and 13.4 is obtained. A smoke consisting of uniform particles should have a right-angled activity curve just touching this curve. Any real burner having a finite spread of particle-sizes should have an activity spectrum lying below the maximum curve.

Examination of figs. 13.3 and 13.4 shows that the behaviour of real burners is described well, both qualitatively and quantitatively, by this theory. The two string-burners in fig. 13.4 behave as if they have fairly uniform smoke particle-size, while the solution burners behave as though producing a wide range of particle-sizes, as in fact electronmicrographs substantiate. A wide log-normal distribution of particle-sizes is found to produce an activity curve very similar in shape to that possessed by solution burners.

The agreement of this simple theory with experiment is of great assistance in the interpretation of the results of tests on solution burners. It should not, however, be allowed to blind us to our ignorance of the exact mechanism of nucleation in the case of mixed smokes.

DEACTIVATION

Because the nucleation efficiency of a crystalline substance depends critically upon its surface-structure, it is to be expected that under certain environmental conditions its activity may be impaired. This is indeed the case with silver iodide and a good deal of effort has gone into investigation of various aspects of the decay of AgI aerosols.

Some of the decay processes are straightforward and apply to most aerosols. When a dense smoke is contained in an enclosure its particle concentration decays through exposure to the walls (approximately exponential) and by coagulation (reciprocal of concentration increasing linearly with time). There may also be sedimentation effects but these are small in the usual size range of AgI smoke particles.

Similarly, many compounds with high vapour-pressures may inhibit nucleation by becoming adsorbed on the surfaces of the AgI particles (Birstein, 1957). This effect is important chiefly as a trap to avoid when designing experimental apparatus for nucleation studies. The mechanism is straightforward in outline, but the whole nucleation process is insufficiently understood in detail for any fundamental knowledge to have been gained from contamination studies.

An interesting phenomenon has been reported by several workers (Picca, 1956; Jaffray and Montmory, 1956b), who find that the formation of an ice-crystal upon a particle or upon an active site on a larger crystal reduces or destroys its activity. Whilst this could conceivably be the result of deformation of the substrate in the region of epitaxy, the most likely explanation seems to be that a growing ice-crystal collects impurities upon its surface, and on evaporation these are deposited upon the active nucleating site, thus lowering its efficiency.

PHOTO-DEACTIVATION

The silver halides are not very stable compounds and may be decomposed by light of sufficiently short wavelength, an effect made use of in the photographic process. Since silver iodide shares in this susceptibility to photo-decomposition, and since its nucleating properties depend critically upon its surface structure it is not surprising to find that its ice-nucleating properties are adversely affected by exposure to ultraviolet light.

This effect was first investigated by Reynolds, Hume, Vonnegut and Schaefer (1951) who stored nuclei from a propane burner in a steel tank of 46 m³ capacity, fitted with either a cellophane or an opaque lid. With the opaque cover in place the concentration of active nuclei, as measured by withdrawing a small known volume and injecting it into a cold-box at $-20°$ C, decreased by an order of magnitude in 24 hr due to the effects of sedimentation, coagulation and diffusion to the walls. When, however, the smoke was irradiated by strong sunlight through the cellophane cover, the concentration of nuclei active above $-20°$ C was found to decrease by two orders of magnitude per hour of exposure. Very similar behaviour was observed when the smoke was irradiated with light from an ultraviolet lamp.

Similar experiments by other workers confirmed this finding, but the observed decay rate varied from more than a factor of 10^6 per hr (Inn, 1951) to as little as a factor between 2 and 5 per hr (Vonnegut and Neubauer, 1951) under similar conditions of irradiation.

Because of the importance of this effect in cloud seeding-operations, field measurements of the decay rate of smoke released from various operational burners were made by Smith *et al.* in Australia (Smith, Heffernan and Seely, 1955; Smith and Heffernan, 1956; Smith, Heffernan and Thomson, 1958). Their technique was to release a fine powder of zinc sulphide from a point very close to the burner output, and to sample the smoke plume downwind. The zinc sulphide particles were counted by a scintillation method under ultraviolet light and gave a measure of the dilution of the plume. The ratio of ice-crystal counts to ZnS particle counts gave a direct measure of the decay of the silver iodide activity.

For a generator burning acetone solution in a hydrogen flame they found an initial decay rate of 10^6 per hour. For the other burners tested, however, burning acetone solution either by itself or in a kerosene flame, the decay rate was between 10 and 10^2 per hour.

The results of all these experiments suggest that there is a real variation in the decay behaviour of different smokes, related, it seems, to the way in which they are prepared. A descriptive theory of this deactivation process based on the simplest possible model has recently been put forward by Fletcher (1959*b*). He considers the details of the mechanism of photolysis in spherical particles of silver iodide active as sublimation nuclei, and traces the decline in their activity under irradiation. For a given smoke the apparent decay rate depends primarily upon the size-distribution of the smoke particles and may vary within wide limits. Chemical composition also plays a part in determining the quantum efficiency of the photolysis process, but this is not yet understood in detail. Broadly speaking, fine smokes are found to decay more quickly than coarse smokes, in agreement with experiment.

Prevention of decay

Because the photo-decay of silver iodide places severe limitations on its use in large-scale cloud seeding, several empirical attempts have been made to find some means of preventing the decay, or at least of decreasing the decay rate. The theory suggests that two approaches may be possible—the addition of an impurity to the AgI to catalyse recombination of electrons and holes without any formation of photolytic silver, or the adsorption of some surface-film which might inactivate electron traps on the crystal surface without inhibiting its nucleation behaviour. Experiments have followed this second approach. Reynolds, Hume and McWhirter (1952) found that a small quantity of ammonia appeared to reactivate to some extent nuclei which had been partially photolysed, though after treatment the decay continued at a slightly accelerated rate. Similar results were found by Sano and Fukuta (1956a), who also investigated the effects of hydrogen sulphide.

Birstein (1952) and later Sano and Fukuta (1956a) investigated the dependence of decay rate upon humidity and found that at relative humidities greater than 60% the decay rate was considerably decreased. Adsorption isotherms showed that at this vapour concentration extraordinarily large water-vapour adsorption occurs (Birstein, 1955, 1956; Sano and Fukuta, 1956a) and the protection from decay was considered to be associated with this effect. The adsorption isotherms are themselves the subject of controversy, however (Karasz *et al.* 1956), and no firm conclusion has yet been reached.

Bolton and Qureshi (1954) have reported a dependence of decay rate upon air-temperature and pressure, but their results have been shown by Warner and Bigg (1956) to be due to the immense initial concentration of their smoke (10^8 particles/cm^3) and to have no relevance to behaviour in the atmosphere.

Whilst no practically useful results have yet come from these studies, the subject is of sufficient importance to warrant increased attention. The almost unexplored possibility of adding small amounts of other substances to the silver iodide solution before burning may also prove worth investigation. In the meantime, if the theory of Fletcher (1959b) describing the deactivation process proves correct, it may at least prove possible to minimize the decay under particular conditions by variations in the size-distribution of particles in the smoke. Unfortunately, this cannot be accomplished with any degree of precision and in practice may reduce to almost arbitrary burner adjustments followed by evaluation of the properties of the smoke.

CONCLUSION

Silver iodide is an almost ideal nucleating agent from many points of view. It is not very expensive when other costs of cloud-seeding operations are considered, it is easily dispersed, and its nucleating behaviour approaches quite closely that of an ideal nucleus (which would follow the broken curve in fig. 13.6). It seems unlikely, therefore, that any more suitable cloud-seeding agent will be found, though the possibility always remains.

The one disadvantage of silver iodide from a practical point of view is the decay in sunlight, although even this may sometimes be an advantage as it prevents large concentrations of nuclei building up which might prove more of an embarrassment than their original scarcity. Nevertheless, it appears possible in principle to decrease the decay rate of silver iodide by appropriate admixtures to its bulk or adsorption on its surface, although little progress has yet been made in this direction.

Whilst it may seem strange that so long after its discovery no better nucleating agent has been devised, this should perhaps be regarded as a confirmation of the essential correctness of our ideas of the mechanics of the nucleation process.

CHAPTER 14

LARGE-SCALE RAIN-MAKING EXPERIMENTS

INTRODUCTION

Of all the methods of cloud modification discussed in chapter 12, the only one which has received really large-scale trials is the seeding of sub-freezing clouds with silver iodide smoke. Some of this seeding has been in an attempt to minimize hailstorms, because of their effects on crops, or to reduce the number of lightning strokes, because of their responsibility for forest fires, but the great majority of experiments have been conducted in an effort to increase rainfall, and it is upon this aspect that we shall concentrate.

Before we discuss any experimental details it is as well to state clearly the question involved: to what extent is it possible to increase by cloud seeding the amount of rain which falls in a specified region? Specification of the region involved is thus an important part of the question, and the result may be reasonably expected to depend greatly upon the geographical situation involved. More important still, however, is a phrase which has not been made explicit. Any talk of an 'increase' implies knowledge of the 'normal', or amount of rain which would have fallen in the absence of any seeding experiments. Most of the difficulty in assessing the results of cloud-seeding experiments springs from our lack of knowledge of this 'normal' amount of precipitation, and skilful statistical design is required if significant answers are to be obtained despite this gap in our basic knowledge.

There are two possible approaches to the problem of determining the efficacy of cloud seeding in a given situation. The first, which is the more appealing from a fundamental point of view, involves three phases—a knowledge of the natural evolution of clouds and in particular the precipitation produced, a knowledge of the way in which individual clouds of various types are affected by seeding, and a detailed cloud census for the region in which seeding is contemplated. Armed with this information it should be possible to make a reliable estimate of the overall results of cloud seeding in the area, and this is one of the aims of the detailed study of cloud physics, whose present status we have outlined in this book. Unfortunately, it will be abundantly clear to the reader that our understanding is as yet far too meagre to allow us to predict in any real case the evolution of a cloud with the precision necessary for a rainfall estimate, though it is not too much to hope that this will one day be possible. Similarly, the behaviour of a seeded cloud is equally uncertain, and no cloud-census programmes appear to be in existence of sufficient comprehensiveness to give the information necessary to the third phase of our hypothetical project.

A few authors have ventured estimates or general comments founded on such basic principles, but of necessity these are as yet limited to order-of-magnitude calculations. Bergeron (1949) examined the problem in general terms and concluded that orographic cloud systems in which systematic condensation is caused by geographical features were most suitable for seeding. Braham et al. (1951) made a radar census of clouds in New Mexico and found a considerable proportion of clouds which appeared suitable for seeding and which did not develop natural precipitation. They were not able to estimate the probable effects of seeding in any quantitative way, however. Ludlam (1955) has analysed orographic clouds in Scandinavia and ventured a conclusion that an ideal seeding-operation might increase snowfall by as much as 100%, though a practical experiment would probably give a much smaller increase.

Valuable though these studies are, they cannot as yet give information of practical value. We are forced, therefore, to abandon this basic approach and to seek instead a direct answer by a statistically designed experiment conducted in the area concerned. Whilst this method gives a result restricted to a particular locality, or at best to a particular type of locality, it has the advantage of directness—the rainfall is its own evidence and uncertainties in our understanding of natural processes are bypassed by experiment.

Unfortunately, most of the experiments which have so far

Introduction

been conducted have been lacking in proper design, the object having been to increase the rainfall if possible, rather than to investigate the possibility of such an increase. In times of water shortage there are strong arguments for such an approach, but fortunately the results of some longer range well-planned experiments are now becoming available, and upon these we can base our discussion.

STATISTICAL DESIGN

Before any of the physical details of an evaluation experiment are considered, it is as well to examine closely the fundamental influences which affect the situation. With these in mind it is then possible to devise an experiment which will yield a significant result with the least amount of effort. In practice this sequence has rarely been followed; experiments have been performed over prolonged periods without any well thought out design and when the results are analysed it is often found that any effects produced by seeding are masked by random fluctuations produced by natural causes.

In this section we shall examine some typical experimental designs and assess their effectiveness.

Serial experiments

The simplest type of experiment to design and perform is what we might call a simple serial experiment. A region of country is selected whose promise from a cloud-seeding point of view is to be assessed and the available historical data on its rainfall is examined. In many cases of interest such rainfall data is available for a period of more than 50 years and allows a reliable value of mean rainfall and standard deviation about this mean, for a selected time unit, to be determined. From this information it is then possible to make certain predictions about the natural rainfall to be expected in the future. Cloud-seeding operations are then commenced in the area and the actual rainfall compared with that expected on the basis of the historical data.

In order to evaluate the effectiveness of this method we require an estimate of the accuracy of the predicted value of the 'unseeded' rainfall in the target area. Some idea of this quantity can be obtained by examining the historical rainfall records.

A convenient survey of a large quantity of rainfall data from the United States has been given by Foster (1948), and this is useful in drawing general conclusions about the reliability of predictions based on historical records. The first thing to be decided in a simple experiment such as we are discussing is the period over which seeding is to take place and over which rainfall is to be evaluated. It is immediately clear that a period of a few days is much too short because the variability of rainfall from day to day is too great. On the other hand, a period of tens of years must be ruled out for practical reasons. Whilst it is clear that the larger the period chosen the less the inherent variation due to natural fluctuations, explicit figures must be examined if a suitable averaging period is to be found.

Table 14.1. *Natural variation of monthly precipitation at Omaha, Nebraska, 1871–1940 (from data of Foster, 1948)*

Month	Maximum (in.)	Minimum (in.)	Mean (in.)	Standard deviation	Coefficient of variation
January	2·80	0·01	0·71	0·54	0·76
February	3·09	0·03	0·86	0·61	0·71
March	4·91	—	1·32	0·98	0·74
April	6·34	0·23	2·46	1·45	0·59
May	11·29	0·55	3·60	2·18	0·61
June	12·70	0·25	4·52	2·64	0·58
July	10·35	0·45	3·69	2·61	0·71
August	12·50	0·18	3·21	1·65	0·51
September	9·32	0·24	3·06	2·01	0·66
October	5·86	0·07	2·15	1·43	0·66
November	6·24	0·03	1·23	1·01	1·01
December	3·33	0·07	0·86	0·71	0·83

Table 14.1 shows, in modified form, data given by Foster (1948) for monthly rainfall in Omaha, Nebraska. These may be taken as fairly typical of continental situations of moderate rainfall (28 in. per year). Means and standard deviations have been calculated from data for the 70-year period from 1871 to 1940. Also shown is the coefficient of variation (standard deviation/mean).

Statistical Design

Now let us examine the implications of this table. Though the rainfall totals for any single day are very far from normally distributed, when sufficient days are added together into groups the total rainfall in these groups becomes more nearly normally distributed. Thus the rainfall totals for the same calendar month of each year are to a first approximation normally distributed, and the approximation becomes quite good for yearly rainfall totals.

If we make this approximation for monthly totals then it is possible to predict from table 14.1 the expected precipitation for some future month and to assign confidence limits to this prediction. Putting this in another and perhaps a better way, if the clouds over Omaha were to be seeded for one particular month and the rainfall found to exceed the expected or mean value for that month by a certain amount, it is possible to state the probability that an increase as large or larger than this should occur merely because of a random fluctuation in the natural rainfall.

In cloud-seeding experiments it is usual to regard a rainfall increase as significant if the probability of an increase as large or larger occurring by chance is less than 5%. Examination of tables of the normal distribution function shows that for an increase to be significant at this level it must exceed 1·65 standard deviations. Table 14.1 now suggests that the best month for cloud-seeding experiments is August when the variability of rainfall is least and, applying our criterion to this month, we find that only precipitation increases greater than 84% of the mean are statistically significant. For other months even larger increases are required.

Experience with cloud-seeding experiments suggests that the increase in precipitation is probably only of the order of 10%, and certainly does not approach 100% as required for significance in an experiment of this kind. We conclude, therefore, that a month is too short an averaging time for a simple experiment of this type.

The year is the next convenient time unit to consider. McDonald (1958), in the course of an excellent review of this topic, gives figures for yearly rainfall at various American stations. These are reproduced in table 14.2. The coefficient of variation for each station is much less than for the monthly totals, but even in the most favourable case (Iowa City) an increase in rainfall of 23% is required for significance, while 40–50% increase is required in other areas. An experiment designed on such a simple basis is thus very insensitive for the detection of small precipitation increases and does not represent a satisfactory approach to the evaluation of cloud-seeding programmes.

Table 14.2. *Natural variability of annual rainfall* (*McDonald*, 1958)

Station	Years of record	Mean annual precipitation (in.)	Coefficient of variation
Iowa City	70	35·2	0·14
Boston, Massachusetts	60	41·0	0·16
Cleveland, Ohio	60	34·0	0·16
Kansas City	63	36·1	0·18
Portland, Oregon	60	42·2	0·19
Bismarck, N. Dakota	66	16·3	0·25
Cheyenne, Wyoming	70	14·6	0·25
Ogden, Utah	60	16·2	0·25
Wichita, Kansas	52	29·2	0·26
Sheridan, Wyoming	47	15·0	0·27
Tucson, Arizona	87	11·2	0·30
Phoenix, Arizona	75	7·6	0·40
San Diego, California	60	9·9	0·44
Yuma, Arizona	83	3·2	0·62

Control-area experiments

The principal obstacle to the use of a simple serial experiment to determine the effects of seeding is, as we have seen, the great variability of natural precipitation even over periods as long as a year. Control-area experiments attempt to reduce the uncertainty involved in predicting the 'unseeded' or natural rainfall in the target area by using data from other stations to aid with the prediction. Thus, instead of the predicted rainfall y in the target area being given by

$$y = a, \qquad (14.1)$$

where a is the historical mean value for the area, we have

$$y = a + \Sigma b_1 x_1, \qquad (14.2)$$

where the x_1 are meteorological variables (rainfall, cloud cover, etc.) associated with one or more control areas during the seeded period, and the coefficients b_1 are determined from analysis of the historical data. If the rainfall in the target area is normally closely related to that in the control areas, this procedure may be expected to increase greatly the accuracy of the predicted rainfall. Thus the degree of variation of actual from predicted rainfall should be much less when (14.2) is used rather than (14.1).

An analysis of commercial rain-making experiments carried on in 1950–4 in Oregon has been made by Decker, Lincoln and Day (1957) using this method. They employed eleven control variables x_1 in their analysis and evaluated the coefficients b_1 by a study of weather records for the previous four-year period. This reduced the coefficient of variation to the extent that a 15·5% increase in rainfall would have been regarded as significant. Since the observed increase was only 6% no conclusions could be drawn. A detailed analysis of the statistical techniques involved in such regression analysis has been given by Thom (1958).

A simpler version of (14.2) often used in practice (MacCready, 1952; Ballay, 1955) takes the form

$$y = bx, \qquad (14.3)$$

where y is target area rainfall, x is control area rainfall, and the constant b is determined from historical rainfall records. For this form to be useful the two areas must be physically close together so that their rainfalls are well correlated, but must be sufficiently separated that seeding-effects do not spread into the control area. The attractiveness of this method lies in its use of only a single constant, and with careful selection of control area the experiment should be quite sensitive.

One defect inherent in all methods which rely upon historical records to evaluate the coefficients in equations like (14.1) or (14.2) is that they neglect the effects of any long-term climatic change. Whilst in principle the coefficients could be made time dependent and extrapolated to new values for the test period, the amount of data available is not usually sufficient to make this feasible, so that the defect remains.

A related defect is of a practical nature and relates to the measurement of precipitation. Since this measurement usually involves interpolation from inadequate rain-gauge data it is not really accurate and it becomes necessary to ensure that the same rain-gauges and evaluation procedures are used during the seeded period as were used for the historical records. This precludes the installations of special rain-gauge networks and usually results in sparse and inadequate precipitation data.

Randomized experiments

A method which at once eliminates both the troubles just mentioned is to seed on only some of the possible opportunities and to compare the rainfall in the target area during seeded and unseeded periods. Since no historical rainfall data is involved, special dense networks of rain-gauges can be installed over the target area and the measurements should be much more representative than those obtained from existing rain-gauge stations.

It is clearly necessary that no bias be introduced in the selection of seeded and unseeded situations, and several means of ensuring this have been adopted. Langmuir (1950) adopted a seeding schedule having a seven-day periodicity, in experiments conducted in New Mexico, and sought a resulting seven-day periodicity in rainfall figures. Such a periodicity was in fact found, commencing at about the same time as the seeding-operation, and was hailed as a substantial effect due to cloud seeding. However, Lewis (1951 b) and Wahl (1951) found similar seven-day periodicities in earlier weather records, and Brier (1954) showed that these could probably be explained by the orderly progress of circumpolar pressure patterns. Because such real or apparent weather periodicities are so common, seeding programmes on a simple alternating basis are unsuitable, and a randomized method must be adopted.

This method amounts to characterizing certain situations as seedable, and then after this has been done deciding on a completely random basis whether or not seeding is to be carried out. Equivalent seedable situations may be simply successive weeks or other convenient short-time units; they may be individual storms possessing certain suitable characteristics, or they may be determined in a more complicated manner. The important thing is that the situation be characterized as seedable before a decision is made as to whether or not it is to be seeded.

Statistical Design

Randomized experiments of this type have only recently been undertaken because of an understandable reluctance on the part of commercial operators and their clients to let half of the seedable situations pass without any action.

An improvement of the simple one-area randomized experiment which is usually employed in practice involves comparison with a neighbouring control area. Once again the control area must be chosen to be sufficiently close to the target area that rainfall in the two is closely correlated; on the other hand, the separation between the two areas must be sufficient so that seeding in one area does not partially seed the other area.

Such a large-scale randomized experiment was commenced in the Snowy Mountains area in Australia in 1955 (Adderley and Twomey, 1958), and since that time similar experiments have been initiated in California, Arizona and Illinois in the United States as well as in several other regions in Australia.

Several procedures are possible. The control area may be left untouched and the target area seeded on a random basis. This method was adopted in the Snowy Mountains experiment for practical reasons associated with works in the control area and has the disadvantage that the seeding-equipment is idle for half of the time. On the other hand, if ground generators are used only one set is required rather than two. Another possibility is to drop the distinction between control and target areas and to seed one area or the other on a random basis. In this way seeding is being carried out continuously and a significant result should be obtained more quickly than if only one area is seeded (Moran, 1959). This technique is being used in some of the other Australian experiments.

The necessity for correct and careful statistical design in experiments of this type can hardly be over emphasized. The claimed results of any rain-making programme will rightly be subjected to searching evaluation by professional statisticians, and it is only by proper statistical design in its initial stages that an experimental programme can hope to produce a convincing result.

EXPERIMENTAL DESIGN

Along with purely statistical considerations in the design of large-scale cloud-seeding experiments, many matters of experimental technique and general organization are also involved. Some of these factors are made more clear if the experimental question is rephrased: 'To what extent is it possible to increase the rainfall in a given region by an economical cloud-seeding programme?' The answer in any given case will thus depend upon the economic value of any precipitation increase, and upon the complexity and cost of the seeding programme involved. There can, therefore, be no unique answer and any results obtained are valid only for a particular region, using a specified seeding-technique and using the precipitation in a particular way.

Thus more expensive techniques may reasonably be used in an operation designed to produce water for hydro-electric power than would be justified merely for irrigation purposes, though in times of critical drought almost any method capable of producing rainfall may be justified.

For such reasons as these the experimental design may vary greatly from one case to another. Seeding from aircraft is ideal in experimental programmes because its efficiency and ease of control lead more quickly to significant results than is the case for seeding from the ground. In cases where rainfall of high economic value is involved or where weather and terrain make ground-seeding impracticable it may also be the best method for routine, as distinct from experimental seeding. In other cases, and where the terrain is suitable, routine seeding might best be carried out with ground-based burners, provided that their lower efficiency is more than compensated for by reduced cost.

Seeding from aircraft

The object of cloud seeding in the type of experiment we are discussing can be stated in quite definite terms. Clouds must be seeded in such a way that as much precipitation as possible develops and falls upon the target area. Now, even when a cloud system as a whole is stationary, as with orographic clouds forming over a mountain chain, the air and water-vapour within the clouds have a general drift characteristic of the prevailing wind, and individual clouds may drift in this way too. Thus, since precipitation takes perhaps 20 min to develop by the Bergeron mechanism, the seeding must be carried out upwind of the target area by a distance equal to the wind-travel in this time.

Here the basic advantages of seeding from aircraft become apparent. In the first place the nuclei can be released directly into the clouds, thus minimizing difficulties due to diffusion and photo-deactivation. Secondly, the region in which the aircraft operates is continuously variable so that smoke may be released at the place and time best calculated to produce rain over the target area. In the case of ground burners such optimum seeding can only be approached by use of a large network of generators in conjunction with a precise knowledge of smoke trajectories, which is ordinarily unavailable.

In a typical experimental situation such as described by Adderley and Twomey (1958) a single aircraft fitted with two generators of the type developed by Warren and Nesbitt and described in chapter 13 is used. The flight path is a line about 50 km in length, approximately normal to the wind direction, and a distance upwind from the target area. This distance is determined by multiplying the wind velocity at the seeding level by 30 min in the case of cumulus clouds with bases at or below 2000 m, or by 60 min in the case of layer-clouds; when the cloud base exceeds 2000 m, $1\frac{1}{2}$ min is added for each 300 m by which the base exceeds this level. Whilst this is purely an empirical procedure and cannot be ideal under all circumstances it appears to give good results.

In seeding from aircraft, the seeding-level, too, is under control. Adderley and Twomey recommend seeding at the $-6°$ C isotherm, which appeared justified on the basis of the activity curve for silver iodide. Later work indicates that $-10°$ C or colder is the best seeding-level for non-turbulent stratus clouds, but that seeding at cloud base is satisfactory for turbulent cumuli.

In this experiment the burners use 500–800 g of AgI per hour and at $-10°$ C produce about 10^{13} active nuclei per gram. If the flight path is traversed about five times per hour and the wind-speed is about 50 km/hr then it is reasonable to approximate the nucleus concentration by assuming them to become uniformly distributed over a depth of about 2 km. With these assumptions the average AgI nucleus concentration is 1–2 per litre. At colder temperatures more AgI nuclei will become active, and there should be some tens of nuclei per litre at $-15°$ C. Because only short exposure-times are involved, and those mostly in cloud, photolytic decay effects should reduce these figures only slightly.

A seeding-technique such as described thus produces an adequate supply of nuclei distributed in a very efficient and flexible manner. Any objections to the method for widespread use are based upon the cost of operating aircraft almost continuously for long periods. These economic considerations must of course be evaluated for each situation. In a situation like the Snowy Mountains experiment, where the rainfall is ultimately used to generate electricity, the economic value of increased rainfall is easily assessed and it is estimated that a rainfall increase of as little as $\frac{1}{2}\%$ in the seeded area would cover the cost of the seeding-programme, though not of the current measuring and evaluation work. If an increase considerably larger than this can be achieved then the use of aircraft is justified. In this particular case the rugged and inaccessible nature of much of the terrain is an additional factor favouring aircraft seeding, because of the difficulty of supplying and servicing ground-based burners.

Ground-based generators

The use of silver iodide smoke-generators from fixed stations on the ground is at first sight a very simple and convenient means of cloud modification. Natural convection is almost certain to carry the smoke up to the clouds, particularly when they are large and conditions are turbulent, and a distribution of burners should allow coverage of any desired area.

Whilst this is true in general substance, it must be remembered that the present objective of rain-making experiments is to determine the effectiveness of silver iodide seeding, and to do this accurately defined experimental areas are required. When ground burners are used, therefore, considerable effort must be expended to determine the trajectory and diffusion behaviour of the smoke after it leaves the burner.

Braham, Seely and Crozier (1952; also Crozier and Seely, 1955) developed a convenient and sensitive technique for use

Experimental Design

in this study. They dispersed a fine powder of zinc sulphide having particle diameters 1–3 μ from a smoke-generator or a centrifugal duster. These particles were then sampled in the plume downwind with an impactor carried in an aircraft. The impaction collections were then examined under ultraviolet light and the ZnS particles recognized by their fluorescence.

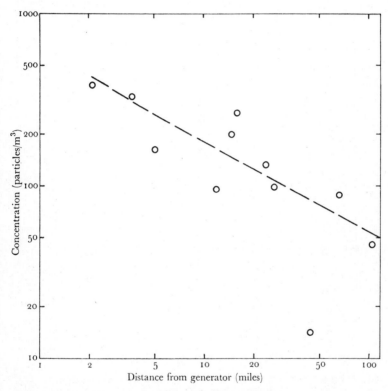

Fig. 14.1. Relation of average particle concentration to distance from source (corrected to 25 miles/h wind). Generator output 3×10^{10} particles/sec (Braham, Seely and Crozier, 1952).

The sensitivity of the method was such that plumes were traced as far as 160 km, and the measurements could be carried out rapidly and accurately. Smith and Heffernan (1954) used an airborne cold-box to trace the silver iodide directly, but because of the background of natural nuclei the plume could only be followed for about 20 km. In addition the measurements took so long that the results of many day's flights had to be combined to obtain any idea of the plume shape.

Several factors will affect the behaviour of the smoke plume, of which the most important are the nature of the terrain, the wind distribution and convective activity in the air. Because these influences can be very complicated all measurements

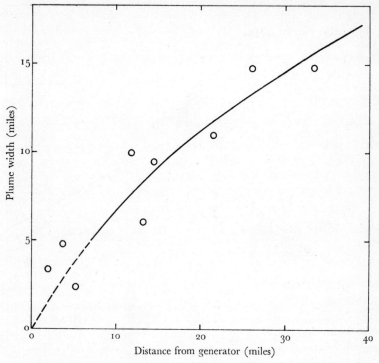

Fig. 14.2. Plume width as a function of distance over flat terrain (Braham, Seely and Crozier, 1952).

have been made under the simplest possible conditions—over flat land with relatively constant wind and only minor convective activity.

Under such relatively simple conditions Braham et al. (1952) found that the smoke plume widened and diluted in a fairly regular manner as shown in figs. 14.1 and 14.2. In the transverse horizontal direction the concentration within the plume was found to vary approximately as $A \exp(-Bx^n)$, where x

Experimental Design

is the distance from the plume centre, A and B are constants, and n is typically between 1·5 and 2·5 (Crozier and Seely, 1955). In the vertical direction, however, the concentration was fairly uniform from ground-level to a height of 1000–3000 m, presumably because of convective mixing.

In their more limited measurements Smith and Heffernan (1954) found a greater variation of smoke concentration with height and also with distance from the burner, as shown in fig. 14.3. Whilst part of this difference may be due to the decay

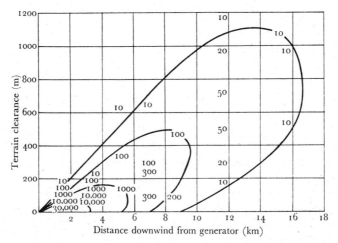

Fig. 14.3. Distribution of ice-forming nuclei in a vertical plane downwind from silver iodide smoke-generator. Figures give number of active nuclei/litre at $-17°$ C. Burner output at this temperature is approximately 3×10^{12} nuclei/sec (Smith and Heffernan, 1954).

of silver iodide activity in sunlight, much may also be due to differing convective conditions.

Serra (1954) has given an analytical discussion of some of the features of the behaviour of smoke plumes which are qualitatively in agreement with experiment, though his calculated plumes are rather narrow. He makes no explicit calculation of mixing in a vertical direction.

When we come to apply these results to seeding experiments however, their limitations immediately become obvious. Rain-making experiments are seldom if ever carried out in plain country under conditions of low convective activity. More typical sites are regions where strong orographic lifting of the airstream takes place, and times most suitable for cloud seeding are often characterized by considerable convective activity. Thus the path of the smoke plume in a mountainous region may be determined very largely by the topography and may differ greatly from the simple behaviour characteristic of flat country. Except on days when low-level inversions restrain convection it seems likely that the vertical- and horizontal-mixing figures given above may be considerably less than those typical of mountainous country. When large clouds are forming much of the smoke probably rises high enough to enter the cloud bases, though predictions of where and when this may happen are difficult to make.

Because of the large variations found from one area to another the location of burners for a seeding project may vary greatly. The same fundamental requirements must, however, be kept in view. The burners must be located sufficiently far upwind from the target area that the smoke is able to reach cloud base and the required precipitation-delay elapse before the cloud has arrived over the target. This will be a matter for compromise, bearing in mind different wind-speeds, cloud types and base heights.

Ground burners may conveniently be located along a line upwind from the seeding-area and normal to the wind direction to give optimum coverage, but other arrangements may prove better in particular cases. Considerably more silver iodide is required than for aircraft burners because of photo-decay in the longer smoke path and because some of the smoke may not reach cloud base at the appropriate time. The difficulty of servicing burners at remote locations in rugged country must also be taken into account, and the relative economics of ground-based and aircraft-seeding must be considered separately for each seeding-project.

When seeding over a very large accessible area is contemplated, and the economic value of the rainfall is not high, ground-based burners are probably the best solution. On the other hand, when optimum results are required with cost being less important, aircraft seeding is a better method. From the point of view of cloud-seeding experiments where the end-product is really information rather than rain, aircraft seeding

Experimental Design

has the additional advantage of leading more quickly to a significant answer because of its greater precision and efficiency. The answer, however, applies only to aircraft seeding, and any extrapolation to other seeding-techniques is dangerous.

EXPERIMENTAL RESULTS

Any assessment of the practical results of cloud-seeding experiments is at the present stage somewhat premature, as no final figures have been published from the large-scale randomized experiments, and we must rely upon interim figures and upon analysis of earlier non-random seeding-programmes.

Results and discussions of various European projects have been given from time to time in the *Bulletin de l'Observatoire du Puy de Dôme*, while the earlier Japanese experiments are summarized in the Report of Rainmaking in Japan (1954). The best analysis, however, is contained in the 1958 Report of the U.S. Advisory Committee for Weather Control. Here the large mass of experimental data available from the results of commercial seeding-operations, mostly using ground-burner techniques, is examined and such statistical conclusions as possible are drawn. The findings of the committee are summarized in the following words (*Report*, vol. 1, p. vi):

On the basis of its statistical evaluation of wintertime cloud seeding using silver iodide as the seeding agent, the Committee concluded that:

(1) The statistical procedures employed indicated that the seeding of winter-type storm clouds in mountainous areas in western United States produced an average increase in precipitation of 10 to 15 per cent from seeded storms with heavy odds that this increase was not the result of natural variations in the amount of rainfall.

(2) In non-mountainous areas, the same statistical procedures did not detect any increase in precipitation that could be attributed to cloud seeding. This does not mean that effects may not have been produced. The greater variability of rainfall patterns in non-mountainous areas made the techniques less sensitive for picking up small changes which might have occurred there than when applied to the mountainous regions.

(3) No evidence was found in the evaluation of any project which was intended to increase precipitation that cloud seeding had produced a detectable negative effect on precipitation.

(4) Available hail frequency data were completely inadequate for evaluation purposes and no conclusions as to the effectiveness of hail suppression projects could be reached.

These findings are in agreement with the opinions of most workers in the field today, though recent Australian results suggest that negative effects may be produced in some situations, in contradiction to (3) above. Early optimistic predictions of immense influences on the weather, caused by seeding, have been abandoned and the view now prevails that seeding will

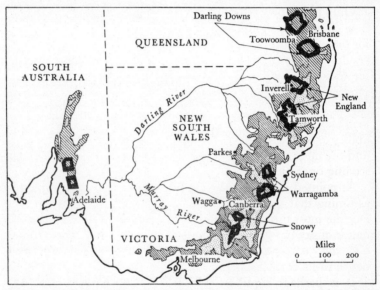

Fig. 14.4. South-eastern Australia showing the location of large-scale cloud-seeding experiments being carried out by the C.S.I.R.O. Mountains shown shaded.

only slightly affect general precipitation patterns. This must not be taken as belittling the value of the increases which seem to have been achieved. On the contrary, if they are substantiated they will represent a very valuable degree of control of precipitation for many purposes.

As examples of current randomized projects we refer to those being conducted by the Commonwealth Scientific and Industrial Research Organization in south-eastern Australia. It can be seen from fig. 14.4 that they are located in more or less mountainous regions where orographic clouds are usual. Special

Experimental Results

rain-gauge networks have been installed in each region and in many cases run-off, stream-flow and other records are also kept.

In the Snowy Mountains experiment, begun in 1955, only the southern area is seeded, the northern area being maintained as a control. In the other experiments, the areas for which are grouped in pairs as shown, both areas are seeded on a random cross-over basis. The experiment located in South Australia was abandoned after three years of operation when it became clear that supercooled clouds were rare in the region so that it was not suitable for silver iodide seeding.

Only the Snowy Mountains experiment has been in progress long enough for significant conclusions to be drawn, and except for the first year's operations (Adderley and Twomey, 1958), no results have been published. Unpublished preliminary analysis for four years' operations, however, indicates that there is a significant difference in precipitation in the seeded area during seeded and unseeded periods, the rainfall during seeded periods being about 20% greater than in unseeded ones. The probability of such a difference occurring by chance is less than 5% and the inference is strong that the increase is due to the cloud-seeding operations.

As pointed out by Adderley, however, one is not justified in assuming that the increase produced is 20%, though this is in fact the most probable value. One can only state with a confidence of say 95% that the rainfall increase lies between about 0% and 50%, and any closer specification reduces the confidence in its correctness.

It is perhaps significant that the most probable value of increase found from the Snowy Mountains experiment is similar to that found for experiments in comparable situations in the U.S. Such agreement cannot, however, be regarded as conclusive, and the final results of properly conducted randomized experiments must be awaited before any really reliable quantitative conclusions can be drawn.

PROSPECTS

We have just seen that as yet we have not been able to obtain an accurate answer to the primary problem of weather control: 'To what extent can the rainfall of a given region be increased by cloud seeding?' In the face of this lack of knowledge it would be rash in the extreme to make any predictions of a definite nature concerning the wider problems of weather control. It is, however, well to recognize that these problems exist, and we shall therefore give some of them a brief discussion.

Present experiments tend to point to the conclusion that cloud seeding using present techniques may be capable of increasing the rainfall over regions where orographic clouds occur, and that the amount of this increase may be perhaps 10 % of the mean rainfall. One question which immediately occurs concerns the origin of this extra moisture. Is it merely robbed from the rainfall of adjoining regions, or is a net increase in rainfall obtained at the expense, say, of increased evaporation from the oceans?

Existing experiments may throw some light upon this question, although the region of depleted rainfall, if it exists, might be expected to lie downwind from the target area, and for obvious reasons normal control areas are not located in this position.

A related question concerns the possibility of increasing rainfall over an area approaching continental dimensions. If cloud seeding merely produces a small-scale redistribution of rainfall, then widespread seeding would not be expected to be very effective, whilst if the redistribution is on a larger scale it may prove possible to augment continental rainfall at the expense of rainfall over the oceans which may normally be regarded as wasted. No answer can as yet be given to these questions.

We have already mentioned programmes designed to reduce the incidence of hailstorms and found that the difficulty in obtaining statistically significant results is such that not even tentative conclusions can yet be drawn. The same is likely to prove the case with experiments designed to reduce precipitation in a given area. This might be done by massive overseeding in the area concerned, which is probably not feasible for economic reasons, or by constant seeding over some orographic features upwind of the area concerned. Though we are not sure on what scale such redistribution of precipitation occurs, the area of reduced precipitation, if any, is certainly many times the area of the region of enhanced rainfall. The amount of the decrease will thus probably be very much less than the 10% characteristic of attainable increases and may well be completely concealed by the natural variability of precipitation.

Another effect of large-scale seeding of quite a different nature has also been discussed from time to time. Much of the driving force for cloud formation comes from the latent heat released when vapour condenses to liquid. At temperatures near $0°$ C this release amounts to approximately 600 cal/g. If widespread cloud seeding takes place so that cloud-tops become glaciated then the additional heat released in the cloud-tops amounts to about 80 cal/g, or about 12% of that released on condensation. In cases of marginal stability this heating may be sufficient to cause spectacular growth of the cloud-tops. This is, however, somewhat unusual, and a more restrained general uplift is probably more likely to occur. The effects of such an uplift over a very large area are difficult to estimate, but it has been suggested that the behaviour of storm systems such as tropical cyclones might be significantly influenced by this means. Experiments are proceeding, particularly in the United States, to investigate the effects of large-scale seeding on the hurricanes which frequently menace the eastern states.

Almost all the rain-making programmes in operation at the present time make use of silver iodide and rely upon the production of ice-crystals for their effect. We must not overlook the possibility that some of the other techniques which were discussed in chapter 12 may be developed sufficiently that their application to large-scale operations becomes feasible.

Some even more ambitious plans to control weather by modifying the balance of solar radiation at the earth's surface have been proposed from time to time. A recent discussion of some of these proposals has been given by Wexler (1958), who concludes that though some of the methods could conceivably produce significant effects, those produced indirectly may be of more importance than the direct effect envisaged. It will probably be as well to leave such ambitious schemes until such time as our knowledge of the earth's weather matches the technological progress achieved in other fields.

APPENDIX—RECENT ADVANCES

SINCE the writing of this book in 1960-61, work has continued on all branches of cloud physics but it seems fair to say that most of the major advances have occurred in fields not discussed in detail in the book—the study of severe storms and of electrification processes in thunder clouds. The first of these topics has been extensively dealt with in a monograph published by the American Meteorological Society (1963) while the two opposed theories of thunderstorm electrification currently receiving most consideration are those of Latham and Mason (1961 a,b) and of Vonnegut (1963). The theory of Latham and Mason attributes the charge separation to thermo-electric effects associated with temperature gradients in soft hail while Vonnegut's mechanism is directly related to the convective activity in the cloud in transporting space charge, thus not requiring the presence of any precipitation elements for its initiation.

In the field of general cloud physics several surveys of Russian work have recently become available in translation (Borovikov, Khrgian *et al*, 1963; Vul'fson and Levin 1963; Vul'fson 1964) and much other work has been published. Generally speaking this does little to change the picture presented in the book as a whole but serves to fill in some previously vague details. The survey below gives brief information and references to some of the more important contributions but is not meant to be in any way exhaustive.

Condensation of water vapour

Until recently the classical theory of homogeneous nucleation from the vapour, as outlined at the beginning of Chapter 3, had been regarded as well founded and substantially verified by experiment. Lothe and Pound (1962), however, pointed out that this treatment considers the embryo droplets to be at rest while in reality they form a macromolecular gas so that their translation and rotation must be considered. This correction increases the concentration of critical embryos by a factor of 10^{17} and destroys the previous good agreement between theory and experiment. Whilst there may be several ways out of this dilemma (Oriani and Sundquist 1963), the matter has not yet been resolved.

Fortunately homogeneous condensation is of little consequence in cloud physics and this correction factor is not involved in the theory of heterogeneous nucleation. The treatment given for this is therefore unchanged. A recent review of many aspects of theory and experiment in condensation and evaporation has been given by Hirth and Pound (1963).

Droplet coalescence

Some attempts have been made to verify Hocking's calculations of collision efficiencies for small droplets (Telford and Thorndike 1961, Woods and Mason 1964) and seem generally to confirm the theory, although a careful model experiment (Telford and Cottis 1964) suggests that some re-examination may be necessary. In larger size ranges the results reported by Woods and Mason (1965) differ from those found by Telford, Thorndike and Bowen which were discussed in Chapter 6 but agree generally with a new calculation by Shafrir and Neiburger (1963).

Radar meteorology

A fairly recent review of this subject has been given by Battan (1959). A very significant advance has, however, occurred quite recently in the development of pulsed-Doppler techniques. By measuring the frequency shift of radar echoes, looking vertically, it is possible to determine the fall-velocity of precipitation elements and hence, with the aid of some additional assumptions, the sizes of these elements (Battan 1964).

Ice-crystal nucleation and growth

Recent reviews on this subject have been given by Fletcher (1965) and by Mason (1965). Whilst it is possible to derive further useful information by application of the classical theory (Fletcher 1963), progress seems most likely through a study of nucleation processes at a molecular level. New calculations of the structure of the interface between two solids (van der Merwe 1963, Fletcher 1964) and speculation on the structure of ice surfaces (Fletcher 1962) is gradually giving more information in this direction, while better understanding of the structure of liquid water in terms of a model involving flickering hydrogen-bonded clusters (Némethy and Scheraga 1962, Kavanau 1964) is also clearly valuable. Careful experiments (Edwards and Evans 1960, 1961, 1962, 1965, Evans 1965) have established fairly closely the nucleation properties of silver iodide while interest has also been shown in the activity of certain organic materials as ice nuclei. (Head 1962, Fukuta 1963, Garten and Head 1965). Some questions have been raised about the basic mechanism of ice nucleation from the vapour by the experiments of Zettlemoyer *et al* (1963) on the ice nucleating properties of silica particles made hydrophobic by various means and this behaviour is not yet understood. Experiments using nuclear magnetic resonance techniques to study the behaviour of water adsorbed on nucleating particles (Barnes 1963) give some promise of progress in this direction.

Experimental studies on ice crystal growth (Mason, Bryant and Van den Heuvel 1963) have revealed a pronounced temperature dependence of the rate of motion of steps across the basal plane of a growing crystal and seem on the verge of explaining the observed habit variations, but no completely satisfactory detailed treatment has yet been put forward.

A review of knowledge of ice-forming nuclei in the atmosphere has been given by Mossop (1963) and this fills out some of the discussion of Chapter 9. New methods of sampling larger air volumes, particularly the technique of collecting nucleating particles on millipore filters (Bigg *et al* 1963, Bigg and Miles 1964), offer the promise of increased sampling reliability. No conclusive evidence has yet come forward either for or against the Meteor Hypothesis and recently attention has been directed to apparent lunar periodicities in rainfall (Bradley *et al* 1962, Brier and Bradley 1964) for which there is no obvious explanation.

Cloud Glaciation

An important advance has recently been made in actually observing the process of glaciation inside clouds and in trying to relate this to the properties of the cloud and the concentration of ice nuclei in the air (Koenig 1963, Braham 1964). The measurements show a strong dependence of glaciation rate on the liquid water content of the cloud and suggest some sort of chain-reaction mechanism for glaciation which is relatively independent of the initial ice-nucleus concentration.

Cloud-modification experiments

Large-scale rain-making experiments have continued to show disappointing results when compared with single-cloud studies, and the result of the Snowy Mountains Experiment, now published (Smith, Adderley and Walsh 1963), can only be described as 'encouraging but inconclusive'. It has been suggested that there is some phenomenon which causes a persistence of the seeding effect for a time of the order of a month or more, which would explain the gradual deterioration of initially promising results (Bowen 1965). Future experiments are being designed to test for and eliminate this effect.

From a different point of view, Malkus and Simpson (1964 a,b) have used silver iodide seeding techniques to release latent heat in a cumulus cloud and to compare the results with predictions based on an assumed model for the development of the cloud. The potentialities of this approach seem considerable.

REFERENCES AND AUTHOR INDEX

Numerals in brackets refer to pages of this book.

Ackerman, B. (1958). Turbulence around tropical cumuli. *J. Met.* **15**, 69.
[26]
Ackerman, B. (1959). The variability of water content of tropical cumuli. *J. Met.* **16**, 191. [11, 14, 21, 26, 28]
Adderley, E. E. (1953). The growth of raindrops in cloud. *Quart. J. R. Met. Soc.* **79**, 380. [191]
Adderley, E. E. and Twomey, S. (1958). An experiment on artificial stimulation of precipitation in the Snowy Mountains region of Australia. *Tellus*, **10**, 275. [335, 337, 345]
Aitken, J. (1883–84). On the formation of small clear spaces in dusty air. *Trans. Roy. Soc. Edinb.* **32**, 239. [86]
Aitken, J. (1887–88). On the number of dust particles in the atmosphere. *Trans. Roy. Soc. Edinb.* **35**, 1. [76]
Aitken, J. (1888–89). On improvements in the apparatus for counting the dust particles in the atmosphere. *Proc. Roy. Soc. Edinb.* **16**, 135. [76]
Aitken, J. (1890–91). On a simple pocket dust counter. *Proc. Roy. Soc. Edinb.* **18**, 39. [76]
Aitken, J. (1910–11). On some nuclei of cloudy condensation. *Proc. Roy. Soc. Edinb.* **31**, 478. [71]
Aitken, J. (1911–12). The sun as a fog producer. *Proc. Roy. Soc. Edinb.* **32**, 183. [66]
Aitken, J. (1923). *Collected Scientific Papers*, ed. C. G. Knott (Cambridge).
[76]
Aliverti, G. and Lovera, G. (1939). I fenomeni meteorologici sull'Oceano e il canto elettrico terrestre. *Atti Accad. Torino*, **74**, 573. [70]
Atlas, D. (1954). The estimation of cloud parameters by radar. *J. Met.* **11**, 309. [185]
Atlas, D. and Banks, H. C. (1951). The interpretation of microwave reflections from rainfall. *J. Met.* **8**, 271. [185]
Atlas, D. and Bartnoff, S. (1953). Cloud visibility, radar reflectivity, and drop-size distribution. *J. Met.* **10**, 143. [107, 185]
Atlas, D., Kerker, M. and Hitschfeld, W. (1953). Scattering and attenuation by non-spherical atmospheric particles. *J. Atmos. Terr. Phys.* **3**, 108. [185]
Atlas, D. and Plank, V. G. (1953). Drop-size history during a shower. *J. Met.* **10**, 291. [192]
Aufm Kampe, H. J. (1950). Visibility and liquid–water content in clouds in the free atmosphere. *J. Met.* **7**, 54. [107]

Aufm Kampe, H. J. and Weickmann, H. K. (1951). The effectiveness of natural and artificial aerosols as freezing nuclei. *J. Met.* **8**, 283.
[231, 239, 241]
Aufm Kampe, H. J. and Weickmann, H. K. (1952). Trabert's formula and the determination of the water content in clouds. *J. Met.* **9**, 167.
[107]
Aufm Kampe, H. J., Weickmann, H. K. and Kedesdy, H. H. (1952). Remarks on 'Electron-microscope study of snow-crystal nuclei'. *J. Met.* **9**, 374. [248, 274]
Aufm Kampe, H. J., Weickmann, H. K. and Kelly, J. J. (1951). The influence of temperature on the shape of ice crystals growing at water saturation. *J. Met.* **8**, 168. [261–2]
Augustin, H. (1954). Rapport technique sur les opérations de pluie artificielle à Madagascar en 1953. *Madagascar, Service Mét.*, Pub. No. 23.
[296]
Austin, J. M. (1948). A note on cumulus growth in a non-saturated environment. *J. Met.* **5**, 103. [23]
Austin, J. M. and Fleisher, A. (1948). A thermodynamic analysis of cumulus convection. *J. Met.* **5**, 240. [23]
Ballay, E. (1955). Methods and results of large scale applications of cloud seeding in the western United States. *Arch. Met., Wien*, **8**, 299. [333]
Bangham, D. H. and Razouk, R. I. (1937a). Adsorption and the wettability of solid surfaces. *Trans. Faraday Soc.* **33**, 1459. [219]
Bangham, D. H. and Razouk, R. I. (1937b). The wetting of charcoal and the nature of the adsorbed phase formed from saturated vapours. *Trans. Faraday Soc.* **33**, 1463. [219]
Barnes, W. H. (1929). The crystal structure of ice between 0° C and −183° C. *Proc. Roy. Soc.* A, **125**, 670. [199]
Barnett, E. W. (1958). A comparative flight test of three instruments for measuring cloud water content. *Univ. of Chicago, Cloud Phys. Lab. Tech. Note*, No. 14. [108]
Barrows, J. S. et al. (1958). Project Skyfire. *U.S. Advisory Committee on Weather Control, Final Report*, **2**, 105. [307]
Bartnoff, S. and Atlas, D. (1951). Microwave determination of particle size distribution. *J. Met.* **8**, 130. [185]
Battan, L. J. and Braham, R. R. (1956). Study of convective precipitation based on cloud and radar observations. *J. Met.* **13**, 587.
[11, 120, 121, 175]
Battan, L. J. and Reitan, C. H. (1957). Droplet size measurements in convective clouds. In *Artificial Stimulation of Rain*, ed. H. Weickmann and W. Smith (London: Pergamon Press), p. 184.
[112, 114, 116]
Bayardelle, M. (1955). Influence des dimensions des noyaux de congélation sur la temperature de congélation de l'eau. *C.R. Acad. Sci., Paris*, **241**, 232. [227, 321]
Becker, R. and Döring, W. (1935). Kinetische Behandlung der Keimbildung in übersättigen Dämpfen. *Ann. Phys., Lpz.*, **24**, 719.
[43, 44]

Benson, G. C. and Shuttleworth, R. (1951). The surface energy of small nuclei. *J. Chem. Phys.* **19**, 350. [45, 47]
Bentley, W. A. (1904). Studies of raindrops and raindrop phenomena. *Weath. Rev.* **32**, 450. [183]
Bentley, W. A. and Humphreys, W. J. (1931). *Snow Crystals* (London and New York: McGraw-Hill). [260]
Bergeron, T. (1935). On the physics of clouds and precipitation. *Proc. 5th Assembly U.G.G.I.*, Lisbon, **2**, 156. [229, 283]
Bergeron, T. (1949). The problem of artificial control of rainfall on the globe. I. *Tellus*, **1**, 32. [328]
Bernal, J. D. (1959). A geometrical approach to the structure of liquids. *Nature, Lond.*, **183**, 141. [202]
Bernal, J. D. and Fowler, R. H. (1933). Theory of water and ionic solution. *J. Chem. Phys.* **1**, 515. [200, 201]
Best, A. C. (1950). The size distribution of rain drops. *Quart. J. R. Met. Soc.* **76**, 16. [187]
Best, A. C. (1951). Drop-size distribution in cloud and fog. *Quart. J. R. Met. Soc.* **77**, 418. [119]
Best, A. C. (1952a). Effect on turbulence and condensation on drop-size distribution in cloud. *Quart. J. R. Met. Soc.* **78**, 28. [137]
Best, A. C. (1952b). The evaporation of raindrops. *Quart. J. R. Met. Soc.* **78**, 200. [192–3]
Bigg, E. K. (1953a). The supercooling of water. *Proc. Phys. Soc.* B, **66**, 688. [206–7, 242]
Bigg, E. K. (1953b). The supercooling of water. Thesis, University of London. [221]
Bigg, E. K. (1953c). The formation of atmospheric ice crystals by the freezing of droplets. *Quart. J. R. Met. Soc.* **79**, 510. [240]
Bigg, E. K. (1956a). Counts of atmospheric freezing nuclei at Carnarvon, Western Australia, Jan. 1956. *Aust. J. Phys.* **9**, 561. [243–4, 249, 256]
Bigg, E. K. (1956b). The detection of atmospheric dust and temperature inversions by twilight scattering. *J. Met.* **13**, 262. [254]
Bigg, E. K. (1957a). January anomalies in cirriform cloud coverage over Australia. *J. Met.* **14**, 524. [254]
Bigg, E. K. (1957b). A new technique for counting ice-forming nuclei in aerosols. *Tellus*, **9**, 394. [231, 233, 243, 249]
Bigg, E. K. and Meade, R. T. (1959). Continuous automatic recording of ice nuclei. *Bull. Obs. Puy de Dome*, No. 4, 125. [236, 244]
Birstein, S. J. (1952). The effect of relative humidity on the nucleating properties of photolysed silver iodide. *Bull. Amer. Met. Soc.* **33**, 431. [325]
Birstein, S. J. (1955). The role of adsorption in heterogeneous nucleation. I. Adsorption of water vapour on silver iodide and lead iodide. *J. Met.* **12**, 324. [217, 316]
Birstein, S. J. (1956). The role of adsorption in hetereogeneous nucleation. II. The adsorption of water vapour on photolysed silver iodide. *J. Met.* **13**, 395. [316]

Birstein, S. J. (1957). Studies on the effects of certain chemicals on the inhibition of nucleation. In *Artificial Stimulation of Rain*, ed. H. Weickmann and W. Smith (London: Pergamon Press), p. 376. [323]

Birstein, S. J. and Anderson, C. E. (1955). The mechanism of atmospheric ice formation. I. The chemical composition of nucleating agents. *J. Met.* **12**, 68. [224]

Blackman, M. and Lisgarten, N. D. (1958). Electron diffraction investigations into the cubic and other structural forms of ice. *Advanc. Phys.* **7**, 189. [198, 200]

Blanchard, D. C. (1949). The use of sooted screens for determining raindrop size and distribution. *G.E. Res. Lab. Project Cirrus Occ. Rep.*, No. 16. [182]

Blanchard, D. C. (1950). The behaviour of water drops at terminal velocity in air. *Trans. Amer. Geophys. Un.* **31**, 836. [163, 194–5]

Blanchard, D. C. (1953). Raindrop size-distribution in Hawaiian rains. *J. Met.* **10**, 457. [188–9]

Bolton, J. G. and Qureshi, N. A. (1954). The effects of air temperature and pressure on the decay of silver iodide. *Bull. Amer. Met. Soc.* **35**, 395. [325]

Bowen, E. G. (1950). The formation of rain by coalescence. *Aust. J. Sci. Res.* A, **3**, 193. [161, 163–7]

Bowen, E. G. (1951). Radar observations of rain and their relation to mechanisms of rain formation. *J. Atmos. Terr. Phys.* **1**, 125. [291]

Bowen, E. G. (1952a). A new method of stimulating convective clouds to produce rain and hail. *Quart. J. R. Met. Soc.* **78**, 37. [296–7]

Bowen, E. G. (1952b). Australian experiments in artificial rainmaking. *Bull. Amer. Met. Soc.* **33**, 244. [300]

Bowen, E. G. (1953). The influence of meteoritic dust on rainfall. *Aust. J. Phys.* **6**, 490. [250–1, 253, 255–6]

Bowen, E. G. (1956a). The relation between rainfall and meteor showers. *J. Met.* **13**, 142. [251, 253]

Bowen, E. G. (1956b). An unorthodox view of the weather. *Nature, Lond.*, **177**, 1121. [251–2]

Bowen, E. G. (1956c). A relation between meteor showers and the rainfall of November and December. *Tellus*, **8**, 394. [253]

Bowen, E. G. (1956d). A relation between snow cover, cirrus cloud, and freezing nuclei in the atmosphere. *Aust. J. Phys.* **9**, 545. [253]

Bowen, E. G. (1956e). January freezing nucleus measurements. *Aust. J. Phys.* **9**, 552. [255]

Bowen, E. G. (1957). Relation between meteor showers and the rainfall of August, September and October. *Aust. J. Phys.* **10**, 412. [253]

Bowen, E. G. (1958). Freezing nucleus measurements in January 1957. *Aust: J. Phys.* **11**, 452. [243, 255]

Bowen, E. G. and Davidson, K. A. (1951). A raindrop spectrograph. *Quart. J. R. Met. Soc.* **77**, 445. [179, 183, 192]

Boylan, R. K. (1926). Atmospheric dust and condensation nuclei. *Proc. R. Irish Acad.* A, **37**, 58. [92]

Bracewell, R. N. (1956). Counts of atmospheric freezing nuclei at Palo Alto, California, Jan. 1956. *Aust. J. Phys.* **9**, 566. [243]

Bragg, W. H. (1922). The crystal structure of ice. *Proc. Phys. Soc.* **34**, 98.
[199]
Braham, R. R. (1960). Physical properties of clouds. *Proc. Amer. Soc. Civ. Engrs*, **86** (IR 1), 111. [302]
Braham, R. R. and Battan, L. J. (1958). Effects of seeding cumulus clouds. *J. Amer. Wat. Wks Ass.* **50**, 185. [297]
Braham, R. R., Battan, L. J. and Byers, H. R. (1957). Artificial nucleation of cumulus clouds. *Met. Monogr.* **11**, 47. [297]
Braham, R. R., Reynolds, S. E. and Harrell, J. H. (1951). Possibilities for cloud seeding as determined by a study of cloud height versus precipitation. *J. Met.* **8**, 416. [121, 328]
Braham, R. R., Seely, B. K. and Crozier, W. D. (1952). A technique for tagging and tracing air parcels. *Trans. Amer. Geophys. Un.* **33**, 825.
[338–40]
Bricard, J. (1953). *Physique des nuages* (Paris: Presses Universitaires de France). [12]
Bridgman, P. W. (1912). Water, in the liquid and five solid forms under pressure. *Proc. Amer. Acad. Arts Sci.* **47**, 441. [198]
Bridgman, P. W. (1935). The pressure–volume–temperature relations of the liquid and the phase diagram of heavy water. *J. Chem. Phys.* **3**, 597.
[198]
Bridgman, P. W. (1937). The phase diagram of water to 45,000 kg/cm^2. *J. Chem. Phys.* **5**, 964. [198]
Brier, G. W. (1954). 7-day periodicities in May 1952. *Bull. Amer. Met. Soc.* **35**, 118. [334]
Brown, E. N. and Willett, J. H. (1955). A three-slide cloud droplet sampler. *Bull. Amer. Met. Soc.* **36**, 123. [106]
Browne, I. C. and Wexler, R. (1953). The collection efficiency of ice crystals. *Quart. J. R. Met. Soc.* **79**, 549. [282]
Brun, R. J. *et al.* (1955). Impingement of cloud droplets on a cylinder and procedure for measuring liquid-water content and droplet sizes in supercooled clouds by rotating multicylinder method. *U.S. N.A.C.A. Rep.*, No. 1215. [106]
Bryant, G. W., Hallett, J. and Mason, B. J. (1960). The epitaxial growth of ice on single-crystalline substrates. *J. Phys. Chem. Solids*, **12**, 189.
[221–2, 226]
Buff, F. P. and Kirkwood, J. G. (1950). Remarks on the surface tension of small droplets. *J. Chem. Phys.* **18**, 991. [45]
Bunker, A. F., Haurwitz, B., Malkus, J. S. and Stommel, H. (1949). Vertical distribution of temperature and humidity over the Caribbean Sea. *Papers in Phys. Ocean. and Met., MIT and Woods Hole Ocean. Inst.* **11**, No. 1. [16, 19, 22]
Burton, W. K. and Cabrera, N. (1949). Crystal growth and surface structure. *Disc. Faraday Soc.* No. 5, 33. [269]
Burton, W. K., Cabrera, N. and Frank, F. C. (1951). The growth of crystals and the equilibrium structure of their surfaces. *Phil. Trans.* **243**, 300.
[270–1]
Buswell, A. M. and Rodebush, W. H. (1956). Water. *Sci. Amer.* **194**, 76. [199]

Byers, H. R. and Braham, R. R. (1948). Thunderstorm structure and circulation. *J. Met.* **5**, 71. [19, 25]

Byers, H. R. and Braham, R. R. (1949). *The Thunderstorm: Report of the Thunderstorm Project* (Washington: U.S. Govt. Printing Office). [19, 26, 27, 292]

Byers, H. R. and Hall, R. K. (1955). A census of cumulus-cloud height versus precipitation in the vicinity of Puerto Rico, Winter–Spring 1953–54. *J. Met.* **12**, 176. [121]

Byers, H. R. and Hull, E. C. (1949). Inflow patterns of thunderstorms as shown by winds aloft. *Bull. Amer. Met. Soc.* **30**, 90. [25]

Byers, H. R., Moses, H. and Harney, P. J. (1949). Measurement of rain temperature. *J. Met.* **6**, 51. [196]

Byers, H. R., Sievers, J. R. and Tufts, B. J. (1957). Distribution in the atmosphere of certain particles capable of serving as condensation nuclei. In *Artificial Stimulation of Rain*, ed. H. Weickmann and W. Smith (London: Pergamon Press), p. 47. [247]

Carte, A. E. (1956). The freezing of water droplets. *Proc. Phys. Soc.* B, **69**, 1028. [207]

Cauer, H. (1951). Some problems of atmospheric chemistry. In *Compendium of Meteorology*, ed. T. F. Malone (Boston: Amer. Met. Soc.), p. 1126. [65, 66]

Cochet, R. (1951). Evolution d'une gouttelette d'eau chargée dans un nouage ou un brouillard à temperature positive. *C.R. Acad. Sci., Paris*, **233**, 190. [148]

Cooper, B. F. (1951). A balloon-borne instrument for telemetering raindrop-size distribution and rainwater content of cloud. *Aust. J. Appl. Sci.* **2**, 43. [183, 191]

Coste, J. H. and Wright, H. L. (1935). The nature of the nucleus in hygroscopic droplets. *Phil. Mag.* Series 7, **20**, 209. [67, 92]

Crozier, W. D. and Seely, B. K. (1955). Concentration distributions in aerosol plumes three to twenty-two miles from a point source. *Trans. Amer. Geophys. Un.* **36**, 42. [338, 341]

Cunningham, R. M., Plank, V. G. and Campen, C. M. (1956). Cloud refractive index studies. *U.S. Air Force Geophys. Res. Directorate, Bedford, Mass., Geophys. Res. Rep. Pap.*, No. 51. [22]

Cwilong, B. M. (1947). Sublimation in a Wilson chamber. *Proc. Roy. Soc.* A, **190**, 137. [206, 230, 235]

Dady, G. (1947). Contribution à l'étude de precipitation. *C.R. Acad. Sci., Paris*, **225**, 1349. [145]

Das, P. K. (1956). The change in the distribution of drops in a cloud. *Indian J. Met. Geophys.* **7**, 55. [137]

Davies, D. A. (1954). Experiments on artificial stimulation of rain in East Africa. *Nature, Lond.*, **174**, 256. [295]

Day, G. A. (1953). Radar observations of rain at Sydney, New South Wales. *Aust. J. Phys.* **6**, 229. [186, 291–2]

Day, G. A. (1958). Sublimation nuclei. *Proc. Phys. Soc.* B, **72**, 296. [227]

Day, G. J. (1955). A refrigerated disc icing meter. *G.B. Air Min. MRP*, 916. [11, 109, 257]

Day, G. J. (1956). Further observations of large cumuliform clouds by the Meteorological Research Flight. *G.B. Air Min. MRP*, 980. [11]
Day, G. J. and Murgatroyd, R. J. (1953). The cumulus cloud investigations made by Meteorological Research Flight during the period August 4th–15th 1952. *G.B. Air Min. MRP*, 826. [10]
Day, J. A. (1958). On the freezing of three to nine micron water droplets. *J. Met.* **15**, 226. [207]
Debye, P. (1945). *Polar Molecules* (New York: Dover Publications), p. 181. [51]
Decker, F. W., Lincoln, R. L. and Day, J. A. (1957). Analysis of cloud seeding efforts in the Tri-County area, Oregon, 1950–54. *Bull. Amer. Met. Soc.* **38**, 134. [333]
Defant, A. (1905). Gesetzmassigkeiten in der Verteilung der verschiedenen Tropfengrössen bei Regenfällen. *Ber. Akad. Wiss. Wien*, **114**, 585. [182, 189–90]
De Reuck, A. V. S. (1957). The surface free energy of ice. *Nature, Lond.*, **179**, 1119. [209]
Dessens, H. (1946). La brume et le brouillard étudiés à l'aide des fils d'araignées. *Ann. Géophys.* **2**, 276. [88]
Dessens, H. (1949). The use of spiders' threads in the study of condensation nuclei. *Quart. J. R. Met. Soc.* **75**, 23. [61, 70, 88]
Diem, M. (1942). Messungen der Grösse von Wolkenelementen. I. *Ann. Hydrogr. Berl.* **70**, 142. [7, 9, 10, 104, 105, 106]
Diem, M. (1948). Messungen der Grösse von Wolkenelementen. II. *Met. Rdsch.* **1**, 261. [104, 111, 116]
Dmitriev, A. A. and Chile, A. V. (1955). On meteor showers and precipitation. *Trudy Morskogo gidrofizicheskogo instituta*, **12**, 181. [251, 253]
Dorsch, R. G. and Boyd, B. (1951). X-ray diffraction study of the internal structure of supercooled water. *U.S. Nat. Adv. Comm. Aero. Tech. Note*, No. 2532. [201]
Dorsch, R. G. and Hacker, P. T. (1950). Photomicrographic investigation of spontaneous freezing temperatures of supercooled water droplets. *U.S. Nat. Adv. Comm. Aero. Tech. Note*, No. 2142. [206–7, 242]
Draginis, M. (1958). Liquid water within convective clouds. *J. Met.* **15**, 481. [11, 21]
Driving, A. Ia., Mironov, A. V., Morozov, V. M. and Khvostiknov, I. A. (1943). The study of optical and physical properties of natural fogs. *Izv. Akad. Nauk SSSR, Ser. Geograf. i. Geofiz.* **2**, 70. *Trans. Nat. Res. Council Canada Tech.* TT-102. [107, 117]
Dufour, L. and Defay, R. (1953). Étude thermodynamique de la pression de saturation par rapport à une gouttelette en suspension dans l'atmosphère. *Inst. Roy. Mét. Belgique*, Pub. Series B, No. 9. [59]
Durbin, W. G. (1956). Droplet sampling in cumulus clouds. *G.B. Air Min. MRP*, 991. [116]
Durbin, W. G. (1959). Droplet sampling in cumulus clouds. *Tellus*, **11**, 202. [11]
East, T. W. R. (1957). An inherent precipitation mechanism in cumulus clouds. *Quart. J. R. Met. Soc.* **83**, 61. [135–6, 137]

Edwards, G. R. and Evans, L. F. (1960). Ice nucleation by silver iodide.
I. Freezing vs. sublimation. *J. Met.* **17**, 627. [321]
Elton, G. A. H., Mason, B. J. and Picknett, R. G. (1958). The relative importance of condensation and coalescence processes on the stability of a water fog. *Trans. Faraday Soc.* **54**, 1724. [137]
Facy, L. (1951). Eclatment des lames mince et noyaux de condensation. *J. Sci. Mét.* **3**, 86. [70]
Facy, L. (1957). Sur le deplacement des aerosols dans un gradient de tension de vapeur. *Met. Soc. Japan*, 75th Anniv. Vol., p. 15.
[71, 224, 319]
Fenn, R. W. and Weickmann, H. K. (1959). Some results of aerosol measurements. *Geofis. pur. appl.* **42**, 53. [236]
Findeisen, W. (1932). Messungen der Grösse und Anzahl der Nebeltropfen zum Studium der Koagulation inhomogenen Nebels. *Beitr. Geophys.* **35**, 295. [144]
Findeisen, W. (1938). Die kolloid-meteorologischen Vorgänge bei der Niederschlagsbildung. *Met. Z.* **55**, 121. [299]
Findeisen, W. (1939). Zur Frage der Regentropfenbildung in reinen Wasserwolken. *Met. Z.* **56**, 365. [160]
Findeisen, W. and Schulz, G. (1944). Experimentelle Untersuchungen über die atmospharisch Eisteilchenbildung. I. *Forsch.-u. ErfahrBer. Reichsamt Wetterdienst* A, 27. [231, 239–40]
Fletcher, N. H. (1958a). Size effect in heterogeneous nucleation. *J. Chem. Phys.* **29**, 572; **31**, 1136. [55, 214–15]
Fletcher, N. H. (1958b). Time lag in ice crystal nucleation in the atmosphere. *Bull. Obs. Puy de Dome*, p. 11. [235]
Fletcher, N. H. (1959a). On ice crystal production by aerosol particles. *J. Met.* **16**, 173. [57, 218–19, 318–20]
Fletcher, N. H. (1959b). A descriptive theory of the photodeactivation of silver iodide as an ice-crystal nucleus. *J. Met.* **16**, 249. [324–6]
Fletcher, N. H. (1959c). The optimum performance of silver iodide smoke generators. *J. Met.* **16**, 385. [317–18, 322]
Fletcher, N. H. (1959d). Entropy effect in ice crystal nucleation. *J. Chem. Phys.* **30**, 1476. [212, 217, 222–3]
Fletcher, N. H. (1960). Nucleation and growth of ice crystals on crystalline substrates. *Aust. J. Phys.* **13**, 408. [217]
Foster, E. E. (1948). *Rainfall and Runoff* (New York: Macmillan Co.), chs. 5–6. [330]
Fournier d'Albe, E. M. (1949). Some experiments on the condensation of water vapour at temperatures below 0° C. *Quart. J. R. Met. Soc.* **75**, 1.
[206, 226]
Fournier d'Albe, E. M. (1957). Some observations of the geographical distribution of giant hygroscopic nuclei. In *Artificial Stimulation of Rain*, ed. H. Weickmann and W. Smith (London: Pergamon Press), p. 73.
[95]
Fournier d'Albe, E. M., Lateef, A. M. A., Rassool, S. I. and Zaidi, I. H. (1955). The cloud seeding trials in the central Punjab, July–September 1954. *Quart. J. R. Met. Soc.* **81**, 574. [296]

References and Author Index

Frank, F. C. (1949). The influence of dislocations on crystal growth. *Disc. Faraday Soc.* **5**, 48. [270]
Frank, F. C. and van der Merwe, J. W. (1949). One dimensional dislocations. *Proc. Roy. Soc.* A, **198**, 205. [212]
Frank, S. R. (1958). Survey and history of hail-suppression operations in the United States. *U.S. Advisory Committee for Weather Control, Report*, Vol. **2**, 264. [307]
Frenkel, J. (1939). A general theory of heterophase fluctuations and pretransition phenomena. *J. Chem. Phys.* **7**, 538. [39, 205]
Frenkel, J. (1946). *Kinetic Theory of Liquids* (Oxford), ch. 7. [39, 201]
Frith, R. (1951). The size of cloud particles in strato-cumulus cloud. *Quart. J. R. Met. Soc.* **77**, 441. [9, 22]
Fritz, S. (1954). Small drops, liquid–water content and transmission in clouds: with replies by Neiburger and by aufm Kampe and Weickmann. *J. Met.* **11**, 428. [107, 117]
Frössling, N. (1938). Über die Verdunstung fallender Tropfen. *Beitr. Geophys.* **52**, 170. [124]
Fuchs, N. and Petrjanoff, J. (1937). Microscopic examination of fog-, cloud- and rain-droplets. *Nature, Lond.*, **139**, 111. [104]
Fukuta, N. (1958). Experimental investigations on the ice forming ability of various chemical substances. *J. Met.* **15**, 17. [224, 309]
Fuquay, D. M. and Wells, H. J. (1957). The Project Skyfire cloud seeding generator. *U.S. Advisory Committee on Weather Control, Final Report*, Vol. **2**, p. 273. [313–14, 318]
Georgii, H. W. and Metnieks, A. L. (1958). An investigation into the properties of atmospheric freezing nuclei and sea-salt nuclei under maritime conditions at the west coast of Ireland. *Geofis. pur. appl.* **41**, 159. [249]
Gibbs, J. W. (1928). *Collected Works*, vol. **1** (New York: Longmans Green and Co.), especially pp. 94, 252–8, 367–8. [42]
Gish, O. H. (1951). Universal aspects of atmospheric electricity. In *Compendium of Meteorology*, ed. T. F. Malone (Boston: Amer. Met. Soc.), p. 101. [152]
Glasstone, S. (1947). *Thermodynamics for Chemists* (New York: D. Van Nostrand Co.). [38, 125, 204]
Goldschlak, L. (1957). The aggregation of ice crystals. Thesis, Pennsylvania State University, Dept. of Met. [279]
Goldstein, S. (ed.) (1938). *Modern Developments in Fluid Dynamics* (Oxford: Clarendon Press), p. 593. [22]
Good, R. J. (1957). Surface entropy and surface orientation of polar liquids. *J. Phys. Chem.* **61**, 810. [51, 202]
Gorham, E. (1958). The influence and importance of daily weather conditions in the supply of chloride, sulphate and other ions to fresh waters from atmospheric precipitation. *Phil. Trans.* B, **241**, 147. [96]
Gourley, M. F. and Crozier, W. D. (1955). Evaporation of sub-microscopic ice crystals. *J. Chem. Phys.* **23**, 1298. [226]
Gunn, K. and Hitschfeld, W. (1951). A laboratory investigation of the coalescence between large and small water drops. *J. Met.* **8**, 7. [145, 149]

Gunn, R. (1951). Precipitation electricity. In *Compendium of Meteorology*, ed. T. F. Malone (Boston: Amer. Met. Soc.), p. 128. [153]
Gunn, R. (1954). Diffusion charging of atmospheric droplets and the resulting combination coefficients. *J. Met.* **11**, 339. [152]
Gunn, R. (1955a). Droplet electrification processes and coagulation in stable and unstable clouds. *J. Met.* **12**, 511. [152, 153]
Gunn, R. (1955b). Raindrop electrification by the association of randomly charged cloud droplets. *J. Met.* **12**, 562. [152]
Gunn, R. (1956). The hyperelectrification of raindrops by atmospheric electric fields. *J. Met.* **13**, 283. [152]
Gunn, R. and Kinzer, G. D. (1949). The terminal velocity of fall for water droplets in stagnant air. *J. Met.* **6**, 243. [194]
Hallett, J. and Mason, B. J. (1958). The influence of temperature and supersaturation on the habit of ice crystals grown from the vapour. *Proc. Roy. Soc.* A, **247**, 440. [262–3]
Haltiner, G. J. (1959). On the theory of convective currents. *Tellus*, **11**, 4. [24]
Hann, J. (1899). Wassergehalt der Wolken und Nebel-Luft. *Met. Z.* **6**, 303. [7]
Helmholtz, L. (1953). The crystal structure of hexagonal silver iodide. *J. Chem. Phys.* **3**, 740. [310]
Henniker, J. C. (1949). The depth of a surface zone of a liquid. *Rev. Mod. Phys.* **21**, 322. [51, 202]
Herdan, G. (1953). *Small-particle Statistics* (Houston: Elsevier Pub. Co.), p. 113. [68]
Heverley, J. R. (1949). Supercooling and crystallization. *Trans. Amer. Geophys. Un.* **30**, 205. [207, 242]
Hocking, L. M. (1958). Three dimensional viscous flow problems solved by the Stokes and Oseen approximations. Thesis, University of London. [142–4]
Hocking, L. M. (1959). The collision efficiency of small drops. *Quart. J. R. Met. Soc.* **85**, 44. [142–4]
Holl, W. and Muhleisen, R. (1955). A new condensation nuclei counter with continuous oversaturation. *Geofis. pur. appl.* **31**, 21. [83]
Horton, R. E. (1948). Statistical distribution of drop sizes and the occurrence of dominant drop sizes in rain. *Trans. Amer. Geophys. Un.* **29**, 624. [190]
Hoshino, S. (1957). Crystal structure and phase transition of some metallic halides. IV. On the anomalous structure of $\alpha - \mathrm{AgI}$. *J. Phys. Soc. Japan*, **12**, 315. [310]
Hosler, C. L. (1950). Preliminary investigation of condensation nuclei under the electron microscope. *Trans. Amer. Geophys. Un.* **31**, 707. [93]
Hosler, C. L. (1951). On the crystallization of supercooled clouds. *J. Met.* **8**, 326. [224]
Hosler, C. L. (1954). Ice crystal formation. *Proc. Toronto Met. Conf.* (1953), p. 253. [206–7, 242]
Houghton, H. G. (1950). A preliminary quantitative estimate of precipitation mechanisms. *J. Met.* **7**, 363. [120, 161, 267, 284–5]

References and Author Index

Houghton, H. G. (1955). On the chemical composition of fog and cloud water. *J. Met.* **12**, 355. [96]

Houghton, H. G. and Cramer, H. E. (1951). A theory of entrainment in convective currents. *J. Met.* **8**, 95. [24]

Houghton, H. G. and Radford, W. H. (1938). On the measurement of drop size and liquid-water content in fogs and clouds. *M.I.T. Papers in Phys. Oceanog. Met.* **6**, No. 4. [114]

Howell, W. E. (1949). The growth of cloud drops in uniformly cooled air. *J. Met.* **6**, 134. [60, 123, 128, 129–32]

Howell, W. E., Wexler, R. and Braun, S. (1949). Contributions to the theory of the constitution of clouds. *Mt. Wash. Obs. Res. Rep.* [136]

Inn, E. C. Y. (1951). Photolytic inactivation of ice-forming silver iodide nuclei. *Bull. Amer. Met. Soc.* **32**, 132. [324]

International Meteorological Organization (1938). *Über aerologische Diagrammpapiere* (Berlin). [3]

Isono, K. (1955). On ice-crystal nuclei and other substances found in snow crystals. *J. Met.* **12**, 456. [248]

Isono, K. and Komabayasi, M. (1954). The influence of volcanic dust on precipitation. *J. Met. Soc. Japan*, Series 2, **32**, 345. [246]

Israel, H. and Schulz, L. (1932). Über die Grossenverteilung der atmosphaerischen Ionen. *Met. Z.* **49**, 226. [93]

Itoo, K. and Hama, K. (1956). Freezing of supercooled water-droplets. *Pap. Met. Geophys., Tokyo*, **6**, 247. [227]

Jacobi, W. (1955). Homogeneous nucleation in supercooled water. *J. Met.* **12**, 408. [206–7]

Jaffray, J. and Montmory, R. (1956a). Congélation orientée de l'eau surfondue sur des surfaces cristallines: 1° Cas du mica muscovite. *C.R. Acad. Sci., Paris*, **243**, 126. [221]

Jaffray, J. and Montmory, R. (1956b). Congélation orientée de l'eau surfondue sur des surfaces cristallines. Cas de l'iodure de plomb PbI_2. *C.R. Acad. Sci., Paris*, **243**, 891. [221, 323]

Jaffray, J. and Montmory, R. (1957). Congélation orientée de l'eau surfondu sur une surface cristalline. Cas de l'argent et de l'iodure d'argent. *C.R. Acad. Sci., Paris*, **244**, 859. [221]

James, D. G. (1954). Fine scale characteristics of the lowest 5000 ft. of the atmosphere as observed from an aircraft. *G.B. Air Min. MRP*, 875. [16]

Jeffreys, H. (1918). Some problems of evaporation. *Phil. Mag.* **35**, 270. [266]

Johnson, J. C. (1950). Measurement of the surface temperature of evaporating waterdrops. *J. Appl. Phys.* **21**, 22. [196]

Junge, C. (1952). Die Konstitution des atmosphärischen Aerosols. *Ann. Met.* (Beiheft). [61, 62, 63, 89, 90, 93, 94]

Junge, C. (1953). Die Rolle der Aerosole und gasformigen Beimengungen der Luft im Spurenstoffhaushalt der Troposphäre. *Tellus*, **5**, 1. [89, 90, 93]

Junge, C. (1955). The size distribution and ageing of natural aerosols as determined from electrical and optical data on the atmosphere. *J. Met.* **12**, 13. [74, 93]

Junge, C. (1957a). Remarks about the size distribution of natural aerosols. In *Artificial Stimulation of Rain*, ed. H. Weickmann and W. Smith (London: Pergamon Press), p. 3. [74, 75]

Junge, C. (1957b). The vertical distribution of aerosols over the ocean. In *Artificial Stimulation of Rain*, ed. H. Weickmann and W. Smith (London: Pergamon Press), p. 89. [97, 247]

Junge, C. (1957c). Some facts about meteoritic dust. In *Artificial Stimulation of Rain*, ed. H. Weickmann and W. Smith (London: Pergamon Press), p. 24. [254, 256]

Karasz, F. E., Champion, W. M. and Halsey, G. D. (1956). The growth of ice layers on the surfaces of anatase and silver iodide. *J. Phys. Chem.* **60**, 376. [217, 325]

Kassander, A. R., Sims, L. L. and McDonald, J. E. (1957). Observations of freezing nuclei over the South Western U.S. In *Artificial Stimulation of Rain*, ed. H. Weickmann and W. Smith (London: Pergamon Press), p. 393. [243-4]

Keily, D. P. (1954). Measurement of drop size distribution and liquid water content in natural clouds. *M.I.T. Dept. of Met.*, *Final Rep.* Contract AF 19 (122)-245. [107, 117]

Keith, C. H. and Arons, A. B. (1953). The growth of sea salt particles by condensation of atmospheric water vapour. *Woods Hole Oceanog. Inst. Tech. Rep.*, No. 5, Ref. 53-48. [178]

Kelvin, Lord (1870). On the equilibrium of vapour at a curved surface of a liquid. *Proc. Roy. Soc. Edinb.* **7**, 63. [42]

Khrgian, A. Kh. and Mazin, I. P. (1952). Distribution of drops according to size in clouds. *Tr. Central Aerolog. Obs. (USSR)*, No. 7, 56. [188]

Kinzer, G. D. and Cobb, W. E. (1958). Laboratory measurements and analysis of the growth and collection efficiency of cloud droplets. *J. Met.* **15**, 138. [146, 149]

Kinzer, G. D. and Gunn, R. (1951). The evaporation, temperature and thermal relaxation-time of freely falling waterdrops. *J. Met.* **8**, 71. [124, 192, 196]

Kirkwood, J. G. and Buff, F. P. (1949). The statistical mechanical theory of surface tension. *J. Chem. Phys.* **17**, 338. [47]

Kleber, W. and Weis, J. (1958). Keimbildung und Epitaxie von Eis (I). *Z. Kristallogr.* **110**, 30. [221]

Kline, D. B. (1949). Investigation of meteorological conditions associated with aircraft icing in layer-type clouds for 1947-48 Winter. *U.S. N.A.C.A. Tech. Note*, No. 1793. [9]

Kline, D. B. and Walker, J. A. (1951). Meteorological analysis of icing conditions encountered in low-altitude stratiform clouds. *U.S. N.A.C.A. Tech. Note*, No. 2306. [9, 10]

Knelman, F., Dombrowski, N. and Newitt, D. M. (1954). Mechanism of the bursting of bubbles. *Nature, Lond.*, **173**, 261. [70]

Kobayashi, T. (1957). Experimental researches on the snow crystal habit and growth by means of a diffusion cloud chamber. *Met. Soc. Japan, 75th Anniv. Vol.*, p. 38. [262]

Kobayashi, T. (1958). On the habit of snow crystals artificially produced at low pressures. *J. Met. Soc. Japan*, Series 2, **36**, 193. [264–5]

Köhler, H. (1921*a*, *b*). Zur Kondensation des Wasserdampfes in der Atmosphäre. *Geofys. Publ.* **2**, No. 3 and No. 6. [58]

Köhler, H. (1925*a*). Untersuchungen über die Elemente des Nebels und der Wolken. *Meddel. Met.-Hydr. Anst. Stockholm*, **2**, No. 5. [58, 96, 106, 189]

Köhler, H. (1925*b*). Über Tropfengruppen und einige Bemerkungen zur Genauigkeit der Tropfenmessungen besonders mit Rücksicht auf Untersuchungen von Richardson. *Met. Z.* **42**, 463. [189]

Köhler, H. (1926). Zur Thermodynamik der Kondensation an hygroskopischen Kernen und Bermerkungen über das Zussammenfliessen der Tropfen. *Meddel. Met.-Hydr. Anst. Stockholm*, **3**, No. 8. [58, 96]

Köhler, H. (1941). An experimental investigation on sea water nuclei. *Nova Acta Soc. Sci. Upsal.* **12**, No. 6. [69]

Komabayasi, M. (1957). Some aspects of rain formation in warm clouds. I. Salinity of individual raindrops and other quantities concerning rainfall. *J. Met. Soc. Japan*, **35**, 205. [178]

Kramers, H. (1946). Heat transfer from spheres to flowing media. *Physica*, **12**, 61. [124]

Krastanow, L. (1940). Über die Bildung der unterkuhlten Wassertropfen und der Eiskristalle in der freien Atmosphäre. *Met. Z.* **57**, 357. [210]

Krastanow, L. (1941). Beitrag zur Theorie der Tropfen- und Kristallbildung in der Atmosphäre. *Met. Z.* **58**, 37. [204]

Krastanow, L. (1943). Über die Bildung und das Wachstum der Eiskristalle in der Atmosphäre. *Met. Z.* **60**, 15. [260]

Kraus, E. B. and Smith, B. (1949). Theoretical aspects of cloud drop distributions. *Aust. J. Sci. Res.* Series A, **2**, 376. [128]

Kraus, E. B. and Squires, P. (1947). Experiments on the stimulation of clouds to produce rain. *Nature, Lond.*, **159**, 489. [300]

Kumai, M. (1951). Electron-microscope study of snow crystal nuclei. *J. Met.* **8**, 151. [248, 274]

Kuroiwa, D. (1955). Growth of snow crystals in a supercooled cloud. *Low Temp. Sci.* A, **14**, 1. Trans. T58J by D.R.B., Canada. [274]

Labrum, N. R. (1952). The scattering of radio waves by meteorological particles. *J. Appl. Phys.* **23**, 1324. [185, 291]

La Mer, V. K. and Gruen, R. (1952). A direct test of Kelvin's equation connecting vapour pressure and radius of curvature. *Trans. Faraday Soc.* **48**, 410. [57]

Landsberg, H. (1938). Atmospheric condensation nuclei. *Ergebn. kosm. Phys.* **3**, 155. [77, 79, 90, 91]

Landsberg, H. and Neuberger, H. (1938). On the frequency distribution of drop sizes in a sleet storm. *Bull. Amer. Met. Soc.* **19**, 354. [189]

Langham, E. J. and Mason, B. J. (1958). The heterogeneous and homogeneous nucleation of supercooled water. *Proc. Roy. Soc.* A, **247**, 493. [206–7, 242]

Langleben, M. P. (1954). The terminal velocity of snowflakes. *Quart. J. R. Met. Soc.* **80**, 174. [277]

Langmuir, I. (1944). Supercooled water droplets in rising currents of cold saturated air. *G.E. Res. Lab. Rep.* W-33-106-SC-65. [88, 123, 127]

Langmuir, I. (1947). *G.E. Res. Lab., First Quart. Prog. Rep. Met. Res.* [300]

Langmuir, I. (1948). The production of rain by a chain reaction in cumulus clouds at temperatures above freezing. *J. Met.* **5**, 175. [139–40, 144, 161–3]

Langmuir, I. (1950). A seven-day periodicity in weather in United States during April 1950. *Bull. Amer. Met. Soc.* **31**, 386. [334]

Langmuir, I. and Blodgett, K. B. (1946). A mathematical investigation of water droplet trajectories. *U.S. Army Air Forces Tech. Rep.*, No. 5418. [87, 104]

Langsdorf, A. (1936). A continuously sensitive cloud chamber. *Phys. Rev.* **49**, 422. [81]

Laws, J. O. (1941). Measurements of the fall-velocity of water-drops and raindrops. *Trans. Amer. Geophys. Un.* **22**, 709. [194]

Laws, J. O. and Parsons, D. A. (1943). The relation of raindrop size to intensity. *Trans. Amer. Geophys. Un.* **24**, 452. [183, 187–8]

Learnard, R. B. (1953). A summary of the icing conditions found on Mount Washington N.H. during the years from 1943 to 1952. *Mt. Wash. Obs. Sci. Rep.*, No. 4. [12]

Lenard, P. (1904). Ueber Regen. *Met. Z.* **21**, 248. [163, 182, 186, 194]

Levin, L. M. (1954*a*). The coagulation of charged cloud droplets. *Dok. Akad. Nauk SSSR*, **94**, 467. Trans. T263R of D.R.B., Canada. [149]

Levin, L. M. (1954*b*). Size distribution function for cloud droplets and rain drops. *Dok. Akad. Nauk SSSR*, **94**, 1045. Trans. T263r of D.R.B., Canada. [119, 188]

Levin, L. M. and Starostina, R. F. (1953). Some results from the investigation of cloud structure. *Dok. Akad. Nauk. SSSR*, **93**, 253. [12, 104]

Levine, J. (1959). Spherical vortex theory of bubble-like motion in cumulus clouds. *J. Met.* **16**, 653. [29]

Lewis, W. (1947). A flight investigation of the meteorological conditions conducive to the formation of ice on airplanes. *U.S. N.A.C.A. Tech. Note*, No. 1393. [9, 10]

Lewis, W. (1951*a*). Meteorological aspects of aircraft icing. In *Compendium of Meteorology*, ed. T. F. Malone (Boston: Amer. Met. Soc.), p. 1197. [8]

Lewis, W. (1951*b*). On a seven-day periodicity. *Bull. Amer. Met. Soc.* **32**, 192. [334]

Lewis, W. and Hoecker, W. H. (1949). Observations of icing conditions encountered in flight during 1948. *U.S. N.A.C.A. Tech. Note*, No. 1904. [9, 10]

Lewis, W., Kline, D. B. and Steinmetz, C. P. (1947). A further investigation of the meteorological conditions conducive to aircraft icing. *U.S. N.A.C.A. Tech. Note*, No. 1424. [9, 10]

Ligda, M. G. H. (1951). Radar storm observation. In *Compendium of Meteorology*, ed. T. F. Malone (Boston: Amer. Met. Soc.), p. 1265. [185]

References and Author Index

Linke, F. (1943). Kondensationskerne im Elektronmikroskop sichtbar gemacht. *Naturwissenschaften*, **19/20**, 230. [93]

Litvinov, I. V. (1956). Determination of the steady-state velocity of falling snow particles. *Izv. Akad. Nauk SSSR, Geophys. Series*, **7**, 853. Trans. T238R by D.R.B., Canada. [278]

Lodge, J. P. and Baer, F. (1954). An experimental investigation of the shatter of salt particles on crystallization. *J. Met.* **11**, 420. [70]

Loeb, L. B., Kip, A. F. and Einarsson, A. W. (1938). On the nature of ionic sign preference in Wilson cloud chamber condensation experiments. *J. Chem. Phys.* **6**, 264. [52]

Lonsdale, Dame K. (1958). The structure of ice. *Proc. Roy. Soc.* A, **247**, 424. [198, 200]

Ludlam, F. H. (1950). The composition of coagulation elements. *Quart. J. R. Met. Soc.* **76**, 52. [287–9]

Ludlam, F. H. (1951a). The heat economy of a rimed cylinder. *Quart. J. R. Met. Soc.* **77**, 663. [8]

Ludlam, F. H. (1951b). Structure of shower clouds. *Nature, Lond.*, **167**, 254. [296]

Ludlam, F. H. (1951c). The production of showers by the coalescence of cloud droplets. *Quart. J. R. Met. Soc.* **77**, 402. [161, 168]

Ludlam, F. H. (1952). The production of showers by the growth of ice particles. *Quart. J. R. Met. Soc.* **78**, 543. [289]

Ludlam, F. H. (1955). Artificial snowfall from mountain clouds. *Tellus*, **7**, 277. [328]

Ludlam, F. H. (1958). The hail problem. *Nubila*, **1**, 12. [29]

Ludlam, F. H. and Saunders, P. M. (1956). Shower formation in large cumuli. *Tellus*, **8**, 424. [19]

Ludlam, F. H. and Scorer, R. S. (1953). Convection in the atmosphere. *Quart. J. R. Met. Soc.* **79**, 317. [15]

Ludlam, F. H. and Scorer, R. S. (1957). *Cloud Study: A Pictorial Guide* (London: Murray). [4]

Luttermoser, R. I. and Brown, W. J. (1959). Freezing nuclei counter. *B.J. Electronics, Borg-Warner Corp. Status Rep. No. 2 to Nat. Sci. Found.*, Contract NSF C41. [236]

MacCready, P. B. (1952). Results of cloud seeding in central Arizona, Winter 1951. *Bull. Amer. Met. Soc.* **33**, 48. [333]

McDonald, J. E. (1953a). Homogeneous nucleation of supercooled water drops. *J. Met.* **10**, 416. [204–5, 209]

McDonald, J. E. (1953b). Erroneous cloud physics use of Raoult's law. *J. Met.* **10**, 68. [58]

McDonald, J. E. (1954). The shape and aerodynamics of large raindrops. *J. Met.* **11**, 478. [194]

McDonald, J. E. (1958). The physics of cloud modification. *Advanc. Geophys.* **5**, 223. [331–2]

Magono, C. (1951). On the fall velocity of snowflakes. *J. Met.* **8**, 199. [277–8]

Magono, C. (1953). On the growth of snow flake and graupel. *Sci. Rep. Yokohama Nat. Univ.*, Sec. 1, No. 2, 18. [277–82]

Magono, C. (1954). On the falling velocity of solid precipitation elements. *Sci. Rep. Yokohama Nat. Univ.*, Sec. 1, No. 3, 33. [277]

Malkus, J. S. (1949). Effects of wind shear on some aspects of convection. *Trans. Amer. Geophys. Un.* **30**, 19. [25, 26, 27, 28]

Malkus, J. S. (1952a). The slopes of cumulus clouds in relation to external wind shear. *Quart. J. R. Met. Soc.* **78**, 530. [27, 28]

Malkus, J. S. (1952b). Recent advances in the study of convective clouds and their interaction with the environment. *Tellus*, **4**, 71. [24, 28]

Malkus, J. S. (1954). Some results of a trade-cumulus cloud investigation. *J. Met.* **11**, 220. [27, 28]

Malkus, J. S. and Ronne, C. (1954). On the structure of some cumulonimbus clouds which penetrated the high tropical troposphere. *Tellus*, **6**, 351. [29]

Malkus, J. S. and Scorer, R. S. (1955). The erosion of cumulus towers. *J. Met.* **12**, 43. [18, 29]

Manson, J. E. (1955). X-ray diffraction study of silver iodide aerosols. *J. Appl. Phys.* **26**, 423. [315]

Manson, J. E. (1957). Calcium carbonate as an ice nucleus. *J. Met.* **14**, 85. [246]

Marshall, J. S., Langille, R. C. and Palmer, W. McK. (1947). Measurement of rainfall by radar. *J. Met.* **4**, 186. [277]

Marshall, J. S. and Langleben, M. P. (1954). A theory of snow crystal habit and growth. *J. Met.* **11**, 104. [262-4, 274]

Marshall, J. S. and Palmer, W. McK. (1948). The distribution of raindrops with size. *J. Met.* **5**, 165. [187-8]

Martyn, D. F. (1954). Comments on a paper by E. G. Bowen entitled 'The influence of meteoritic dust on rainfall'. *Aust. J. Phys.* **7**, 358. [255]

Mason, B. J. (1952a). The spontaneous crystallization of supercooled water. *Quart. J. R. Met. Soc.* **78**, 22. [204-5, 209]

Mason, B. J. (1952b). Production of rain and drizzle by coalescence in stratiform clouds. *Quart. J. R. Met. Soc.* **78**, 275. [137, 161, 169]

Mason, B. J. (1953). The growth of ice crystals in a supercooled water cloud. *Quart. J. R. Met. Soc.* **79**, 104. [261-3, 267, 268]

Mason, B. J. (1957). *The Physics of Clouds* (Oxford). [151, 261, 291]

Mason, B. J. (1958). The supercooling and nucleation of water. *Advanc. Phys.* **7**, 221. [222]

Mason, B. J. (1959). Recent developments in the physics of rain and rain making. *Weather*, **14**, 81. [320]

Mason, B. J. and Hallett, J. (1956). Artificial ice-forming nuclei. *Nature, Lond.*, **177**, 681. [225, 315]

Mason, B. J. and Maybank, J. (1958). Ice nucleating properties of some natural mineral dusts. *Quart. J. R. Met. Soc.* **84**, 235. [226, 246, 257]

Mason, B. J. and Ramanadham, R. (1953). A photoelectric raindrop spectrometer. *Quart. J. R. Met. Soc.* **79**, 490. [184]

Mason, B. J. and Ramanadham, R. (1954). Modification of the size distribution of falling raindrops by coalescence. *Quart. J. R. Met. Soc.* **80**, 388. [193]

Maulaud, J. (1950). Mésure de la température de la pluie. *J. Sci. Mét.* **2**, 75. [196]
May, K. R. (1945). The cascade impactor. An instrument for sampling coarse aerosols. *J. Sci. Instrum.* **22**, 187. [89, 105]
May, K. R. (1947). The development of an instrument for measuring droplet size and concentration in natural clouds. *G.B. Chem. Def. Exp. Station, Porton, Tech. Paper*, No. 30. [105]
May, K. R. (1950). The measurement of airborne droplets by the magnesium oxide method. *J. Sci. Instrum.* **27**, 128. [105]
Mazur, J. (1943). The number and size distribution of water particles in natural clouds. *Met. Off., London, Met. Res. Paper*, No. 109. [7, 9, 10, 104, 105]
Meteorology Research Inc. (1957). Report on the Pasadena cooperative program of ice nuclei measuring techniques. *Advisory Committee on Weather Control, U.S.A.* [238, 239, 241]
Meyer, J. and Pfaff, W. (1935). Zur Kenntnis der Kristallisation von Schmelzen III. *Zeit. Anorg. Chem.* **224**, 305. [206]
Millman, P. M. (1954). Meteor showers and rainfall. *J. R. Astr. Soc. Can.* **48**, 226. [256]
Montmory, R. (1955). L'epitaxie et la nucléation de l'eau surfondue. *Bull. Obs. Puy de Dome*, p. 108. [222, 225]
Montmory, R. (1956). La congélation de l'eau surfondue sur une surface cristalline. *Bull. Obs. Puy de Dome*, p. 126. [221]
Montmory, R. and Jaffray, J. (1957). Epitaxies de la glace sur l'iodure d'argent. *C.R. Acad. Sci., Paris*, **245**, 2221. [221]
Moore, D. J. and Mason, B. J. (1954). The concentration, size distribution and production rate of large salt nuclei over the oceans. *Quart. J. R. Met. Soc.* **80**, 583. [70]
Moran, P. A. P. (1959). The power of a cross-over test for the artificial stimulation of rain. *Aust. J. Stat.* **1**, 47. [335]
Morton, B. R. (1957). Buoyant plumes in a moist atmosphere. *J. Fluid Mech.* **2**, 127. [16, 24]
Morton, B. R., Taylor, G. I. and Turner, J. S. (1956). Turbulent gravitational convection from maintained and instantaneous sources. *Proc. Roy. Soc. A*, **234**, 1. [16]
Mossop, S. C. (1956a). Sublimation nuclei. *Proc. Phys. Soc. B*, **69**, 161. [226]
Mossop, S. C. (1956b). The nucleation of supercooled water by various chemicals. *Proc. Phys. Soc. B*, **69**, 165. [207, 224]
Mossop, S. C., Carte, A. E. and Hefferman, K. J. (1956). Counts of atmospheric freezing nuclei at Pretoria, January 1956. *Aust. J. Phys.* **9**, 556. [243, 256]
Mount Washington Observatory (1946). The multicylinder method. *Mon. Res. Bull.* **2**, No. 6. [106]
Murgatroyd, R. J. and Garrod, M. P. (1956). The measurements of natural freezing nuclei made by the Meteorological Research Flight during January 1956. *G.B. Air Min. MRP*, No. 998. [239-40, 244]
Naito, H. and Sugawara, K. (1954). Determination of chemical composition of silver iodide smoke for rain making. *Rep. Rain Making Japan*, **1**, 77. [315]

Nakaya, U. (1951). The formation of ice crystals. In *Compendium of Meteorology*, ed. T. F. Malone (Boston: Amer. Met. Soc.), p. 207.
[260, 262]
Nakaya, U. (1954). *Snow Crystals: Natural and Artificial* (Cambridge, U.S.A.: Harvard University Press). [260, 262]
Nakaya, U. (1955). Snow crystals and aerosols. *J. Fac. Sci. Hokkaido Univ.* Ser. 2, **4**, 341. [265]
Nakaya, U., Hanaxima, M. and Dezuno, K. (1939). Experimental researches on window hoar crystals, a general survey. *J. Fac. Sci. Hokkaido Univ.* Ser. 2, **3**, 1. [221]
Nakaya, U. and Terada, T. (1934). Simultaneous observations of the mass, falling velocity and form of individual snow crystals. *J. Fac. Sci. Hokkaido Univ.* Ser. 2, **1**, 191. [275–7, 282]
Nathan, A. M. and Hill, D. (1955). Design plan for a freezing nuclei meter. *New York University Res. Div.*, Tech. Rep. 389.01. [236]
Nathan, A. M. and Hill, D. (1957). Freezing nuclei meter. *New York University Res. Div.*, Tech. Rep. 389.02. [236]
Neel, C. B. and Steinmetz, C. P. (1952). A heated-wire liquid-water content meter for measuring icing severity. *U.S. N.A.C.A. Washington*, Tech. Note, No. 2615. [108]
Neiburger, M. (1949). Reflection, absorption and transmission of insolation by stratus cloud. *J. Met.* **6**, 98. [9]
Newkirk, J. B. and Turnbull, D. (1955). Nucleation of ammonium iodide crystals from aqueous solutions. *J. Appl. Phys.* **26**, 579. [216]
Newmann, J. (1954). Fluctuations of long-period accumulations of daily rainfall amounts. *Aust. J. Phys.* **7**, 522. [255]
Niederdorfer, E. (1932). Messungen der Grösse der Regentropfen. *Met. Z.* **49**, 1. [182, 189]
Nolan, P. J. and Pollak, L. W. (1946). The calibration of a photo-electric nucleus counter. *Proc. R. Irish Acad.* **51**A, 9. [79, 80]
Ockman, N. and Sutherland, G. B. B. M. (1958). Infra-red and Raman spectra of single crystals of ice. *Proc. Roy. Soc.* A, **247**, 434. [200]
Okita, T. (1958a). Water-blue film method for measurement of cloud and fog droplets. *J. Met. Soc. Japan*, **36**, 164. [105]
Okita, T. (1958b). Observations on vertical change of raindrop size distribution. *Sci. Rep. Tohoku Univ.* Ser. 5, **10**, 1. [191, 193]
Oliver, M. B. and Oliver, V. J. (1955). Rainfall and stardust. *Bull. Amer. Met. Soc.* **36**, 147. [255]
Öpik, E. J. (1951). Astronomy and the bottom of the sea. *Irish Astr. J.* **1**, 145. [254]
Orr, C., Hurd, F. K., Hendrix, W. P. and Junge, C. (1958). The behaviour of condensation nuclei under changing humidities. *J. Met.* **15**, 240. [61]
Ouchi, K. (1954). Freezing mechanism of supercooled water. *Sci. Rep. Tohoku Univ.* Ser. 5, **6**, 43. [205]
Oura, H. and Hori, J. (1953). Optical method of measuring drop-size distribution of fog. In *Studies in Fogs* (Hokkaido University), p. 327.
[107]

Owens, G. V. (1957). A pneumatically operated cloud droplet sampler. *University of Chicago Cloud Phys. Lab. Tech. Note*, No. 7. [106]
Owston, P. G. (1958). The structure of ice-I, as determined by X-ray and neutron diffraction analysis. *Advanc. Phys.* **7**, 171. [200]
Palmer, H. P. (1949). Natural ice-particle nuclei. *Quart. J. R. Met. Soc.* **75**, 15. [239]
Papée, H. M. (1959). Microcalorimetry of adsorption of water vapour on lead di-iodide. *Canad. J. Chem.* **37**, 375. [217]
Pauling, L. (1935). Structure and entropy of ice and of other crystals with randomness of atomic arrangement. *J. Amer. Chem. Soc.* **57**, 2680. [200]
Pauthenier, M. and Brun, E. (1940). Dénombrement des gouttelettes d'un brouillard au moyen d'un champ électrique ionise. *C.R. Acad. Sci., Paris*, **211**, 295. [104]
Pauthenier, M. and Cochet, R. (1950). Influence de la charge électrique de l'obstacle dans le méchanisme de captation de particules en suspension dans un fluide en mouvement. *C.R. Acad. Sci., Paris*, **231**, 213. [148]
Pauthenier, M. and Loutfoullah, N. (1950). Le balayage électrique des brouillards. *C.R. Acad. Sci., Paris*, **231**, 953. [148]
Pearcey, T. and Hill, G. W. (1956). The accelerated motion of droplets and bubbles. *Aust. J. Phys.* **9**, 19. [146]
Pearcey, T. and Hill, G. W. (1957). A theoretical estimate of the collection efficiencies of small droplets. *Quart. J. R. Met. Soc.* **83**, 77.
[141–2, 144, 145, 190, 281]
Perkins, P. J. (1959). Summary of statistical icing cloud data measured over United States and North Atlantic, Pacific and Arctic Oceans during routine aircraft operations. *N.A.S.A. Memo* 1-19-59 E. [12]
Pettersson, H. (1958). Rate of accretion of cosmic dust on the earth. *Nature, Lond.*, **181**, 330. [256]
Phillips, B. B. and Kinzer, G. D. (1958). Measurements of the size and electrification of droplets in cumuliform clouds. *J. Met.* **15**, 369. [151]
Picca, R. (1956). Nouveaux essais de désactivation de l'iodure d'argent en taut qu'agent glacogène. *C.R. Acad. Sci., Paris*, **242**, 489. [323]
Pollak, L. W. (1952). A condensation nuclei counter with photographic recording. *Geofis. pur. appl.* **22**, 75. [79]
Pollak, L. W. (1957). Methods of measuring condensation nuclei. *Geofis. pur. appl.* **36**, 21. [76, 80]
Pollak, L. W. and Murphy, T. (1953). Comparison of photoelectric nuclei counters. *Geofis. pur. appl.* **25**, 44. [80]
Pollak, L. W. and O'Connor, T. C. (1955). A photo-electric condensation nucleus counter of high precision. *Geofis. pur. appl.* **32**, 139. [80]
Pound, G. M., Madonna, L. A. and Peake, S. L. (1953). Critical supercooling of pure water droplets by a new microscopic technique. *J. Colloid Sci.* **8**, 187. [207]
Pound, G. M., Madonna, L. A. and Sciulli, C. (1951). Kinetics of nucleation in atmospheric phase transitions. *Carnegie Inst. of Tech., Metals Res. Lab. Quart. Rep.*, No. 5. [47, 210]

Pound, G. M., Simnad, M. T. and Yang, L. (1954). Heterogeneous nucleation of crystals from vapour. *J. Chem. Phys.* **22**, 1215. [54]

Powell, C. F. (1928). Condensation phenomena at different temperatures. *Proc. Roy. Soc.* A, **119**, 553. [47]

Priestley, C. H. B. and Ball, F. K. (1955). Continuous convection from an isolated source of heat. *Quart. J. R. Met. Soc.* **81**, 144. [16]

Pruppacher, H. R. and Sänger, R. (1955a). Mechanismus der Vereisung unterkühlter Wassertropfen durch disperse Keimsubstanzen. *Zeit. Angew. Math. Phys.* **6**, 407. [224, 309]

Pruppacher, H. R. and Sänger, R. (1955b). Mechanismus der Vereisung unterkühlter Wassertropfen durch disperse Keimsubstanzen. *Zeit. Angew. Math. Phys.* **6**, 485. [225]

Ranz, W. E. and Wong, J. B. (1952). Impaction of dust and smoke particles on surface and body collectors. *Industr. Engng Chem. (Industr.)*, **44**, 1371. [87, 88, 89, 281–2]

Rau, W. (1950). Über die Wirkungsweise der Gefrierkerne in unterkühlten Wasser. *Z. Naturf.* **5a**, 667. [242]

Rau, W. (1953). Über die Einfluss des Tropfenvolumens auf die Unterkühlbarkeit von Wassertropfen und die Bedeutung des Gefrierkernspectrums. *Z. Naturf.* **8a**, 197. [242]

Rau, W. (1954). Die Gefrierkerngehalte der Verschiedenen Luftmassen. *Met. Rdsch.* **7**, 205. [239, 241, 244, 249]

Reiss, H. (1952). The statistical mechanical theory of irreversible condensation. *J. Chem. Phys.* **20**, 1216. [45]

Reynolds, S. E., Hume, W. and McWhirter, M. (1952). Effects of sunlight and ammonia on the action of silver iodide particles as sublimation nuclei. *Bull. Amer. Met. Soc.* **33**, 26. [325]

Reynolds, S. E., Hume, W., Vonnegut, B. and Schaefer, V. J. (1951). Effect of sunlight on the action of silver iodide particles as sublimation nuclei. *Bull. Amer. Met. Soc.* **32**, 47. [323]

Rich, T. A. (1955). A photo-electric nucleus counter with size discrimination. *Geofis. pur. appl.* **31**, 60. [80]

Richardson, E. G. (1953). Processes of convection and evaporation. *Brit. J. Appl. Phys.* **4**, 65. [124]

Rigby, E. C., Marshall, J. S. and Hitschfeld, W. (1954). The development of the size distribution of raindrops during their fall. *J. Met.* **11**, 362. [193]

Rittenberger, W. (1959). Zur Struktur der Wolken. *Arch. Met., Wien*, A, **11**, 333. [12]

Rooth, C. (1957). On a special aspect of the condensation process and its importance in the treatment of cloud particle growth. *Tellus*, **9**, 372. [122]

Rosenblatt, P. and La Mer, V. K. (1946). Motion of a particle in a temperature gradient; thermal repulsion as a radiometer phenomenon. *Phys. Rev.* **70**, 385. [86]

Rossby, C. G. (1936). Dynamics of steady ocean currents in the light of experimental fluid mechanics. *Papers in Phys. Ocean. and Met., MIT and Woods Hole Ocean. Inst.* **5**, No. 1. [22]

References and Author Index 371

Roulleau, M. (1957). Méchanismes de la congélation de l'eau. *J. Sci. Mét.* **9**, 127. [227, 321]

Roulleau, M. (1958). Méchanisme de la congélation de l'eau. *J. Sci. Mét.* **10**, 10. [205, 207]

Ryde, J. W. (1947). The attenuation of radar echoes produced at centimetre wavelengths by various meteorological phenomena. In *Meteorological Factors in Radio Wave Propagation* (Phys. Soc. and Roy. Met. Soc.), p. 169. [184, 290]

Sander, A. and Damköhler, G. (1943). Übersättigung bei spontanen Keimbildung in Wasserdampf. *Naturwissenschaften*, **31**, 460. [47, 210]

Sano, I., Fujitani, Y. and Maena, Y. (1956). An experimental investigation on ice nucleating properties of some chemical substances. *J. Met. Soc. Japan*, Ser. 3, **34**, 104. [226, 316, 320]

Sano, I. and Fukuta, N. (1956a). Effects of water vapour, ammonia and hydrogen sulphide against the decay of silver iodide smoke under irradiation of ultraviolet light. *J. Met. Soc. Japan*, Ser. 2, **34**, 34. [217, 316, 325]

Sano, I. and Fukuta, N. (1956b). Observations of ice-nucleating temperature of some chemical substances. *J. Met. Soc. Japan*, Ser. 2, **34**, 293. [224]

Sartor, D. (1954). A laboratory investigation of collision efficiencies, coalescence and electrical charging of simulated cloud droplets. *J. Met.* **11**, 91. [146, 150, 153]

Sawyer, K. F. and Walton, W. H. (1950). The 'Conifuge'—a size-separating sampling device for airborne particles. *J. Sci. Instrum.* **27**, 372. [87]

Saxton, R. L. and Ranz, W. E. (1952). Thermal force on an aerosol particle in a temperature gradient. *J. Appl. Phys.* **23**, 917. [86]

Schaefer, V. J. (1945). Demountable rotating multicylinders for measuring liquid water content and particle size of clouds in above or below freezing temperatures. *G.E. Res. Lab. Rep.*, Contract W33-038-AC-9151. [106]

Schaefer, V. J. (1946). The production of ice crystals in a cloud of supercooled water droplets. *Science*, **104**, 457. [299]

Schaefer, V. J. (1948a). The production of clouds containing supercooled water droplets or ice crystals under laboratory conditions. *Bull. Amer. Met. Soc.* **29**, 175. [206]

Schaefer, V. J. (1948b). The detection of ice nuclei in the free atmosphere. Project Cirrus, Occasional Rep. No. 9, Gen. Elec. Co. [230, 233]

Schaefer, V. J. (1949). The formation of ice crystals in the laboratory and the atmosphere. *Chem. Rev.* **44**, 291. [245]

Schaefer, V. J. (1953). Project Cirrus final rep. under contract No. DA-36-039-SC-15345, G.E. Res. Lab., Schenectady, R.L.-785, p. 77. [311]

Schaefer, V. J. (1954a). The concentration of ice nuclei in air passing the summit of Mount Washington. *Bull. Amer. Met. Soc.* **35**, 310. [244, 249]

Schaefer, V. J. (1954b). Silver and lead iodides as ice-crystal nuclei. *J. Met.* **11**, 417. [222, 226]

Schaefer, V. J. (1954c). Properties of particles of snow and the electrical effects they produce in storms. *Trans. Amer. Geophys. Un.* **28**, 587.
[277]

Schaefer, V. J. (1957). The question of meteoritic dust in the atmosphere. In *Artificial Stimulation of Rain*, ed. H. Weickmann and W. Smith (London: Pergamon Press), p. 18. [257]

Schindelhauer, F. (1925). Versuch einer Registrierung der Tropfenzahl bei Regenfällen. *Met. Z.* **42**, 25. [183]

Schmidt, F. H. (1947). Some speculations on the resistance to the motion of cumuliform clouds. *Koninklijk Nederlands Met. Inst. de Bilt*, No. 125, Meded. Verhand Serie B, **1**, (8). [22]

Scholz, J. (1931). Ein neuer Apparat zur Bestimmung der Zahl der geladenen und ungeladenen Kerne. *Z. InstrumKde.* **51**, 505. [78]

Scholz, J. (1932). Vereinfachter Bau eines Kernzählers. *Met. Z.* **49**, 381.
[78]

Schotland, R. M. (1957a). The collision efficiency of cloud drops. In *Artificial Stimulation of Rain*, ed. H. Weickmann and W. Smith (London: Pergamon Press), p. 170. [147]

Schotland, R. M. (1957b). The collision efficiency of cloud drops of equal size. *J. Met.* **14**, 381. [147]

Schotland, R. M. (1957c). Droplet collision efficiency. *Res. Div., New York University College of Eng.*, Final Rep., Contract No. AF 19(604)-993.
[147]

Schotland, R. M. and Kaplin, E. J. (1956). The collision efficiency of cloud droplets. *Res. Div., New York University Coll. of Eng. Sci. Rep.*, No. 1, Contract No. AF 19(604)-993. [147]

Schrage, R. W. (1953). *A Theoretical Study of Interphase Mass Transfer* (New York: Columbia University Press). [122]

Scorer, R. S. (1957). Experiment on the convection of isolated masses of buoyant fluid. *J. Fluid Mech.* **2**, 583. [18, 19]

Scorer, R. S. and Ludlam, F. H. (1953). The bubble theory of penetrative convection. *Quart. J. R. Met. Soc.* **79**, 96. [17, 23, 28]

Scorer, R. S. and Ronne, C. (1956). Experiments with convection bubbles. *Weather*, **11**, 151. [18, 19]

Sears, G. W. (1958). Effect of poisons on crystal growth. *J. Chem. Phys.* **29**, 1045. [265]

Serra, L. (1954). Diffusion dans l'atmosphère des noyaux d'iodure d'argent. *Bull. Obs. Puy de Dome*, p. 65. [341]

Shaw, D. and Mason, B. J. (1955). The growth of ice crystals from the vapour. *Phil. Mag.* **46**, 249. [221, 262, 272]

Sidgwick, N. V. (1950). *Chemical Elements and their Compounds* (Oxford), vol. I, p. 123. [309]

Simpson, G. C. (1941). Sea-salt and condensation nuclei. *Quart. J. R. Met. Soc.* **67**, 163. [176]

Smith, E. J. (1949). Experiments in seeding cumuliform cloud layers with dry ice. *Aust. J. Sci. Res.* A, **2**, 78. [300]

Smith, E. J. and Heffernan, K. J. (1954). Airborne measurements of the concentration of natural and artificial freezing nuclei. *Quart. J. R. Met. Soc.* **80**, 182. [231–2, 239–40, 243–4, 316–17, 339, 341]
Smith, E. J. and Heffernan, K. J. (1956). The decay of the ice nucleating properties of silver iodide released from a mountain top. *Quart. J. R. Met. Soc.* **82**, 301. [324]
Smith, E. J., Heffernan, K. J. and Thompson, W. J. (1958). The decay of the ice-nucleating properties of silver iodide smoke released from an aircraft. *Quart. J. R. Met. Soc.* **84**, 162. [324]
Smith, E. J., Heffernan, K. J. and Seely, B. K. (1955). The decay of ice-nucleating properties of silver iodide in the atmosphere. *J. Met.* **12**, 379. [324]
Smith, E. J., Kassander, A. R. and Twomey, S. (1956). Measurements of natural freezing nuclei at high altitudes. *Nature, Lond.*, **177**, 82. [243]
Soulage, G. (1955a). Résultats preliminaires d'une étude expérimentale des noyaux glaçogènes naturels. *Arch. Met. Wien*, **8**, 211. [248]
Soulage, G. (1955b). Étude de générateurs de fumées d'iodure d'argent. *Bull. Obs. Puy de Dome*, p. 1. [317]
Soulage, G. (1957). Les noyaux de congélation de l'atmosphere. *Ann. Géophys.* **13**, 103. [248]
Soulage, G. (1958a). Contribution des fumées industrielles à l'enrichissement de l'atmosphère en noyaux glacogènes. *Bull. Obs. Puy de Dome*, No. 4, 121. [246]
Soulage, G. (1958b). Influence des émissions d'iodure d'argent de l'Association d'Études sur le pouvoir glaçogène de l'air en periode grêligène. *Assoc. d'Études de moyens de lutte contre les fleaux atmospheriques*, Toulouse, Report No. 6, p. 24. [306]
Spilhaus, A. F. (1948). Raindrop size, shape and falling speed. *J. Met.* **5**, 108. [194]
Squires, P. (1952a). The growth of cloud drops by condensation. I. General characteristics. *Aust. J. Sci. Res.* A, **5**, 59. [123, 124, 128, 132]
Squires, P. (1952b). The growth of cloud drops by condensation. II. The formation of large cloud drops. *Aust. J. Sci. Res.* A, **5**, 473. [136]
Squires, P. (1956). The micro-structure of cumuli in maritime and continental air. *Tellus*, **8**, 443. [110, 121]
Squires, P. (1958a). The microstructure and colloidal stability of warm clouds. I. The relation between structure and stability. *Tellus*, **10**, 256. [110–15, 117, 119, 121]
Squires, P. (1958b). The microstructure and colloidal stability of warm clouds. II. The causes of variations in microstructure. *Tellus*, **10**, 262. [110, 112, 133]
Squires, P. (1958c). The spatial variation of liquid water and droplet concentration in cumuli. *Tellus*, **10**, 372. [11, 20, 21, 22, 30, 105, 112]
Squires, P. (1958d). Penetrative downdraughts in cumuli. *Tellus*, **10**, 381. [30]
Squires, P. and Gillespie, C. A. (1952). A cloud-droplet sampler for use on aircraft. *Quart. J. R. Met. Soc.* **78**, 387. [106]

Squires, P. and Twomey, S. (1957). Some observations of sea-salt nuclei in Hawaii during Project Shower. *Tellus*, **9**, 538. [179]

Squires, P. and Twomey, S. (1958). Some observations relating to the stability of warm cumuli. *Tellus*, **10**, 272. [121, 180, 295]

Squires, P. and Warner, J. (1957). Some measurements in the orographic cloud of the island of Hawaii and in trade wind cumuli. *Tellus*, **9**, 475. [12, 110]

Stommel, H. (1947). Entrainment of air into a cumulus cloud. I. *J. Met.* **4**, 91. [22, 28]

Stommel, H. (1951). Entrainment of air into a cumulus cloud. II. *J. Met.* **8**, 127. [23, 28]

Styles, R. S. and Campbell, F. W. (1953). Radar observations of rain from non-freezing clouds. *Aust. J. Phys.* **6**, 73. [186]

Swinbank, W. C. (1947). Collision of cloud droplets. *Nature, Lond.*, **159**, 849. [145]

Swinbank, W. C. (1954). Comments on a paper by E. G. Bowen entitled 'The influence of meteoritic dust on rainfall'. *Aust. J. Phys.* **7**, 354. [255]

Takahasi, Y. (1934). *J. Met. Soc. Japan*, Ser. 2, **12**. [190]

Telford, J. (1955). A new aspect of coalescence theory. *J. Met.* **12**, 436. [171-2]

Telford, J. (1960a). Freezing nuclei above the tropopause. *J. Met.* **17**, 86. [244]

Telford, J. (1960b). Freezing nuclei from industrial processes. *Bull. Amer. Met. Soc.* **16**, 676. [246]

Telford, J. (unpublished). [298-9]

Telford, J. W., Thorndike, N. S. and Bowen, E. G. (1955). The coalescence between small water drops. *Quart. J. R. Met. Soc.* **81**, 241. [145, 150, 153]

Thom, H. C. S. (1958). A statistical method of evaluating augmentation of precipitation by cloud seeding. *U.S. Advisory Committee for Weather Control, Final Rep.* **2**, 5. [333]

Thomson, J. J. (1888). *Application of Dynamics to Physics and Chemistry*, 1st ed. (Cambridge), p. 165. [49]

Tohmfor, G. and Volmer, M. (1938). Die Keimbildung unter dem Einfluss elektrischer Ladungen. *Ann. Phys., Lpz.*, Ser. 5, **33**, 109. [49]

Tolman, R. C. (1949). The effect of droplet size on surface tension. *J. Chem. Phys.* **17**, 333. [47]

Tominaga, H. and Kinumaki, S. (1954). Note on a method of spraying AgI-in-Liq. NH_3 solution for the purpose of the artificial seeding. *Sci. Rep. Tokohu Univ.*, Ser. 5, **6**, 39. [312]

Trabert, W. (1901). Die Extinktion des Lichtes in einem truben Medium (Sehweite in Wolken). *Met. Z.* **18**, 518. [14]

Tsuji, M. (1950). On the growth of cloud particles and the degree of super-saturation in convective clouds. *J. Met. Soc. Japan*, **28**, 129. [128]

Turnbull, D. (1950). Kinetics of heterogeneous nucleation. *J. Chem. Phys.* **18**, 198. [57, 217, 219]

References and Author Index

Turnbull, D. and Fisher, J. C. (1949). Rate of nucleation in condensed systems. *J. Chem. Phys.* **17**, 71. [204]
Turnbull, D. and Vonnegut, B. (1952). Nucleation catalysis. *Industr. Engng Chem. (Industr.)*, **44**, 1292. [212, 216]
Turner, J. S. (1955). The salinity of rainfall as a function of drop size. *Quart. J. R. Met. Soc.* **81**, 418. [179]
Twomey, S. (1953a). On the measurement of precipitation intensity by radar. *J. Met.* **10**, 66. [186]
Twomey, S. (1953b). The identification of individual hygroscopic particles in the atmosphere by a phase transition method. *J. Appl. Phys.* **24**, 1099. [61]
Twomey, S. (1954). The composition of hygroscopic particles in the atmosphere. *J. Met.* **11**, 334. [61, 95]
Twomey, S. (1955). The distribution of sea-salt nuclei in air over land. *J. Met.* **12**, 81. [95, 97, 170, 180, 247]
Twomey, S. (1956). The electrification of individual cloud droplets. *Tellus*, **8**, 445. [151]
Twomey, S. (1958). Quantitative aspects of seeding rates for use in supercooled clouds. *Bull. Obs. Puy de Dome*, p. 333. [303–4]
Twomey, S. (1959a). The nuclei of natural cloud formation. I. The chemical diffusion method and its application to atmospheric nuclei. *Geofis. pur. appl.* **43**, 227. [83–5, 134]
Twomey, S. (1959b). The nuclei of natural cloud formation. II. The supersaturation in natural clouds and the variation of cloud droplet concentration. *Geofis. pur. appl.* **43**, 243. [83, 85, 101, 128, 131–2, 135]
Twomey, S. (1959c). The influence of droplet concentration on rain formation and stability in clouds. *Bull. Obs. Puy de Dome*, p. 1. [157–8]
Twomey, S. (1959d). An experimental test of the Volmer theory of heterogeneous nucleation. *J. Chem. Phys.* **30**, 941. [54, 57]
Twomey, S. and McMaster, K. N. (1955). The production of condensation nuclei by crystallizing salt particles. *Tellus*, **7**, 458. [70]
Van der Merwe, J. H. (1949). Misfitting monolayers and oriented overgrowth. *Disc. Faraday Soc.* **5**, 201. [212]
Verma, A. R. (1953). *Crystal Growth and Dislocations* (London: Butterworth). [271]
Verzàr, F. (1953). Kondensationskernzähler mit automatischer Registrierung. *Arch. Met., Wien*, A, **5**, 372. [79]
Vollrath, R. E. (1936). A continuously active cloud chamber. *Rev. Sci. Instrum.* **7**, 409. [83]
Volmer, M. (1939). *Kinetik der Phasenbildung* (Dresden and Leipzig: Steinkopff), p. 100. [52]
Volmer, M. and Flood, H. (1934). Tröpfchenbildung in Dämpfen. *Z. phys. Chem.* A, **170**, 273. [47–8]
Volmer, M. and Weber, A. (1926). Keimbildung in übersättigen Gebilden. *Z. Phys. Chem.* **119**, 277. [43–4]
Vonnegut, B. (1947). The nucleation of ice formation by silver iodide. *J. Appl. Phys.* **18**, 593. [299, 301, 309]

Vonnegut, B. (1948). Influence of butyl alcohol on shape of snow crystals formed in the laboratory. *Science*, **107**, 621. [265]

Vonnegut, B. (1949a). A capillary collector for measuring the deposition of water drops on a surface moving through clouds. *Rev. Sci. Instrum.* **20**, 110. [8, 107]

Vonnegut, B. (1949b). Nucleation of supercooled water clouds by silver iodide smokes. *Chem. Rev.* **44**, 277. [317]

Vonnegut, B. (1950). Techniques for generating silver iodide smoke. *J. Colloid Sci.* **5**, 37. [312–13]

Vonnegut, B. (1951). Silver iodide smoke. Project Cirrus Occasional Rep. No. 13, in Final Rep. Contract W-36-039-SC-38141, G.E. Res. Lab., Schenectady, RL-566, p. 27. [312–13]

Vonnegut, B. and Moore, C. B. (1958). Preliminary attempts to influence convective electrification in cumulus clouds by the introduction of space charge into the lower atmosphere. *Conf. on Atmos. Elect.*, Wentworth-by-the-sea, N.H., A.D. Little Inc., Cambridge, Mass. [298]

Vonnegut, B. and Neubauer, R. (1951). Recent experiments on the effect of ultraviolet light on silver iodide nuclei. *Bull. Amer. Met. Soc.* **32**, 356. [324]

Von Straten, F. W., Ruskin, R. E. and Mastenbrook, H. J. (1958). Preliminary experiments using carbon black for cloud modification and formation. *U.S. Naval Res. Lab. Rep.*, No. 5235. [306]

Wahl, E. (1951). On a seven-day periodicity in weather in the United States during April, 1960. *Bull. Amer. Met. Soc.* **32**, 193. [334]

Wait, G. R. and Parkinson, W. D. (1951). Ions in the atmosphere. In *Compendium of Meteorology*, ed. T. F. Malone (Boston: Amer. Met. Soc.), p. 120. [152]

Warner, J. (1955). The water content of cumuliform cloud. *Tellus*, **7**, 449. [10, 20]

Warner, J. (1957). An instrument for the measurement of freezing nucleus concentration. *Bull. Obs. Puy de Dome*, p. 33. [235, 239–41, 243]

Warner, J. and Bigg, E. K. (1956). The effects of air temperature and pressure on the decay of silver iodide. *Bull. Amer. Met. Soc.* **37**, 94. [326]

Warner, J. and Newnham, T. D. (1952). A new method of measurement of cloud-water content. *Quart. J. R. Met. Soc.* **78**, 46. [12, 108]

Warner, J. and Newnham, T. D. (1958). Time lag in ice crystal nucleation in the atmosphere. *Bull. Obs. Puy de Dome*, p. 1. [235]

Warner, J. and Squires, P. (1958). Liquid water content and the adiabatic model of cumulus development. *Tellus*, **10**, 390. [11, 14, 108]

Warner, J. and Twomey, S. (1956). The use of silver iodide for seeding individual clouds. *Tellus*, **8**, 453. [302]

Warren, D. R. and Nesbitt, M. V. (1955). An airborne silver iodide dispensing burner. *Aust. Dept. of Supply, Aero. Res. Lab., Mech. Eng. Note*, ARL/ME 200. [313–14, 337]

Watson, H. H. (1936). A system for obtaining dust samples from mine air. *Trans. Instn Min. Metall.* **46**, 155. [86]

Webb, W. L. and Gunn, R. (1955). The net electrification of natural cloud droplets at the earth's surface. *J. Met.* **12**, 211. [150]

References and Author Index 377

Wegener, A. (1911). *Thermodynamik der Atmosphare*, (Leipzig: J. A. Barth). [283]

Weickmann, H. (1947). Die Eisphase in der Atmosphare. *Rep. and Trans.*, 716 (*Volkenrode*), *Min. of Supply*, London. [206, 248]

Weickmann, H. (1951). A theory of the formation of ice crystals. *Arch. Met., Wien*, A, **4**, 309. [222, 225]

Weickmann, H. (1957a). Current understanding of the physical processes associated with cloud nucleation. *Beitr. Phys. Atmos.* **30**, 97. [301, 305]

Weickmann, H. (1957b). Recent measurements of the vertical distribution of Aitken nuclei. In *Artificial Stimulation of Rain*, ed. H. Weickmann and W. Smith (London: Pergamon Press), p. 81. [92]

Weickmann, H. and aufm Kampe, H. J. (1953). Physical properties of cumulus clouds. *J. Met.* **10**, 204. [10, 14, 20, 21, 104, 112, 116]

Went, F. W. (1956). *Some Aspects of Plant Research in Australia* (Melbourne: C.S.I.R.O.). [65]

Wexler, H. (1958). Modifying weather on a large scale. *Science*, **128**, 1059. [347]

Wexler, R. (1948). Rain intensities by radar. *J. Met.* **5**, 171. [186]

Wexler, R. (1951). Theory and observation of radar storm detection. In *Compendium of Meteorology*, ed. T. F. Malone (Boston: Amer. Met. Soc.), p. 1283. [185]

Wexler, R. (1954). Growth of rain in warm clouds. *Woods Hole Oceanog. Inst. Tech. Rep.*, No. 28, Ref. 54–32. [177]

Weyl, W. A. (1951). Surface structure of water and some of its physical and chemical manifestations. *J. Colloid Sci.* **6**, 389. [202, 279]

Whipple, F. L. (1950). The theory of micro-meteorites. I. In an isothermal atmosphere. *Proc. Nat. Acad. Sci., Wash.*, **36**, 687. [254]

Whipple, F. L. (1951). The theory of micro-meteorites. II. In heterothermal atmospheres. *Proc. Nat. Acad. Sci., Wash.*, **37**, 19. [254]

Whipple, F. L. and Hawkins, G. S. (1956). On meteors and rainfall. *J. Met.* **13**, 236. [256]

Whipple, F. J. W. and Chalmers, J. A. (1944). On Wilson's theory of the collection of charge by falling drops. *Quart. J. R. Met. Soc.* **70**, 103. [151]

Whytlaw-Gray, R. and Patterson, H. S. (1932). *Smoke* (London: Arnold and Co.), ch. 6. [71]

Wieland, W. (1955). Eine neue Methode des Kondensationskernzählung. *Eidg. Kom. Stud. Hagelbildung Hagelabwehr, E.T.H. Zurich*, Publ. No. 6. [79]

Wieland, W. (1956). Die Wasserdampfkondensation an naturlichem Aerosol bei geringen Ubersattigungen. *Eidg. Tech. Hochsch. Zurich*, Promotionsarbeit No. 2577. [82, 83, 99, 100]

Wiesner, J. (1895). Beitrage zur Kenntnis des tropischen Regens. *Ber. Akad. Wiss. Wien*, **104**, 1397. [182]

Wigand, A. (1919). Die vertikale Verteilung des Kondensationskerne in der freien Atmosphare. *Ann. Phys., Lpz.*, **59**, 689. [92]

Wilson, C. T. R. (1897). Condensation of water vapour in the presence of dust-free air and other gases. *Phil. Trans.* A, **189**, 265. [47, 48]

Wilson, C. T. R. (1899). On the comparative efficiency as condensation nuclei of positively and negatively charged ions. *Phil. Trans.* A, **193**, 289. [48]

Wilson, C. T. R. (1929). Some thunderstorm problems. *J. Franklin Inst.* **208**, 1. [151]

Woodcock, A. H. (1940). Convection and soaring over the open sea. *J. Mar. Res.* **3**, 248. [16]

Woodcock, A. H. (1950). Sea salt nuclei in a tropical storm. *J. Met.* **7**, 397. [97]

Woodcock, A. H. (1951). Atmospheric salt particles and raindrops. *Woods Hole Ocean. Inst. Contr.* N6 ONR-27702 *Tech. Rep.*, No. 14. [247]

Woodcock, A. H. (1952). Atmospheric salt particles and raindrops. *J. Met.* **9**, 200. [97, 176]

Woodcock, A. H. (1953). Salt nuclei in marine air as a function of altitude and wind force. *J. Met.* **10**, 362. [97, 98, 170]

Woodcock, A. H. and Blanchard, D. C. (1955). Tests of the salt-nuclei hypothesis of rain formation. *Tellus*, **7**, 437. [179]

Woodcock, A. H. and Gifford, M. (1949). Sampling atmospheric sea-salt nuclei over the ocean. *J. Mar. Res.* **8**, 177. [88, 97]

Woodcock, A. H., Kientzler, C. F., Arons, A. B. and Blanchard, D. C. (1953). Giant condensation nuclei from bursting bubbles. *Nature, Lond.*, **172**, 1144. [70]

Woodcock, A. H. and Wyman, J. (1947). Convective motion in air over the sea. *Ann. N.Y. Acad. Sci.* **48**, 749. [16]

Woodward, B. (1959). The motion around isolated thermals. *Quart. J. R. Met. Soc.* **85**, 144. [18, 19]

Workman, E. J. and Reynolds, S. E. (1949). Thunderstorm electricity. *New Mexico Instn Min. Tech., Prog. Rep.*, No. 6. [239, 241]

World Meteorological Organization (1956). *International Cloud Atlas*, vols. 1 and 2. [4]

Wright, H. L. (1936). The size of atmospheric nuclei: some deductions from measurements on the numbers of charged and uncharged nuclei at Kew Observatory. *Proc. Phys. Soc.* **48**, 675. [58]

Wulff, G. (1901). Zur Frage der Geschwindigkeit des Wachstums und der Auflösung der Kristallflächen. *Z. Kristallogr.* **34**, 449. [259–60]

Wylie, R. G. (1953). The freezing of supercooled water in glass. *Proc. Phys. Soc.* B, **66**, 241. [206–7]

Yamamoto, G. and Ohtake, T. (1953). Electron microscope study of cloud and fog nuclei. *Sci. Rep. Tohoku Univ.* Ser. 5, **5**, 141. [94]

Zacharov, I. (1952). Influence des Perseides sur la transparence atmospherique. *Bull. Central Astr. Instn, Czechoslovakia*, **3**, 82. [255]

Zaitsev, V. A. (1950). Liquid water content and distribution of drops in cumulus clouds. *Trudy Glavnoi Geofiz. Obs.* No. 19 (81), 12. Trans. *Nat. Res. Council of Canada* TT 395. [10, 14, 20, 112, 117, 118]

SUPPLEMENTARY REFERENCES

American Meteorological Society (1963). Severe Local Storms. *Met. Monogr.* **5** (No. 27).
Barnes, G. T. (1963). Phase transitions in water sorbed on ice forming nuclei. *Zeits angew. Math. Phys.* (ZAMP) **14**, 510.
Battan, L. J. (1959). Radar meteorology. Univ. of Chicago Press, Chicago.
Battan, L. J. (1964). Some observations of vertical velocities and precipitation sizes in a thunderstorm. *J. Appl. Met.* **3**, 415.
Bigg, E. K. and Miles, G. T. (1964). The results of large-scale measurements of natural ice nuclei. *J. Atmos. Sci.* **21**, 396.
Bigg, E. K., Mossop, S. C., Meade, R. T. and Thorndike, N. S. C. (1963). The measurement of ice nucleus concentration by means of millipore filters. *J. Appl. Met.* **2**, 266.
Borovikov, A. M., Khrgian, A. Kh. *et al* (1963). Cloud Physics. Israel Program for Scientific Translations, Jerusalem.
Bowen, E. G. (1965). Lessons learned from long-term cloud-seeding experiments. *Proc. Int. Conf. on Cloud Physics, Tokyo and Sapporo*, p. 429.
Bradley, D. A., Woodbury, M. A. and Brier, G. W. (1962). Lunar synodical period and widespread precipitation. *Science* **137**, 748.
Braham, R. R. (1964). What is the role of ice in summer rain showers? *J. Atmos. Sci.* **21**, 640.
Brier, G. W. and Bradley, D. A. (1964). The lunar synodical period and precipitation in the United States. *J. Atmos. Sci.* **21**, 386.
Edwards, G. R. and Evans, L. F. (1960, 1961, 1965). Ice nucleation by silver iodide, Parts I, II. *J. Met.* **17**, 627; **18**, 760; Parts III, IV to be published.
Edwards, G. R. and Evans, L. F. (1962). Effect of surface charge on ice nucleation by silver iodide. *Trans. Faraday Soc.* **58**, 1649.
Evans, L. F. (1965). Requirements of an ice nucleus. *Nature* **206**, 822.
Fletcher, N. H. (1962). Surface structure of water and ice. *Phil. Mag.* **7**, 225; (1963) **8**, 1425.
Fletcher, N. H. (1963). Nucleation by crystalline particles. *J. Chem. Phys.* **38**, 237.
Fletcher, N. H. (1964). Crystal interfaces. *J. Appl. Phys.* **35**, 234.
Fletcher, N. H. (1965). The Theory of ice crystal nucleation. *Proc. Internat. Conf. on Cloud Physics, Tokyo and Sapporo*, p. 458.
Fukuta, N. (1963). Ice nucleation by metaldehyde. *Nature* **199**, 475.
Garten, V. A. and Head, R. B. (1965). A theoretical basis of ice nucleation by organic crystals. *Nature* **205**, 160.
Head, R. B. (1962). Ice nucleation by some cyclic compounds. *J. Phys. Chem. Solids* **23**, 1371.
Hirth, J. P. and Pound, G. M. (1963). *Condensation and evaporation.* (Progress in Materials Science, Vol. II) Pergamon Press, London.
Koenig, L. R. (1963). The glaciating behaviour of small cumulonimbus clouds. *J. Atmos. Sci.* **20**, 29.
Kavanau, J. L. (1964). *Water and Solute-water Interactions*, pp. 8-23, Holden-Day Inc., San Francisco.

Latham, J. and Mason, B. J. (1961a). Electric charge transfer associated with temperature gradients in ice. *Proc. Roy. Soc.* **A.260,** 523.
Latham, J. and Mason, B. J. (1961b). Generation of electric charge associated with the formation of soft hail in thunderclouds. *Proc. Roy. Soc.* **A.260,** 537.
Lothe, J. and Pound, G. M. (1962). Reconsiderations of nucleation theory. *J. Chem. Phys.* **36,** 2080.
Malkus, J. S. and Simpson, R. H. (1964a). Modification experiments on tropical cumulus clouds. *Science* **145,** 451.
Malkus, J. S. and Simpson, R. H. (1964b). Notes on the potentialities of cumulonimbus and hurricane seeding experiments. *J. Appl. Met.* **3,** 470.
Mason, B. J. (1965). The nucleation and growth of ice crystals. *Proc. Internat. Conf. on Cloud Physics, Tokyo and Sapporo,* p. 467.
Mason, B. J., Bryant, G. W. and Van den Heuvel, A. P. (1963). *Phil. Mag.* **8,** 505. The growth habits and surface structure of ice crystals.
Mossop, S. C. (1963). Atmospheric ice nuclei. *Zeits. angew Math. Phys.* (ZAMP), **14,** 456.
Némethy, G. and Scheraga, H. A. (1962). Structure of water and hydrophobic bonding in proteins. I. A model for the thermodynamic properties of liquid water. *J. Chem. Phys.* **36,** 3382.
Oriani, R. A. and Sundquist, B. E. (1963). Emendations to nucleation theory and the homogeneous nucleation of water from the vapor. *J. Chem. Phys.* **38,** 2082.
Shafrir, U. and Neiburger, M. (1963). Collision efficiencies of two spheres falling in a viscous medium. *J. Geophys. Res.* **68,** 4141.
Smith, E. J., Adderley, E. E. and Walsh, D. T. (1963). A cloud-seeding experiment in the Snowy Mountains, Australia. *J. Appl. Met.* **2,** 324.
Telford, J. W. and Cottis, R. E. (1964). Cloud droplet collisions. *J. Atmos. Sci.* **21,** 549.
Telford, J. W. and Thorndike, N. S. C. (1961). Observation of small drop collisions. *J. Met.* **18,** 382.
Van der Merwe, J. H. (1963). Crystal Interfaces. I and II. *J. Appl. Phys.* **34,** 117, 123.
Vonnegut, B. (1963). Some facts and speculations concerning the origin and role of thunderstorm electricity. *Met. Monogr.* **5,** 224.
Vul'fson, N. I. (1964). *Convective motions in a free atmosphere.* Israel Program for Scientific Translations, Jerusalem.
Vul'fson, N. I. and Levin, L. M. (eds) (1963). *Studies of Clouds, Precipitation and Thunderstorm electricity.* Amer. Met. Soc., Boston.
Woods, J. D. and Mason, B. J. (1964). Experimental determination of collection efficiencies for small water droplets in air. *Quart. J. Roy. Met. Soc.* **90,** 373.
Woods, J. D. and Mason, B. J. (1965). The wake capture of water drops in air. *Quart. J. Roy. Met. Soc.* **91,** 35.
Zettlemoyer, A. C., Tcheurekdjian, N. and Hosler, C. L. (1963). Ice nucleation by hydrophobic substrates. *Zeits. angew. Math. Phys.* (ZAMP) **14,** 496.

SUBJECT INDEX

adsorption of water-vapour
  on nuclei, 217
  on silver iodide, 217, 325
adsorption of impurities on silver iodide, 323, 325
aerosols
  coagulation of, 71-5
  diffusion of, 72-3
  sources, 64-71
air-mass
  and cloud stability, 109, 156-9
  condensation nuclei in, 93-102, 133
  continental, 93-6, 98-102, 133, 249
  ice-nuclei in, 249
  maritime, 97-102, 133, 249
aircraft
  icing of, 7-8, 12
  in cloud-seeding, 336-8
Aitken nuclei
  and ice-nuclei, 249
  concentration, 91-2
  counting, 76-9
  size, 93
  sources, 64-71
  variation with height, 92
ammonia
  and dispersion of AgI, 311-12
  and photolysis of AgI, 325
ammonium sulphate, 94
atmosphere
  condensation nuclei in, 64-102
  ice-nuclei in, 229-58
  standard I.C.A.N., 192-3
Australia, rain-making experiments in, 344-5

birds, and convection, 16
bubble theory, 17-19
  of cloudy convection, 28-9
burners, silver iodide, 312-14
  ammonia, 312
  charcoal, 312
  cord, 312
  optimum efficiency of, 322
  output characteristic of, 316-18
  solution, 313-14

cadmium iodide, 226
capillary collector, 8, 108
carbon black
  for cloud dissipation, 306
  for cloud formation, 306
  for droplet sampling, 105
  for raindrop sampling, 105
cells
  convection, 16
  in thunderstorms, 18
chain reaction in rain production, 163
charcoal burners for AgI, 312
chemical potential, 39-40
Clausius-Clapeyron equation, 125, 204, 220
clay minerals as ice-nuclei, 247-50
cloud census, 120-1, 302, 327-8
cloud chambers
  chemical gradient, 83-5, 101, 262
  thermal gradient, 81-3, 100, 262-3
cloud modification, 293-308
  dry-ice seeding, 300-1, 305
  electrical techniques, 298-9
  hail suppression, 306-7
  lightning suppression, 306-7
  non-freezing clouds, 294-9, 308
  parameters available, 294
  quantitative aspects of AgI seeding, 303-4
  salt seeding, 295-6
  silver iodide seeding, 301-4
  sub-freezing clouds, 299-308
clouds
  cirrus, and meteor showers, 254
  continental: droplet concentration in, 110-13, 134-5; droplet size in, 114, 157; modification techniques, 293-308; rainfall from, 173-5; stability of, 119-21, 156-9, 293-4
  convective: depth for precipitation, 120, 166-8, 174-5; droplet concentration in, 20, 110-13, 134-5;

clouds (cont.)
  droplet size-distribution, 113–19; downdraughts in, 20, 25; entrainment by, 22–8; formation of, 5, 6, 15–30; liquid water content of, 10–11, 13–15, 109; radar echoes from, 186, 291–2; rain formation in, 160–81; roots of, 15–16; sea-salt nuclei in, 180; seeding of, 337; stability of, 119–21, 156–9; structure of, 19–22; updraughts in, 5, 20
  cumulus, see convective
  detrainment by, 23–4
  dissipation of, 304–6
  downdraughts in, 20, 25, 29
  droplets in: concentration of, 110–13, 134–5; electrical charges on, 147–53; growth of, by coalescence, 138–56, 171–3; growth of, by condensation, 122–38; sampling of, 104–6; size-distribution of, 113–19, 136–8, 171
  dynamics of, 1–30
  entrainment by, 15, 18–19, 22–8
  formation of, 2–3
  forms of, 4–7
  glaciation of, 303–4, 347
  instability of: by coalescence, 34, 293; by ice-crystals, 35, 294; induced, 36, 293–306; parameters governing, 294
  layer: dissipation of, 305; droplet concentration, 110–13, 135; droplet sizes, 113–14; formation of, 5, 7; liquid water content of, 7–9, 12–13; radar echoes from, 186; rainfall from, 5, 161, 169; seeding of, 305
  lifetime of, 5, 7, 15, 20, 27, 137, 302
  liquid water content of: compared with adiabatic value, 14, 15, 21; comparison of types, 7–13; convective cloud, 10–11, 13–15, 109, 293; for hail formation, 288–9; layer-cloud, 7–9, 12–13; measurement of, 8, 12, 14, 107–109
  maritime: droplet concentration, 110–13, 134–5; droplet sizes, 114, 157; Hawaiian orographic, 115, 176–9, 189; modification of, 294; rainfall from, 173–5, 189; stability of 119–21, 158, 293–4
  microphysics of, 31–6

  modification of, 293–308: see cloud modification
  noctilucent, 253, 255
  non-freezing: microstructure of, 103–21; modification of, 294–9, 308; rainfall from, 160–96; stability of, 156–9; theory of development of, 122–59
  orographic: formation of, 6; Hawaiian, 115, 176–9, 189; liquid water content of, 12; raindrop sizes from, 189; rainmaking in, 328, 336; sea-salt nuclei in, 176–9
  stability of: dynamic, 3, 18; microphysical, 34–5, 119–21, 156–9, 173–5, 293–4
  sub-freezing: glaciation of, 304; growth of ice-crystals in, 261, 273–4; instability of, 35; in the atmosphere, 3, 33, 35; in ice-nucleus counters, 230; modification of, 299–308; precipitation from, 283–92
  supercooled, see above, sub-freezing
  updraughts (q.v.) in, 4–7, 20
  wind shear and, 21, 25, 26
coagulation
  of aerosols, 71–5
  of silver iodide smoke, 324
coalescence of cloud droplets, 138–56
  and cloud stability, 156–9
  on collision, 154–5
  critical radius for, 144
  as discrete process, 171–2
  effect on droplet spectra, 137
  electrical effects on, 147–53
  experimental results, 144–7
  and growth of raindrops, 161–75
  prevention of, in sampling, 104
  theory of, 139–44
coalescence of raindrops, 190, 191
cold box, see counters, ice-nucleus
collection efficiency
  definition, 139, 141
  of drops and droplets, 138–56, 158, 173
  of ice-crystals, 281–2
  see also coalescence
collision efficiency, see collection efficiency
collision of cloud droplets, see coalescence
combustion, as source of nuclei, 67

Subject Index

condensation
  in ascending air, 2
  in the atmosphere, 32–3, 122–38
  cloud-droplet growth by, 122–38
  cloud-droplet spectrum by, 136
  heterogeneous, 48–63
  homogeneous, 39–45
  on insoluble particles, 55–7
  on insoluble surfaces, 52–5
  on ions, 48–52
  on mixed particles, 62–3
  on soluble particles, 58–62
  plume theory and, 26
  thermodynamics of, 38–9
condensation nuclei
  ions as, 48–52
  insoluble particles as, 55–8
  insoluble surfaces as, 52–5
  mixed, 62–3
  soluble, 58–62
condensation nuclei in the atmosphere, 64–102
  active at low supersaturations, 98–102
  activity of, 99–101, 130, 133
  capture by droplets, 71
  chemical composition of, 94–5
  and cloud droplets, 129–35
  and cloud modification, 294–5
  coagulation of, 71–5
  concentration of, 91, 95, 98–102
  content of air-masses, 99–102
  continental nuclei, 93–9
  experimental methods, 74–90
  removal of, 76, 295
  sea-salt, 97–8, 159
  sizes of, 67, 70, 74, 75, 93–108
  sources of, 64–71: combustion, 67; dispersion, 68–71; gas reactions, 65–6
  spectrum of, 99–101, 130, 133
contact angle, 52–3
  analogue for solids, 211
  and nucleation efficiency, 55–7, 63, 214–15
continental clouds, *see* clouds, continental
control area, in rain-making experiments, 332–5
convection
  and birds, 16
  and clouds, 5–6, 15–30
  below cloud base, 16

convective instability, 5
correlation, statistical
  in meteor hypothesis, 253–5
  in rain-making experiments, 332
counters
  cloud droplet, 104–7
  condensation nucleus: Aitken, 76–9; chemical gradient, 83–5; photoelectric, 79–80; Scholz, 78; thermal gradient, 81–3
  ice-nucleus: constant temperature, 233–4; continuous, 236; controlled cooling, 231–2; difficulties with, 237–8; expansion chamber, 231, 234–5; thermal precipitation, 236–7
C.S.I.R.O., 344–5
crystals, theory of growth of, 270–1:
  see also ice-crystals
cupric sulphide, 225
cybotactic groups, 201
cylinders, rotating
  droplet size-distribution by, 106
  impaction efficiency of, 88–9
  liquid water measurement by, 12

detrainment by clouds, 23–4
diffusion
  of aerosols, 72–3, 247
  of water-vapour, 1, 122–4, 264–5
  self, in liquids, 204
disk, spinning, 145
dislocations
  and condensation, 58
  and crystal growth, 270–1
  at solid interface, 211–12
dispersion
  of solids, 68–9
  of solutions, 69–71
dissipation of cloud, 304–6
Doppler shift, in radar echo, 185
downdraughts
  in cumuli, 20, 25
  penetrative, 29
droplets, cloud
  concentration of, 20, 110–13, 134–5
  electric charges on, 147–53
  freezing of, 33, 242–3
  growth by coalescence, 137, 138–59, 171–3
  growth by condensation, 122–38
  growth in cloud, 127–56
  sampling techniques, 104–7

droplets, cloud (cont.)
  size-distribution of, 113–19, 136–8, 171
droplets, freezing of
  cloud, 33, 242–3
  AgI suspension, 227
  pure water, 207
dry ice
  cloud seeding with, 300–1, 305
  production of ice-crystals by, 299
dye
  for droplet sampling, 105
  for raindrop sampling, 182

Einstein equation, 73
elastic effects in nucleation, 212, 216
electric charges on cloud droplets
  effect on coalescence, 148–50, 155, 298–9
  experimental, 150–1
  production of, 298–9
  sign of, 151
  theory of, 151–3
electric fields
  in the atmosphere, 153
  effect on coalescence, 149–50, 153, 181
  and condensation on ions, 52
  in thunderstorms, 151, 153
electrical effects
  on droplet coalescence, 147–53, 155, 181
  production of artificial, 298–9
  on raindrop shape, 195
electrostatic precipitation, 104, 316
embryos
  charged, 49
  concentration of, 43
  critical, 42, 53, 60, 203–4
  crystalline, 203
  elastic strain in, 212–13
  free energy of, 41, 49, 53, 55, 204, 208, 213
  in phase changes, 39, 57
  retention of, 57, 226
  size-distribution of, 41
entrainment, 15, 18–28
  and bubble theory, 18–19
  and detrainment, 23–4
  and jet theory, 22
  and plume theory, 28
entropy
  effect in ice-nucleation, 212

and free energy, 38
of fusion, 204
and orientation, 202
epitaxy, 211, 222
evaporation of raindrops, 190–4
expansion, adiabatic
  in the atmosphere, 2–3
  saturation ratio produced by, 77–8
  water released by, 4

filter paper, for drop sampling, 182–3
flames, as source of nuclei, 67, 92
flour, for drop sampling, 182
fluctuations, phase, 39
  and condensation, 42, 50, 58
  and freezing, 201
fluctuation, statistical
  in coalescence events, 171–2
  in rainfall, 330–2
free energy, 38
  of charged embryo, 49
  and condensation, 40–1
  of crystalline embryo, 203, 208
  and elastic strain, 212–13
  of embryo, 41, 53
  of ice, 209, 259–60
  of ice-water interface, 204–5, 207
  surface, 40, 45
  and Wulff's theorem, 259–60
freezing of droplets
  AgI suspension, 227
  cloud, 33, 242–3
  pure water, 207
freezing, nucleation of, see nucleation of freezing
freezing nuclei, see ice-forming nuclei
frost, in ice-nucleus counters, 231, 233, 237

gas reactions, 65–6
glaciation of clouds, 303–4, 347
graupel
  density, 275
  growth of, 286–7
  shape of, 282
  terminal velocity, 276
  time for formation, 287

habit
  of condensation embryo, 41, 52
  of ice-crystals, 260–5, 273
  of ice-embryo, 203, 205, 208, 211, 217

Subject Index

hail
  formation of, 287–9
  suppression of, 306–7, 344
heat flow
  in droplet growth, 125
  in hail growth, 287–8
  in ice-coated cylinders, 8
  in ice-crystal growth, 267
humidity, effect on decay of AgI, 325
hurricanes, effect of seeding, 346–7
hydrogen bonding in ice, 199–200
hygroscopic particles
  and cloud formation, 130–2
  condensation on, 58–62
  from the sea, 69–71
  identification of, 61
  phase-changes of, 61–2
hysteresis in phase-changes, 62

I.C.A.N. standard atmosphere, 192–3
ice
  allotropic forms, 198
  crystal structure, 197–200, 223
  dielectric constant, 291
  entropy, 200
  lattice parameters, 200
  surface free energy, 209, 259–60
  surface free energy against water, 204–5, 207
  surface-structure, 279
ice-crystals
  and cloud instability, 8, 35
  detection of, 230–1, 233
  epitaxial growth, 222
  growth from vapour, 259–73
  growth in cloud, 261, 273–4
  growth rate, 266–73, 284: effect of cloud droplets, 274; effect of habit, 267; effect of supersaturation, 267–72; effect of temperature, 268, 284; electrical analogue, 266–7; molecular theory, 268–72
  habit, 260–5, 273; effect of impurities, 265; effect of supersaturation, 262–5; effect of temperature, 261–5; effect on growth rate, 266–7; effect on terminal velocity, 276; equilibrium, 260; limiting, 273
  relation between mass and size, 275
  terminal velocity, 275–9
  and thunderstorms, 153, 307

ice-forming nuclei
  growth of ice on, 211
  lattice misfit, 225
  nucleation thresholds, 224–6
  surface-structure, 212, 228
ice-forming nuclei, activity
  industrial smokes, 246
  meteor dust, 257
  minerals, 226, 246
  pure chemicals, 254–6
  soils, 245
  volcanic dust, 246
ice-forming nuclei in the atmosphere, 229–58
  above tropopause, 250
  activity spectrum, 239–45
  concentration, 243–5, 255, 283; spatial variations, 243–4, 249–50; temporal variations, 244–5, 255
  correlation: with meteorological variables, 248–9; with meteor showers, 255
  experimental methods, 229–39
  glaciation produced by, 304
  meteor hypothesis, 250–8
  nuclei from snow crystals, 247–8
  origins of, 245–58
  peaks in concentration, 255
  size-distribution, 246, 248, 254
icing of aircraft, 7–8, 12
impaction
  of cloud droplets, 104–9
  of nuclei, 87–9
impaction efficiency
  calculations, 87–9
  of jets, 88
  of obstacles, 89
impactor, cascade, 89
industry
  as source of Aitken nuclei, 66, 91
  as source of ice-nuclei, 246
instability, convective, 5–6
instability of clouds
  by coalescence, 34, 293
  containing ice-crystals, 35, 294
  induced, 36, 293–306
inversions
  and Aitken count, 93
  and ice-nuclei, 244
  and orographic cloud, 177
  and sea-salt, 247
ions
  charging of droplets by, 151–2

ions (cont.)
  condensation on, 48–52, 93
  removal of, 46
  production of, 298–9

jets
  impaction efficiency of, 88
  lateral entrainment by, 22

kaolinite, 246
Kelvin equation, 42, 57
  experimental verification, 57
konimeter, 89–90

lapse-rate
  adiabatic, 2–3, 6
  in clouds, 19, 20
layer-clouds, see clouds, layer
lead iodide
  adsorption of water-vapour, 217
  crystal structure, 223, 225
  ice-epitaxy on, 222
lifetime of clouds, 5, 7, 15, 20, 27, 137, 302
light, ultraviolet, action on AgI, 323–5
lightning
  as source of nuclei, 65
  suppression of, 306–7
liquid water
  and downdraughts, 30
  release by expansion, 3, 4
liquid water in clouds
  compared with adiabatic value, 14, 15, 21
  comparison of cloud types, 7–9, 12–13
  convective cloud, 10–11, 13–15, 109, 293
  for hail formation, 288–9
  layer-cloud, 7–9, 12–13
  measurement of, 8, 12, 14, 107–9
  spatial variation, 20–2
liquids
  self-diffusion in, 204
  structure of, 201–2
log-normal distribution, 69
  condensation nuclei, 94
  raindrops, 188
  silver iodide smoke, 316
Ludlam limit, 8

maritime clouds, see clouds, maritime
melting band, radar, 186
mesh, for drop sampling, 182

meteor dust
  accretion rate, 255–7
  as ice-nucleus: in atmosphere, 252; in laboratory, 257
  melting of, 254
  time of fall, 254
meteor hypothesis, 250–8
  correlations of rainfall with: cirrus clouds, 254; dust clouds, 254; freezing nucleus peaks, 255; meteor showers, 253; noctilucent clouds, 253, 255; snowfall, 253
  criticisms of, 255–7
  rainfall peaks, 251–4
  time-delay, 254
meteor showers
  and noctilucent clouds, 253, 255
  periodic, 253
  and rainfall peaks, 252–3
mica, 221
microphysics of clouds, 31–6
minerals, as ice-nuclei, 245–6
misfit of crystals
  and epitaxy, 222–3
  and ice-nucleation, 212, 216, 225
mixing ratio, adiabatic, 3
model experiments
  cloud growth, 18
  droplet collision, 146–7
  ice-crystal growth, 264
mountains
  cloud observations, 117–18
  orographic cloud, 6, 177

noctilucent clouds, 253, 255
nucleation of condensation
  heterogeneous, 48–63: insoluble particles, 55–8; ions, 48–52; mixed particles, 62–3; rate, 51, 54, 56; size-effect, 55–7; soluble particles, 58–62; surfaces, 52–5
  homogeneous, 39–48: difficulties in theory, 45; experiment, 45–8; rate, 46, 47; theory, 39–45
nucleation of freezing
  heterogeneous, 210–28: experiments, 205–7, 221–8; nuclei, 212; rate, 214; size-effect, 214, 227; and sublimation, 217–19, 224
  homogeneous, 33, 202–7: effect of droplet size, 207; experiment, 205–7; habit of embryo, 203; rate, 204; theory, 202–5

Subject Index

nucleation of freezing (*cont.*)
  solutions, 219–21
nucleation of sublimation, 208–28
  heterogeneous, 210–28: and freezing, 217–19, 224; experiment, 221–8; nuclei, 212; rate, 213; size-effect, 215, 226
  homogeneous, 33, 208–10: and condensation, 209–10; experiment, 210; rate, 208
  nuclei, condensation, freezing, sublimation, *see* condensation nuclei *or* ice-forming nuclei

obstacles, impaction by, 87–90
oil, for drop sampling, 104–5
optical methods
  cloud drop sizes, 107, 117
  ice-crystal detection, 230, 232, 236
  liquid water content, 107
  raindrop sizes, 183–4
orientation of molecules
  oil-water interface, 147
  water surface, 51, 154, 202
orographic clouds, *see* clouds, orographic
overseeding, 303–4, 346

peaks
  in rainfall, 251–3, 255
  in ice-nucleus count, 255
penetrative downdraughts, 29
phase-changes
  hysteresis in, 62
  in salt droplets, 61–2
  nucleation of, 37
  thermodynamics of, 38–9
plume theory, 16–17, 24, 28
plumes
  dilution of, 339
  smoke, 324, 338–42
  vertical extent of, 341
  width of, 340
potential, chemical, 39, 40
Prandtl number, 124
precipitation
  electrostatic, 104, 316
  induced, 36, 293–308, 327–47
  mechanisms, 35, 160–96, 283–92
  thermal, 86, 316
  see also graupel, hail, snow, rain
precipitation intensity
  and radar echo, 186
  and raindrop size-distribution, 187–9

radar
  band echoes, 186, 291–2
  cloud census, 302
  column echoes, 186, 292
  detection of precipitation, 121, 182, 184–6
  echo-intensity and precipitation, 186
  echoes from rain, 184–6
  echoes from snow, 185, 290–1
  melting band, 291
  wind detection, 20
radiation, solar, 2, 324, 347
rain from non-freezing clouds, 160–96, 294–9
  maritime and continental clouds, 173–5
  sea-salt nuclei, 175–81
  stimulation of, 294–9
  time for development, 166, 172, 175
rain from sub-freezing clouds, 283–92, 299–308, 327–47
  coalescence, 290
  graupel, 286–7
  hail, 287–9
  ice-crystal growth, 283–5
  radar studies, 290–2
  rain-making, 327–47
  snowflakes, 285–6
  stimulation of, 299–308
raindrop size-spectrum
  determination of, 182–4
  expressions for, 186–9
  in Hawaiian rain, 189
  modification below cloud base, 190–4
  preferred drop sizes, 189–90
  and radar echo, 186
  variation with height, 191–2
  variation with time, 183, 191–2
raindrop spectrograph, 179, 183
raindrops
  coalescence of, 190
  deformation, 194
  growth, 161–75
  sampling of, 182–4
  sea-salt nuclei, 175–81
  shattering of, 163–95
  size-distribution, 163, 166–7, 174
  temperature, 196
  terminal velocity, 194–5
rainfall
  decrease by seeding, 343–4, 346
  increase by seeding, 327–8, 333, 343–4

rainfall (*cont.*)
  peaks and meteor showers, 251–2, 255
  periodicities, 250–3, 334
  variability, 330–2
rain-making experiments, 327–47
  control area, 332–4
  design of, 335–43
  economics of, 336, 338
  effect of cloud base, 337
  effect of cloud depth, 296–7, 300
  ice-nucleus concentration: by aircraft seeding, 337; by ground burners, 339–41
  randomized, 334–5
  results of 343–5
  seeding: from aircraft, 336–8; from ground, 338–43
  serial, 329–32
  statistical design, 329–35
randomization, in rain-making, 334–5
Raoult's law, 58, 220
replicas
  of cloud droplets, 105
  of ice-crystals, 261
Reynolds number, 124, 142
rime, *see* graupel
roots, of cumuli, 15–16

salinity
  of rain, 179
  of rime, 96
salt
  as cloud seeding agent, 295–6
  concentration in rain, 179
  nuclei from 70, 95–8
saturation ratio, *see* supersaturation
sea, as source of nuclei, 69–71, 95–8
sea-salt nuclei
  and cloud stability, 159
  concentration over continents, 95–7, 247, 295
  concentration over sea, 97–8
  and rainfall, 170, 175–81
  size-distribution, 97–8
sedimentation
  of cloud droplets, 15
  of nuclei, 78
seeding of clouds, *see also* cloud modification
  from aircraft, 336–8
  from ground burners, 338–43
  large-scale effects, 347

seeding-rate
  with dry ice, 301, 305
  with silver iodide, 303–4, 337
shattering of drops, 163
silver iodide
  adsorption of impurities, 323, 325
  adsorption of water, 217, 325
  allotropic forms, 309–10
  as condensation nucleus, 57, 319
  crystal structure, 223, 309–10
  dispersion of, 311–14
  as freezing nucleus, 227, 321
  as ice-forming nucleus, 225, 318–21
  as sublimation nucleus, 319–20
  ice-epitaxy on, 222, 323
  photolysis, 323–5
  solubility, 310, 312
silver iodide smoke
  chemical composition, 315
  in cloud modification, 299, 301–7
  coagulation, 324
  deactivation, 322–5
  mixed smokes, 321–2
  nucleation behaviour, 316–22
  photo-deactivation, 323–5, 338
  production of, 312–14
  size-distribution, 316
  size effect in nucleation, 55–7, 207, 214–15, 216–17, 319–21
smoke
  coagulation of, 71–5
  condensation nuclei in, 66–7, 91
  ice-nuclei in, 246
  plumes of, 338–42
snow crystals, nuclei in, 247–8
snowflakes, *see also* ice-crystals
  aggregation of, 279
  growth of, 279–82, 285–6
  terminal velocity, 275–8
  time for formation, 286
  symmetry, 260–1
Snowy Mountains experiment, 335, 345
sodium iodide, 310, 312–13, 315, 321–2
soil
  condensation nuclei from, 68, 92, 96
solutions
  dispersion of, 69–71
  freezing of, 219–21
  ice-nuclei from, 247–9
  vapour-pressure of, 58–9
spider web, 61, 88

stability of clouds
 dynamic, 3, 18
 microphysical, 34–5, 119–21, 156–9, 173–5, 293–4
statistical design, in cloud seeding, 329–35
statistical significance
 of correlations in meteor hypothesis, 258
 of rainfall increases, 331–3, 345
 of rainfall peaks, 251, 255
stratosphere
 dust clouds in, 254
 transit of meteor dust, 253
sublimation, *see* nucleation of sublimation
sublimation nuclei, *see also* ice-forming nuclei
 decay of, 227
 ideal, 320
 pre-activation of, 226, 257
sugar solution, 231, 233
supercooling, critical
 additional for solutions, 220
 additional for strain, 216
 in ascending air, 128
supersaturation
 over charged droplets, 49
 by chemical gradient, 85
 in clouds, 127, 129–34, 165
 for condensation: homogeneous, 45–7; on insoluble particles, 56; on ions, 48, 51; on soluble particles, 60; on surfaces, 55
 for crystal growth, 270
 over droplets, 42
 by expansion, 77–8
 for freezing, 205–7, 214, 218–19
 and freezing or sublimation, 217–19
 for homogeneous sublimation, 210
 influence on ice-growth, 267–72
 influence on ice-habit, 262–5
 over solution droplets, 59–60
 for sublimation, 208–10, 215, 218–19
 in terms of temperature, 126
 by thermal gradient, 82
surface-structure
 and coalescence, 154
 and condensation, 57–8, 219
 and nucleation, 57–8, 217
 and sign preference, 51
 of ice, 205, 269
 of water, 51, 154

temperature
 and ice-crystal growth, 268
 and ice-crystal habit, 261–5
 of rain, 196
thermal precipitation, 86, 236–7
thermodynamics of phase-changes, 38–9
thunderstorms
 electrical effects and coalescence, 153, 156, 181
 origin of electrification, 151
 radar echoes from, 186
 structure of, 20, 25–6
 suppression of, 307
time delay
 in cloud dispersion, 305
 in cloud seeding, 302, 336–7, 342
 in meteor hypothesis, 254
trajectories of droplets, 139–43
tropopause, 247, 250
turbulence
 and cloud formation, 7, 26
 and drop disintegration, 195
 in plume theory, 17

ultraviolet light
 action on AgI, 323–5
 as source of nuclei, 65
updraughts
 and cloud formation, 4–7
 in cumuli, 20
 and droplet concentration, 134–5
 in layer-clouds, 5
 and raindrop growth, 166

Van't Hoff factor, 58
vapour-pressure
 over curved surface, 42
 of growing droplet, 125
 of solution droplet, 58–60
vegetation as source of nuclei, 65
velocity, terminal
 of cloud droplets, 160
 of graupel, 276–9
 of ice-crystals, 275–7
 of raindrops, 194–5
 of snowflakes, 276–8
ventilation, 124–7, 267, 287
volcanic dust, as ice-nucleus, 246

wake
 of falling droplet, 141, 147
 of falling ice-crystals, 281

Subject Index

water
    dielectric constant, 51, 291
    molecule, 197–8
    structure, 197, 201–2, 207
    surface free energy, 40, 45
    surface-structure, 51, 154, 202
water-cycle, 1–2
weather control
    prospects for, 345–7
    U.S. Advisory Committee on, 343–4

wind shear
    and cloud growth, 25–6
    and cloud structure, 21
    and precipitation, 300
    and turbulence, 26
wind tunnel, 145, 195
wire, hot, 108
world rainfall curve, 251–2
Wulff's theorem, 259–60

zinc sulphide, 324